Finite Mathematics
For the Managerial, Life, and Social Sciences

ELEVENTH EDITION

Soo T. Tan
Stonehill College

Prepared by

Andy Bulman-Fleming

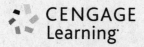

CENGAGE
Learning

Australia • Brazil • Mexico • Singapore • United Kingdom • United States

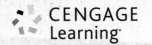
For product information and technology assistance, contact us at **Cengage Learning Customer & Sales Support, 1-800-354-9706**.

For permission to use material from this text or product, submit all requests online at **www.cengage.com/permissions** Further permissions questions can be emailed to **permissionrequest@cengage.com**.

ISBN-13: 978-1-285-84572-2
ISBN-10: 1-285-84572-2

Cengage Learning
200 First Stamford Place, 4th Floor
Stamford, CT 06902
USA

Cengage Learning is a leading provider of customized learning solutions with office locations around the globe, including Singapore, the United Kingdom, Australia, Mexico, Brazil, and Japan. Locate your local office at: **www.cengage.com/global**.

Cengage Learning products are represented in Canada by Nelson Education, Ltd.

To learn more about Cengage Learning Solutions, visit **www.cengage.com**.

Purchase any of our products at your local college store or at our preferred online store **www.cengagebrain.com**.

Printed in the United States of America
1 2 3 4 5 6 7 18 17 16 15 14

CONTENTS

CHAPTER 1   Straight Lines and Linear Functions   1

1.1   The Cartesian Coordinate System   1

1.2   Straight Lines   3

1.3   Linear Functions and Mathematical Models   10

1.4   Intersection of Straight Lines   15

1.5   The Method of Least Squares   19

Chapter 1 Review   23

Chapter 1 Before Moving On   26

CHAPTER 2   Systems of Linear Equations and Matrices   27

2.1   Systems of Linear Equations: An Introduction   27

2.2   Systems of Linear Equations: Unique Solutions   30

2.3   Systems of Linear Equations: Underdetermined and Overdetermined Systems   38

2.4   Matrices   44

2.5   Multiplication of Matrices   48

2.6   The Inverse of a Square Matrix   54

2.7   Leontief Input-Output Model   62

Chapter 2 Review   65

Chapter 2 Before Moving On   69

CHAPTER 3   Linear Programming: A Geometric Approach   73

3.1   Graphing Systems of Linear Inequalities in Two Variables   73

3.2   Linear Programming Problems   78

3.3   Graphical Solutions of Linear Programming Problems   83

3.4   Sensitivity Analysis   94

Chapter 3 Review   99

Chapter 3 Before Moving On   102

CHAPTER 4   Linear Programming: An Algebraic Approach   105

4.1   The Simplex Method: Standard Maximization Problems   105

4.2   The Simplex Method: Standard Minimization Problems   120

4.3 The Simplex Method: Nonstandard Problems (Optional) 132

Chapter 4 Review 142

Chapter 4 Before Moving On 150

CHAPTER 5 **Mathematics of Finance 153**

5.1 Compound Interest 153

5.2 Annuities 157

5.3 Amortization and Sinking Funds 160

5.4 Arithmetic and Geometric Progressions 166

Chapter 5 Review 168

Chapter 5 Before Moving On 170

CHAPTER 6 **Sets and Counting 171**

6.1 Sets and Set Operations 171

6.2 The Number of Elements in a Finite Set 174

6.3 The Multiplication Principle 178

6.4 Permutations and Combinations 179

Chapter 6 Review 183

Chapter 6 Before Moving On 185

CHAPTER 7 **Probability 187**

7.1 Experiments, Sample Spaces, and Events 187

7.2 Definition of Probability 188

7.3 Rules of Probability 191

7.4 Use of Counting Techniques in Probability 195

7.5 Conditional Probability and Independent Events 197

7.6 Bayes' Theorem 202

Chapter 7 Review 207

Chapter 7 Before Moving On 210

CHAPTER 8 **Probability Distributions and Statistics 211**

8.1 Distributions of Random Variables 211

8.2 Expected Value 214

8.3   Variance and Standard Deviation   217

8.4   The Binomial Distribution   221

8.5   The Normal Distribution   224

8.6   Applications of the Normal Distribution   226

     Chapter 8 Review   *228*

     Chapter 8 Before Moving On   *230*

CHAPTER 9   Markov Chains and the Theory of Games   233

9.1   Markov Chains   233

9.2   Regular Markov Chains   236

9.3   Absorbing Markov Chains   239

9.4   Game Theory and Strictly Determined Games   244

9.5   Games with Mixed Strategies   248

     Chapter 9 Review   *252*

     Chapter 9 Before Moving On   *255*

APPENDIX A   Introduction to Logic   257

A.1   Propositions and Connectives   257

A.2   Truth Tables   258

A.3   The Conditional and the Biconditional Connectives   259

A.4   Laws of Logic   261

A.5   Arguments   263

A.6   Applications of Logic to Switching Networks   266

APPENDIX C   Review of Logarithms   268

1 STRAIGHT LINES AND LINEAR FUNCTIONS

1.1 The Cartesian Coordinate System

Problem-Solving Tips

Suppose you are asked to determine whether a given statement is true or false, and you are also asked to explain your answer. How would you answer the question?

If you think the statement is true, then prove it. On the other hand, if you think the statement is false, then give an example that disproves the statement. For example, the statement "If a and b are real numbers, then $a - b = b - a$" is false, and an example that disproves it may be constructed by taking $a = 3$ and $b = 5$. For these values of a and b, we find $a - b = 3 - 5 = -2$, but $b - a = 5 - 3 = 2$, and this shows that $a - b \neq b - a$. Such an example is called a **counterexample**.

Concept Questions   page 6

1. a. $a < 0$ and $b > 0$ **b.** $a < 0$ and $b < 0$ **c.** $a > 0$ and $b < 0$

Exercises   page 7

1. The coordinates of A are $(3, 3)$ and it is located in Quadrant I.

3. The coordinates of C are $(2, -2)$ and it is located in Quadrant IV.

5. The coordinates of E are $(-4, -6)$ and it is located in Quadrant III.

7. A **9.** E, F, and G **11.** F

For Exercises 13–19, refer to the following figure.

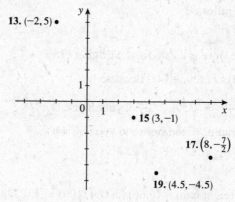

21. Using the distance formula, we find that $\sqrt{(4-1)^2 + (7-3)^2} = \sqrt{3^2 + 4^2} = \sqrt{25} = 5$.

23. Using the distance formula, we find that $\sqrt{[4-(-1)]^2 + (9-3)^2} = \sqrt{5^2 + 6^2} = \sqrt{25 + 36} = \sqrt{61}$.

1

25. The coordinates of the points have the form $(x, -6)$. Because the points are 10 units away from the origin, we have $(x - 0)^2 + (-6 - 0)^2 = 10^2$, $x^2 = 64$, or $x = \pm 8$. Therefore, the required points are $(-8, -6)$ and $(8, -6)$.

27. The points are shown in the diagram. To show that the four sides are equal, we compute

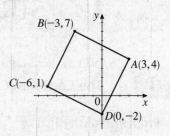

$$d(A, B) = \sqrt{(-3 - 3)^2 + (7 - 4)^2} = \sqrt{(-6)^2 + 3^2} = \sqrt{45},$$
$$d(B, C) = \sqrt{[-6 - (-3)]^2 + (1 - 7)^2} = \sqrt{(-3)^2 + (-6)^2} = \sqrt{45},$$
$$d(C, D) = \sqrt{[0 - (-6)]^2 + [(-2) - 1]^2} = \sqrt{(6)^2 + (-3)^2} = \sqrt{45},$$
and $d(A, D) = \sqrt{(0 - 3)^2 + (-2 - 4)^2} = \sqrt{(3)^2 + (-6)^2} = \sqrt{45}$.

Next, to show that $\triangle ABC$ is a right triangle, we show that it satisfies the Pythagorean Theorem. Thus, $d(A, C) = \sqrt{(-6 - 3)^2 + (1 - 4)^2} = \sqrt{(-9)^2 + (-3)^2} = \sqrt{90} = 3\sqrt{10}$ and $[d(A, B)]^2 + [d(B, C)]^2 = 90 = [d(A, C)]^2$. Similarly, $d(B, D) = \sqrt{90} = 3\sqrt{10}$, so $\triangle BAD$ is a right triangle as well. It follows that $\angle B$ and $\angle D$ are right angles, and we conclude that $ADCB$ is a square.

29. The equation of the circle with radius 5 and center $(2, -3)$ is given by $(x - 2)^2 + [y - (-3)]^2 = 5^2$, or $(x - 2)^2 + (y + 3)^2 = 25$.

31. The equation of the circle with radius 5 and center $(0, 0)$ is given by $(x - 0)^2 + (y - 0)^2 = 5^2$, or $x^2 + y^2 = 25$.

33. The distance between the points $(5, 2)$ and $(2, -3)$ is given by $d = \sqrt{(5 - 2)^2 + [2 - (-3)]^2} = \sqrt{3^2 + 5^2} = \sqrt{34}$. Therefore $r = \sqrt{34}$ and the equation of the circle passing through $(5, 2)$ and $(2, -3)$ is $(x - 2)^2 + [y - (-3)]^2 = 34$, or $(x - 2)^2 + (y + 3)^2 = 34$.

35. a. The coordinates of the suspect's car at its final destination are $x = 4$ and $y = 4$.

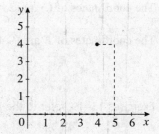

b. The distance traveled by the suspect was $5 + 4 + 1$, or 10 miles.

c. The distance between the original and final positions of the suspect's car was $d = \sqrt{(4 - 0)^2 + (4 - 0)^2} = \sqrt{32} = 4\sqrt{2}$, or approximately 5.66 miles.

37. Suppose that the furniture store is located at the origin O so that your house is located at $A(20, -14)$. Because

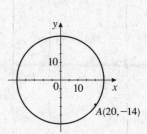

$d(O, A) = \sqrt{20^2 + (-14)^2} = \sqrt{596} \approx 24.4$, your house is located within a 25-mile radius of the store and you will not incur a delivery charge.

39. The cost of shipping by freight train is $(0.66)(2000)(100) = 132,000$, or \$132,000.
The cost of shipping by truck is $(0.62)(2200)(100) = 136,400$, or \$136,400.
Comparing these results, we see that the automobiles should be shipped by freight train. The net savings are $136,400 - 132,000 = 4400$, or \$4400.

41. To determine the VHF requirements, we calculate $d = \sqrt{25^2 + 35^2} = \sqrt{625 + 1225} = \sqrt{1850} \approx 43.01$.

Models B, C, and D satisfy this requirement.

To determine the UHF requirements, we calculate $d = \sqrt{20^2 + 32^2} = \sqrt{400 + 1024} = \sqrt{1424} \approx 37.74$. Models C and D satisfy this requirement.

Therefore, Model C allows him to receive both channels at the least cost.

43. a. Let the positions of ships A and B be $(0, y)$ and $(x, 0)$, respectively. Then

$y = 25\left(t + \frac{1}{2}\right)$ and $x = 20t$. The distance D in miles between the two ships is

$$D = \sqrt{(x - 0)^2 + (0 - y)^2} = \sqrt{x^2 + y^2} = \sqrt{400t^2 + 625\left(t + \frac{1}{2}\right)^2} \quad (1).$$

b. The distance between the ships 2 hours after ship A has left port is obtained by letting $t = \frac{3}{2}$ in Equation (1),

yielding $D = \sqrt{400\left(\frac{3}{2}\right)^2 + 625\left(\frac{3}{2} + \frac{1}{2}\right)^2} = \sqrt{3400}$, or approximately 58.31 miles.

45. a. Suppose that $P = (x_1, y_1)$ and $Q = (x_2, y_2)$ are endpoints of the line segment and that

the point $M = \left(\dfrac{x_1 + x_2}{2}, \dfrac{y_1 + y_2}{2}\right)$ is the midpoint of the line segment PQ. The distance

between P and Q is $\sqrt{(x_2 - x_1)^2 + (y_2 - y_1)^2}$. The distance between P and M is

$$\sqrt{\left(\frac{x_1 + x_2}{2} - x_1\right)^2 + \left(\frac{y_1 + y_2}{2} - y_1\right)^2} = \sqrt{\left(\frac{x_2 - x_1}{2}\right)^2 + \left(\frac{y_2 - y_1}{2}\right)^2} = \frac{1}{2}\sqrt{(x_2 - x_1)^2 + (y_2 - y_1)^2},$$

which is one-half the distance from P to Q. Similarly, we obtain the same expression for the distance from M to P.

b. The midpoint is given by $\left(\dfrac{4 - 3}{2}, \dfrac{-5 + 2}{2}\right)$, or $\left(\dfrac{1}{2}, -\dfrac{3}{2}\right)$.

47. False. The distance between $P_1(a, b)$ and $P_3(kc, kd)$ is

$d = \sqrt{(kc - a)^2 + (kd - b)^2}$

$\neq |k|\, D = |k|\sqrt{(c - a)^2 + (d - b)^2} = \sqrt{k^2(c - a)^2 + k^2(d - b)^2} = \sqrt{[k(c - a)]^2 + [k(d - b)]^2}$.

49. Referring to the figure in the text, we see that the distance between the two points is given by the length of the hypotenuse of the right triangle. That is, $d = \sqrt{(x_2 - x_1)^2 + (y_2 - y_1)^2}$.

1.2   Straight Lines

Problem-Solving Tips

When you solve a problem in the exercises that follow each section, first read the problem. Before you start computing or writing out a solution, try to formulate a strategy for solving the problem. Then proceed by using your strategy to solve the problem.

Here we summarize some general problem-solving techniques that are covered in this section.

1. To show that two lines are parallel, you need to show that the slopes of the two lines are equal or that their slopes are both undefined.

2. **To show that two lines L_1 and L_2 are perpendicular**, you need to show that the slope m_1 of L_1 is the negative reciprocal of the slope m_2 of L_2; that is, $m_1 = -1/m_2$.

3. **To find the equation of a line**, you need the slope of the line and a point lying on the line. You can then find the equation of the line using the point-slope form of the equation of a line: $(y - y_1) = m(x - x_1)$.

Concept Questions   page 19

1. The slope is $m = \dfrac{y_2 - y_1}{x_2 - x_1}$, where $P(x_1, y_1)$ and $P(x_2, y_2)$ are any two distinct points on the nonvertical line. The slope of a vertical line is undefined.

3. **a.** $m_1 = m_2$ **b.** $m_2 = -\dfrac{1}{m_1}$

Exercises   page 19

1. Referring to the figure shown in the text, we see that $m = \dfrac{2 - 0}{0 - (-4)} = \dfrac{1}{2}$.

3. This is a vertical line, and hence its slope is undefined.

5. $m = \dfrac{y_2 - y_1}{x_2 - x_1} = \dfrac{8 - 3}{5 - 4} = 5$.   7. $m = \dfrac{y_2 - y_1}{x_2 - x_1} = \dfrac{8 - 3}{4 - (-2)} = \dfrac{5}{6}$.

9. $m = \dfrac{y_2 - y_1}{x_2 - x_1} = \dfrac{d - b}{c - a}$, provided $a \neq c$.

11. Because the equation is already in slope-intercept form, we read off the slope $m = 4$.

   a. If x increases by 1 unit, then y increases by 4 units.

   b. If x decreases by 2 units, then y decreases by $4(-2) = -8$ units.

13. (e)   15. (a)   17. (f)

19. The slope of the line through A and B is $\dfrac{-10 - (-2)}{-3 - 1} = \dfrac{-8}{-4} = 2$. The slope of the line through C and D is $\dfrac{1 - 5}{-1 - 1} = \dfrac{-4}{-2} = 2$. Because the slopes of these two lines are equal, the lines are parallel.

21. The slope of the line through the point $(1, a)$ and $(4, -2)$ is $m_1 = \dfrac{-2 - a}{4 - 1}$ and the slope of the line through $(2, 8)$ and $(-7, a + 4)$ is $m_2 = \dfrac{a + 4 - 8}{-7 - 2}$. Because these two lines are parallel, m_1 is equal to m_2. Therefore, $\dfrac{-2 - a}{3} = \dfrac{a - 4}{-9}$, $-9(-2 - a) = 3(a - 4)$, $18 + 9a = 3a - 12$, and $6a = -30$, so $a = -5$.

23. We use the point-slope form of an equation of a line with the point $(3, -4)$ and slope $m = 2$. Thus $y - y_1 = m(x - x_1)$ becomes $y - (-4) = 2(x - 3)$. Simplifying, we have $y + 4 = 2x - 6$, or $y = 2x - 10$.

25. Because the slope $m = 0$, we know that the line is a horizontal line of the form $y = b$. Because the line passes through $(-3, 2)$, we see that $b = 2$, and an equation of the line is $y = 2$.

27. We first compute the slope of the line joining the points $(2, 4)$ and $(3, 7)$ to be $m = \dfrac{7 - 4}{3 - 2} = 3$. Using the point-slope form of an equation of a line with the point $(2, 4)$ and slope $m = 3$, we find $y - 4 = 3\,(x - 2)$, or $y = 3x - 2$.

29. We first compute the slope of the line joining the points $(1, 2)$ and $(-3, -2)$ to be $m = \dfrac{-2 - 2}{-3 - 1} = \dfrac{-4}{-4} = 1$. Using the point-slope form of an equation of a line with the point $(1, 2)$ and slope $m = 1$, we find $y - 2 = x - 1$, or $y = x + 1$.

31. The slope of the line through A and B is $\dfrac{2 - 5}{4 - (-2)} = -\dfrac{3}{6} = -\dfrac{1}{2}$. The slope of the line through C and D is $\dfrac{6 - (-2)}{3 - (-1)} = \dfrac{8}{4} = 2$. Because the slopes of these two lines are the negative reciprocals of each other, the lines are perpendicular.

33. We use the slope-intercept form of an equation of a line: $y = mx + b$. Because $m = 3$ and $b = 4$, the equation is $y = 3x + 4$.

35. We use the slope-intercept form of an equation of a line: $y = mx + b$. Because $m = 0$ and $b = 5$, the equation is $y = 5$.

37. We first write the given equation in the slope-intercept form: $x - 2y = 0$, so $-2y = -x$, or $y = \frac{1}{2}x$. From this equation, we see that $m = \frac{1}{2}$ and $b = 0$.

39. We write the equation in slope-intercept form: $2x - 3y - 9 = 0$, $-3y = -2x + 9$, and $y = \frac{2}{3}x - 3$. From this equation, we see that $m = \frac{2}{3}$ and $b = -3$.

41. We write the equation in slope-intercept form: $2x + 4y = 14$, $4y = -2x + 14$, and $y = -\frac{2}{4}x + \frac{14}{4} = -\frac{1}{2}x + \frac{7}{2}$. From this equation, we see that $m = -\frac{1}{2}$ and $b = \frac{7}{2}$.

43. An equation of a horizontal line is of the form $y = b$. In this case $b = -3$, so $y = -3$ is an equation of the line.

45. We first write the equation $2x - 4y - 8 = 0$ in slope-intercept form: $2x - 4y - 8 = 0$, $4y = 2x - 8$, $y = \frac{1}{2}x - 2$. Now the required line is parallel to this line, and hence has the same slope. Using the point-slope form of an equation of a line with $m = \frac{1}{2}$ and the point $(-2, 2)$, we have $y - 2 = \frac{1}{2}\,[x - (-2)]$ or $y = \frac{1}{2}x + 3$.

47. We first write the equation $3x + 4y - 22 = 0$ in slope-intercept form: $3x + 4y - 22 = 0$, so $4y = -3x + 22$ and $y = -\frac{3}{4}x + \frac{11}{2}$. Now the required line is perpendicular to this line, and hence has slope $\frac{4}{3}$ (the negative reciprocal of $-\frac{3}{4}$). Using the point-slope form of an equation of a line with $m = \frac{4}{3}$ and the point $(2, 4)$, we have $y - 4 = \frac{4}{3}\,(x - 2)$, or $y = \frac{4}{3}x + \frac{4}{3}$.

49. The midpoint of the line segment joining $P_1\,(-2, -4)$ and $P_2\,(3, 6)$ is $M\left(\dfrac{-2 + 3}{2}, \dfrac{-4 + 6}{2}\right)$ or $M\left(\frac{1}{2}, 1\right)$.

Using the point-slope form of the equation of a line with $m = -2$, we have $y - 1 = -2\left(x - \frac{1}{2}\right)$ or $y = -2x + 2$.

51. A line parallel to the x-axis has slope 0 and is of the form $y = b$. Because the line is 6 units below the axis, it passes through $(0, -6)$ and its equation is $y = -6$.

53. We use the point-slope form of an equation of a line to obtain $y - b = 0 \, (x - a)$, or $y = b$.

55. Because the required line is parallel to the line joining $(-3, 2)$ and $(6, 8)$, it has slope $m = \dfrac{8 - 2}{6 - (-3)} = \dfrac{6}{9} = \dfrac{2}{3}$. We also know that the required line passes through $(-5, -4)$. Using the point-slope form of an equation of a line, we find $y - (-4) = \frac{2}{3} \, [x - (-5)]$, $y = \frac{2}{3}x + \frac{10}{3} - 4$, and finally $y = \frac{2}{3}x - \frac{2}{3}$.

57. Because the point $(-3, 5)$ lies on the line $kx + 3y + 9 = 0$, it satisfies the equation. Substituting $x = -3$ and $y = 5$ into the equation gives $-3k + 15 + 9 = 0$, or $k = 8$.

59. $3x - 2y + 6 = 0$. Setting $y = 0$, we have $3x + 6 = 0$ or $x = -2$, so the x-intercept is -2. Setting $x = 0$, we have $-2y + 6 = 0$ or $y = 3$, so the y-intercept is 3.

61. $x + 2y - 4 = 0$. Setting $y = 0$, we have $x - 4 = 0$ or $x = 4$, so the x-intercept is 4. Setting $x = 0$, we have $2y - 4 = 0$ or $y = 2$, so the y-intercept is 2.

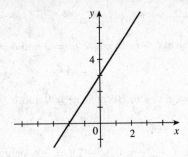

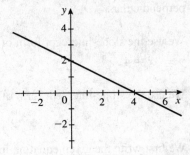

63. $y + 5 = 0$. Setting $y = 0$, we have $0 + 5 = 0$, which has no solution, so there is no x-intercept. Setting $x = 0$, we have $y + 5 = 0$ or $y = -5$, so the y-intercept is -5.

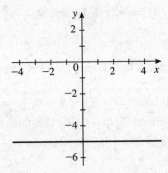

65. Because the line passes through the points $(a, 0)$ and $(0, b)$, its slope is $m = \dfrac{b - 0}{0 - a} = -\dfrac{b}{a}$. Then, using the point-slope form of an equation of a line with the point $(a, 0)$, we have $y - 0 = -\dfrac{b}{a} \, (x - a)$ or $y = -\dfrac{b}{a}x + b$, which may be written in the form $\dfrac{b}{a}x + y = b$. Multiplying this last equation by $\dfrac{1}{b}$, we have $\dfrac{x}{a} + \dfrac{y}{b} = 1$.

67. Using the equation $\dfrac{x}{a} + \dfrac{y}{b} = 1$ with $a = -2$ and $b = -4$, we have $-\dfrac{x}{2} - \dfrac{y}{4} = 1$. Then $-4x - 2y = 8$, $2y = -8 - 4x$, and finally $y = -2x - 4$.

69. Using the equation $\dfrac{x}{a} + \dfrac{y}{b} = 1$ with $a = 4$ and $b = -\frac{1}{2}$, we have $\dfrac{x}{4} + \dfrac{y}{-1/2} = 1$, $-\frac{1}{4}x + 2y = -1$, $2y = \frac{1}{4}x - 1$, and so $y = \frac{1}{8}x - \frac{1}{2}$.

71. The slope of the line passing through A and B is $m = \dfrac{7 - 1}{1 - (-2)} = \dfrac{6}{3} = 2$, and the slope of the line passing through B and C is $m = \dfrac{13 - 7}{4 - 1} = \dfrac{6}{3} = 2$. Because the slopes are equal, the points lie on the same line.

73. The slope of the line L passing through P_1 (1.8, −6.44) and P_2 (2.4, −5.72) is $m = \dfrac{-5.72 - (-6.44)}{2.4 - 1.8} = 1.2$, so an

equation of L is $y - (-6.44) = 1.2 (x - 1.8)$ or $y = 1.2x - 8.6$.

Substituting $x = 5.0$ into this equation gives $y = 1.2 (5) - 8.6 = -2.6$. This shows that the point P_3 (5.0, −2.72) does not lie on L, and we conclude that Alison's claim is not valid.

75. a.

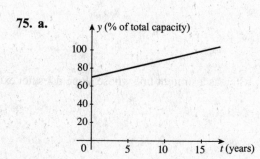

b. The slope is 1.9467 and the y-intercept is 70.082.

c. The output is increasing at the rate of 1.9467% per year. The output at the beginning of 1990 was 70.082%.

d. We solve the equation $1.9467t + 70.082 = 100$, obtaining $t \approx 15.37$. We conclude that the plants were generating at maximum capacity during April 2005.

77. a. $y = 0.55x$ **b.** Solving the equation $1100 = 0.55x$ for x, we have $x = \dfrac{1100}{0.55} = 2000$.

79. Using the points $(0, 0.68)$ and $(10, 0.80)$, we see that the slope of the required line is

$m = \dfrac{0.80 - 0.68}{10 - 0} = \dfrac{0.12}{10} = 0.012$. Next, using the point-slope form of the equation of a line, we have

$y - 0.68 = 0.012 (t - 0)$ or $y = 0.012t + 0.68$. Therefore, when $t = 14$, we have $y = 0.012 (14) + 0.68 = 0.848$, or 84.8%. That is, in 2004 women's wages were 84.8% of men's wages.

81. a, b.

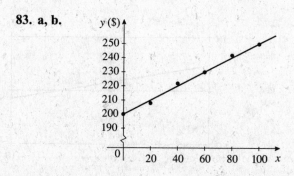

c. The slope of L is $m = \dfrac{8.2 - 1.3}{3 - 0} = 2.3$, so an equation of L is

$y - 1.3 = 2.3 (x - 0)$ or $y = 2.3x + 1.3$.

d. The change in spending in the first quarter of 2014 is estimated to be $2.3 (4) + 1.3$, or 10.5%.

83. a, b.

c. Using the points $(0, 200)$ and $(100, 250)$, we see that the

slope of the required line is $m = \dfrac{250 - 200}{100} = \dfrac{1}{2}$.

Therefore, an equation is $y - 200 = \frac{1}{2}x$ or $y = \frac{1}{2}x + 200$.

d. The approximate cost for producing 54 units of the commodity is $\frac{1}{2} (54) + 200$, or $227.

85. a, b.

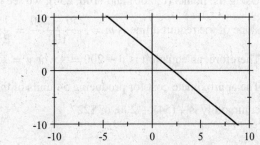

c. The slope of L is $m = \dfrac{9.0 - 5.8}{5 - 1} = \dfrac{3.2}{4} = 0.8$. Using the point-slope form of an equation of a line, we have

$y - 5.8 = 0.8\,(x - 1) = 0.8x - 0.8$, or $y = 0.8x + 5$.

d. Using the equation from part c with $x = 9$, we have

$y = 0.8\,(9) + 5 = 12.2$, or \$12.2 million.

87. Yes. A straight line with slope zero ($m = 0$) is a horizontal line, whereas a straight line whose slope does not exist is a vertical line (m cannot be computed).

89. True. The slope of the line is given by $-\dfrac{2}{4} = -\dfrac{1}{2}$.

91. True. The slope of the line $Ax + By + C = 0$ is $-\dfrac{A}{B}$. (Write it in slope-intercept form.) Similarly, the slope of the line $ax + by + c = 0$ is $-\dfrac{a}{b}$. They are parallel if and only if $-\dfrac{A}{B} = -\dfrac{a}{b}$, that is, if $Ab = aB$, or $Ab - aB = 0$.

93. True. The slope of the line $ax + by + c_1 = 0$ is $m_1 = -\dfrac{a}{b}$. The slope of the line $bx - ay + c_2 = 0$ is $m_2 = \dfrac{b}{a}$. Because $m_1 m_2 = -1$, the straight lines are indeed perpendicular.

95. Writing each equation in the slope-intercept form, we have $y = -\dfrac{a_1}{b_1}x - \dfrac{c_1}{b_1}$ ($b_1 \neq 0$) and $y = -\dfrac{a_2}{b_2}x - \dfrac{c_2}{b_2}$ ($b_2 \neq 0$). Because two lines are parallel if and only if their slopes are equal, we see that the lines are parallel if and only if $-\dfrac{a_1}{b_1} = -\dfrac{a_2}{b_2}$, or $a_1 b_2 - b_1 a_2 = 0$.

Technology Exercises   page 28

Graphing Utility

1.

3.

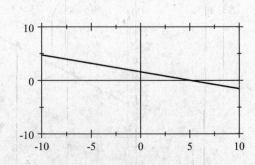

5. a.

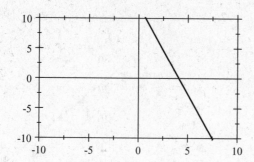

b.

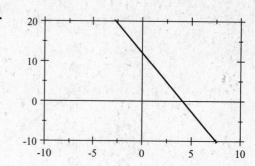

7. a.

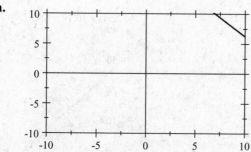

b.

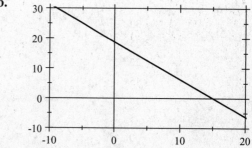

9.

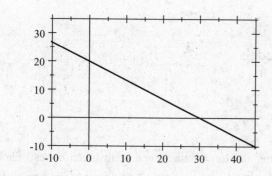

11.

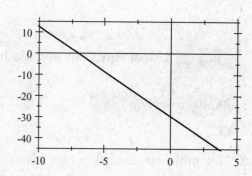

Excel

1.

$3.2x + 2.1y - 6.72 = 0$

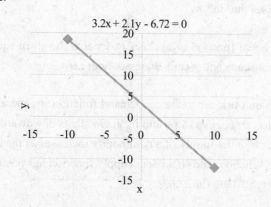

3.

$1.6x + 5.1y = 8.16$

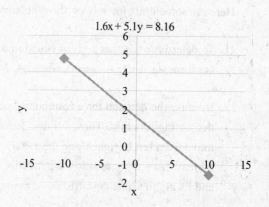

5.

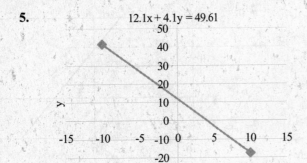

$12.1x + 4.1y = 49.61$

7.

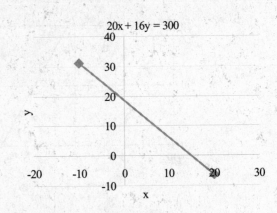

$20x + 16y = 300$

9.

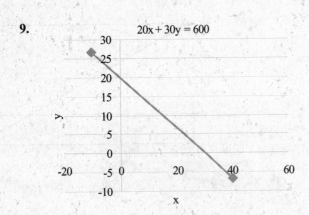

$20x + 30y = 600$

11.

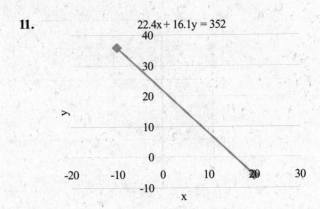

$22.4x + 16.1y = 352$

1.3   Linear Functions and Mathematical Models

Problem-Solving Tips

New mathematical terms in each section are defined either in blue boldface type or in green boxes. Each time you encounter a new term, read through the definition and then try to express the definition in your own words without looking at the book. Once you understand these definitions, it will be easier for you to work the exercise sets that follow each section.

Here are some hints for solving the problems in the exercises that follow:

1. To determine whether a given equation defines y as a linear function of x, check to see that the given equation has the form $Ax + By + C = 0$, where A, B, and C are constants and A and B are not both zero.

2. Because the demand for a commodity decreases as its unit price increases, a demand function is generally a decreasing function. Thus, a linear demand function has a negative slope and its graph slants downwards as we move from left to right along the x-axis. Similarly, because the supply of a commodity increases as the unit price increases, a supply function is generally an increasing function, and so a linear supply function has positive slope and its graph slants upwards as we move from left to right along the x-axis.

Concept Questions   page 36

1. a. A function is a rule that associates with each element in a set A exactly one element in a set B.

b. A linear function is a function of the form $f(x) = mx + b$, where m and b are constants. For example, $f(x) = 2x + 3$ is a linear function.

c. The domain and range of a linear function are both $(-\infty, \infty)$.

d. The graph of a linear function is a straight line.

3. Negative, positive

Exercises   page 36

1. Yes. Solving for y in terms of x, we find $3y = -2x + 6$, or $y = -\frac{2}{3}x + 2$.

3. Yes. Solving for y in terms of x, we find $2y = x + 4$, or $y = \frac{1}{2}x + 2$.

5. Yes. Solving for y in terms of x, we have $4y = 2x + 9$, or $y = \frac{1}{2}x + \frac{9}{4}$.

7. y is not a linear function of x because of the quadratic term $2x^2$.

9. y is not a linear function of x because of the nonlinear term $-3y^2$.

11. a. $C(x) = 8x + 40{,}000$, where x is the number of units produced.

b. $R(x) = 12x$, where x is the number of units sold.

c. $P(x) = R(x) - C(x) = 12x - (8x + 40{,}000) = 4x - 40{,}000$.

d. $P(8000) = 4(8000) - 40{,}000 = -8000$, or a loss of \$8,000. $P(12{,}000) = 4(12{,}000) - 40{,}000 = 8000$, or a profit of \$8000.

13. $f(0) = 2$ gives $m(0) + b = 2$, or $b = 2$. Thus, $f(x) = mx + 2$. Next, $f(3) = -1$ gives $m(3) + 2 = -1$, or $m = -1$.

15. Let V be the book value of the office building after 2008. Since $V = 1{,}000{,}000$ when $t = 0$, the line passes through $(0, 1000000)$. Similarly, when $t = 50$, $V = 0$, so the line passes through $(50, 0)$. Then the slope of the line is given by $m = \dfrac{0 - 1{,}000{,}000}{50 - 0} = -20{,}000$. Using the point-slope form of the equation of a line with the point $(0, 1000000)$, we have $V - 1{,}000{,}000 = -20{,}000(t - 0)$, or $V = -20{,}000t + 1{,}000{,}000$.
In 2013, $t = 5$ and $V = -20{,}000(5) + 1{,}000{,}000 = 900{,}000$, or \$900,000.
In 2018, $t = 10$ and $V = -20{,}000(10) + 1{,}000{,}000 = 800{,}000$, or \$800,000.

17. The consumption function is given by $C(x) = 0.75x + 6$. Thus, $C(0) = 6$, or 6 billion dollars; $C(50) = 0.75(50) + 6 = 43.5$, or 43.5 billion dollars; and $C(100) = 0.75(100) + 6 = 81$, or 81 billion dollars.

19. a. $y = I(x) = 1.033x$, where x is the monthly benefit before adjustment and y is the adjusted monthly benefit.

b. His adjusted monthly benefit is $I(1220) = 1.033(1220) = 1260.26$, or \$1260.26.

21. Let the number of tapes produced and sold be x. Then $C(x) = 12{,}100 + 0.60x$, $R(x) = 1.15x$, and $P(x) = R(x) - C(x) = 1.15x - (12{,}100 + 0.60x) = 0.55x - 12{,}100$.

23. Let the value of the workcenter system after t years be V. When $t = 0$, $V = 60,000$ and when $t = 4$, $V = 12,000$.

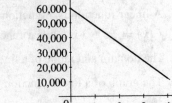

a. Since $m = \dfrac{12,000 - 60,000}{4} = -\dfrac{48,000}{4} = -12,000$, the

rate of depreciation $(-m)$ is $\$12,000/\text{yr}$.

b. Using the point-slope form of the equation of a line with the

point $(4, 12000)$, we have $V - 12,000 = -12,000(t - 4)$,

or $V = -12,000t + 60,000$.

d. When $t = 3$, $V = -12,000(3) + 60,000 = 24,000$, or

$\$24,000$.

25. The formula given in Exercise 24 is $V = C - \dfrac{C - S}{N}t$. When $C = 1,000,000$, $N = 50$, and

$S = 0$, we have $V = 1,000,000 - \dfrac{1,000,000 - 0}{50}t$, or $V = 1,000,000 - 20,000t$. In 2013, $t = 5$ and

$V = 1,000,000 - 20,000(5) = 900,000$, or $\$900,000$. In 2018, $t = 10$ and $V = 1,000,000 - 20,000(10) = 800,000$,

or $\$800,000$.

27. a. $D(S) = \dfrac{Sa}{1.7}$. If we think of D as having the form $D(S) = mS + b$, then $m = \dfrac{a}{1.7}$, $b = 0$, and D is a linear

function of S.

b. $D(0.4) = \dfrac{500(0.4)}{1.7} \approx 117.647$, or approximately 117.65 mg.

29. a. The graph of f passes through the points $P_1 (0, 17.5)$ and $P_2 (10, 10.3)$. Its slope is $\dfrac{10.3 - 17.5}{10 - 0} = -0.72$.

An equation of the line is $y - 17.5 = -0.72(t - 0)$ or $y = -0.72t + 17.5$, so the linear function is

$f(t) = -0.72t + 17.5$.

b. The percentage of high school students who drink and drive at the beginning of 2014 is projected to be

$f(13) = -0.72(13) + 17.5 = 8.14$, or 8.14%.

31. a. The line passing through $P_1 (0, 61)$ and $P_2 (4, 51)$ has slope $m = \dfrac{61 - 51}{0 - 4} = -2.5$, so its equation is

$y - 61 = -2.5(t - 0)$ or $y = -2.5t + 61$. Thus, $f(t) = -2.5t + 61$.

b. The percentage of middle-income adults in 2021 is projected to be $f(t) = -2.5(5) + 61$, or 48.5%.

33. a. Since the relationship is linear, we can write $F = mC + b$, where m and b are constants. Using the condition

$C = 0$ when $F = 32$, we have $32 = b$, and so $F = mC + 32$. Next, using the condition $C = 100$ when $F = 212$,

we have $212 = 100m + 32$, or $m = \frac{9}{5}$. Therefore, $F = \frac{9}{5}C + 32$.

b. From part a, we have $F = \frac{9}{5}C + 32$. When $C = 20$, $F = \frac{9}{5}(20) + 32 = 68$, and so the temperature equivalent to

$20° C$ is $68° F$.

c. Solving for C in terms of F, we find $\frac{9}{5}C = F - 32$, or $C = \frac{5}{9}F - \frac{160}{9}$. When $F = 70$, $C = \frac{5}{9}(70) - \frac{160}{9} = \frac{190}{9}$,

or approximately $21.1° C$.

35. a. $2x + 3p - 18 = 0$, so setting $x = 0$ gives
$3p = 18$, or $p = 6$. Next, setting $p = 0$ gives
$2x = 18$, or $x = 9$.

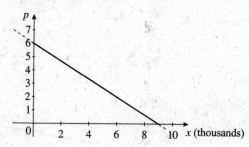

b. If $p = 4$, then $2x + 3(4) - 18 = 0$,
$2x = 18 - 12 = 6$, and $x = 3$. Therefore, the
quantity demanded when $p = 4$ is 3000.
(Remember that x is measured in units of a
1000.)

37. a. $p = -3x + 60$, so when $x = 0$, $p = 60$ and
when $p = 0$, $-3x = -60$, or $x = 20$.

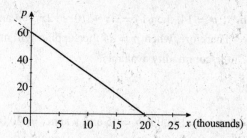

b. When $p = 30$, $30 = -3x + 60$, $3x = 30$, and
$x = 10$. Therefore, the quantity demanded
when $p = 30$ is 10,000 units.

39. When $x = 1000$, $p = 55$, and when $x = 600$, $p = 85$. Therefore, the graph of the linear demand equation is the
straight line passing through the points $(1000, 55)$ and $(600, 85)$. The slope of the line is $\dfrac{85 - 55}{600 - 1000} = -\dfrac{3}{40}$.
Using this slope and the point $(1000, 55)$, we find that the required equation is $p - 55 = -\frac{3}{40}(x - 1000)$, or
$p = -\frac{3}{40}x + 130$. When $x = 0$, $p = 130$, and this means that there will be no demand above \$130. When $p = 0$,
$x = 1733.33$, and this means that 1733 units is the maximum quantity demanded.

41. The demand equation is linear, and we know that the line passes through the points $(1000, 9)$ and $(6000, 4)$.
Therefore, the slope of the line is given by $m = \dfrac{4 - 9}{6000 - 1000} = -\dfrac{5}{5000} = -0.001$. Since the equation of
the line has the form $p = ax + b$, $9 = -0.001(1000) + b$, so $b = 10$. Therefore, an equation of the line is
$p = -0.001x + 10$. If $p = 7.50$, we have $7.50 = -0.001x + 10$, so $0.001x = 2.50$ and $x = 2500$. Thus, the
quantity demanded when the unit price is \$7.50 is 2500 units.

43. a. $3x - 4p + 24 = 0$. Setting $x = 0$, we obtain
$3(0) - 4p + 24 = 0$, so $-4p = -24$ and $p = 6$. Setting
$p = 0$, we obtain $3x - 4(0) + 24 = 0$, so $3x = -24$ and
$x = -8$.

b. When $p = 8$, $3x - 4(8) + 24 = 0$, $3x = 32 - 24 = 8$, and
$x = \frac{8}{3}$. Therefore, 2667 units of the commodity would be
supplied at a unit price of \$8. (Here again x is measured in
units of 1000.)

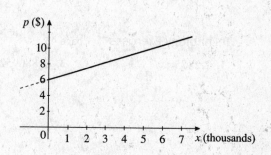

45. a. $p = 2x + 10$, so when $x = 0$, $p = 10$, and when $p = 0$,
  $x = -5$.

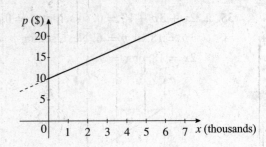

  b. If $p = 14$, then $14 = 2x + 10$, so $2x = 4$ and $x = 2$.
  Therefore, when $p = 14$ the supplier will make 2000 units of
  the commodity available.

47. When $x = 10,000$, $p = 45$ and when $x = 20,000$, $p = 50$. The
  slope of the line passing through (10000, 45) and (20000, 50) is

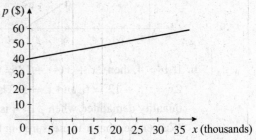

  $$m = \frac{50 - 45}{20,000 - 10,000} = \frac{5}{10,000} = 0.0005,\text{ so using the}$$

  point-slope form of an equation of a line with the point
  (10000, 45), we have $p - 45 = 0.0005\,(x - 10,000)$,
  $p = 0.0005x - 5 + 45$, and $p = 0.0005x + 40$.

  If $p = 70$, then $70 = 0.0005x + 40$ and $0.0005x = 30$, so $x = \frac{30}{0.0005} = 60,000$. (If x is expressed in units of a
  thousand, then the equation may be written in the form $p = \frac{1}{2}x + 40$.)

49.

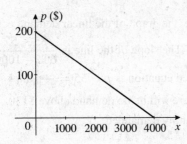

  b. The highest price is $200 per unit.

  c. To find the quantity demanded when $p = 100$, we solve
  $-0.005x + 200 = 100$, obtaining $x = \dfrac{-100}{-0.05} = 2000$, or
  2000 units per month.

51.

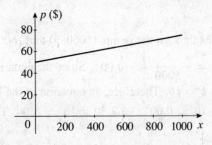

  b. The lowest price is $50 per unit.

  c. We solve the equation $0.025x + 50 = 100$, obtaining
  $x = \dfrac{100 - 50}{0.0025} = 2000$, or 2000 units per month.

53. False. $P(x) = R(x) - C(x) = sx - (cx + F) = (s - c)x - F$. Therefore, the firm is making a profit if
  $P(x) = (s - c)x - F > 0$, or $x > \dfrac{F}{s - c}$.

Technology Exercises   page 43

  1. 2.2875   **3.** 2.880952381   **5.** 7.2851648352   **7.** 2.4680851064

1.4   Intersection of Straight Lines

Problem-Solving Tips

1. **To find the break-even point,** solve the simultaneous equations $p = R(x)$ and $p = C(x)$ for x and p.

2. **To find the market equilibrium for a commodity,** find the point of intersection of the supply and demand equations for the commodity. (Market equilibrium prevails when the quantity produced is equal to the quantity demanded.)

Concept Questions   page 49

1. The intersection must lie in the first quadrant because only the parts of the demand and supply curves in the first quadrant are of interest.

3.

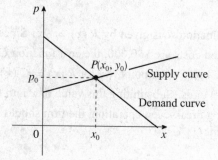

Exercises   page 50

1. We solve the system $y = 3x + 4$, $y = -2x + 14$. Substituting the first equation into the second yields $3x + 4 = -2x + 14$, $5x = 10$, and $x = 2$. Substituting this value of x into the first equation yields $y = 3(2) + 4$, so $y = 10$. Thus, the point of intersection is $(2, 10)$.

3. We solve the system $2x - 3y = 6$, $3x + 6y = 16$. Solving the first equation for y, we obtain $3y = 2x - 6$, so $y = \frac{2}{3}x - 2$. Substituting this value of y into the second equation, we obtain $3x + 6\left(\frac{2}{3}x - 2\right) = 16$, $3x + 4x - 12 = 16$, $7x = 28$, and $x = 4$. Then $y = \frac{2}{3}(4) - 2 = \frac{2}{3}$, so the point of intersection is $\left(4, \frac{2}{3}\right)$.

5. We solve the system $y = \frac{1}{4}x - 5$, $2x - \frac{3}{2}y = 1$. Substituting the value of y given in the first equation into the second equation, we obtain $2x - \frac{3}{2}\left(\frac{1}{4}x - 5\right) = 1$, so $2x - \frac{3}{8}x + \frac{15}{2} = 1$, $16x - 3x + 60 = 8$, $13x = -52$, and $x = -4$. Substituting this value of x into the first equation, we have $y = \frac{1}{4}(-4) - 5 = -1 - 5$, so $y = -6$. Therefore, the point of intersection is $(-4, -6)$.

7. We solve the equation $R(x) = C(x)$, or $15x = 5x + 10{,}000$, obtaining $10x = 10{,}000$, or $x = 1000$. Substituting this value of x into the equation $R(x) = 15x$, we find $R(1000) = 15{,}000$. Therefore, the break-even point is $(1000, 15000)$.

9. We solve the equation $R(x) = C(x)$, or $0.4x = 0.2x + 120$, obtaining $0.2x = 120$, or $x = 600$. Substituting this value of x into the equation $R(x) = 0.4x$, we find $R(600) = 240$. Therefore, the break-even point is $(600, 240)$.

11. a.

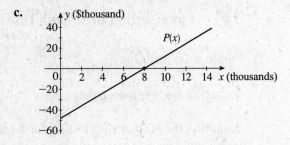

b. We solve the equation $R(x) = C(x)$ or $14x = 8x + 48{,}000$, obtaining $6x = 48{,}000$, so $x = 8000$. Substituting this value of x into the equation $R(x) = 14x$, we find $R(8000) = 14(8000) = 112{,}000$. Therefore, the break-even point is $(8000, 112000)$.

d. $P(x) = R(x) - C(x) = 14x - 8x - 48{,}000 = 6x - 48{,}000$. The graph of the profit function crosses the x-axis when $P(x) = 0$, or $6x = 48{,}000$ and $x = 8000$. This means that the revenue is equal to the cost when 8000 units are produced and consequently the company breaks even at this point.

13. Let x denote the number of units sold. Then, the revenue function R is given by $R(x) = 9x$. Since the variable cost is 40% of the selling price and the monthly fixed costs are $50,000, the cost function C is given by $C(x) = 0.4(9x) + 50{,}000 = 3.6x + 50{,}000$. To find the break-even point, we set $R(x) = C(x)$, obtaining $9x = 3.6x + 50{,}000$, $5.4x = 50{,}000$, and $x \approx 9259$, or 9259 units. Substituting this value of x into the equation $R(x) = 9x$ gives $R(9259) = 9(9259) = 83{,}331$. Thus, for a break-even operation, the firm should manufacture 9259 bicycle pumps, resulting in a break-even revenue of $83,331.

15. a. The cost function associated with using machine I is $C_1(x) = 18{,}000 + 15x$. The cost function associated with using machine II is $C_2(x) = 15{,}000 + 20x$.

c. Comparing the cost of producing 450 units on each machine, we find $C_1(450) = 18{,}000 + 15(450) = 24{,}750$ or $24,750 on machine I, and $C_2(450) = 15{,}000 + 20(450) = 24{,}000$ or $24,000 on machine II. Therefore, machine II should be used in this case. Next, comparing the costs of producing 550 units on each machine, we find $C_1(550) = 18{,}000 + 15(550) = 26{,}250$ or $26,250 on machine I, and $C_2(550) = 15{,}000 + 20(550) = 26{,}000$, or $26,000 on machine II. Therefore, machine II should be used in this instance. Once again, we compare the cost of producing 650 units on each machine and find that $C_1(650) = 18{,}000 + 15(650) = 27{,}750$, or $27,750 on machine I and $C_2(650) = 15{,}000 + 20(650) = 28{,}000$, or $28,000 on machine II. Therefore, machine I should be used in this case.

d. We use the equation $P(x) = R(x) - C(x)$ and find $P(450) = 50(450) - 24{,}000 = -1500$, indicating a loss of $1500 when machine II is used to produce 450 units. Similarly, $P(550) = 50(550) - 26{,}000 = 1500$, indicating a profit of $1500 when machine II is used to produce 550 units. Finally, $P(650) = 50(650) - 27{,}750 = 4750$, for a profit of $4750 when machine I is used to produce 650 units.

17. We solve the two equations simultaneously, obtaining $18t + 13.4 = -12t + 88$, $30t = 74.6$, and $t \approx 2.486$, or approximately 2.5 years. So shipments of LCDs will first overtake shipments of CRTs just before mid-2003.

19. a.

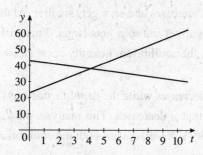

b. We solve the two equations simultaneously, obtaining $\frac{11}{3}t + 23 = -\frac{11}{9}t + 43$, $\frac{44}{9}t = 20$, and $t = 4.09$. Thus, electronic transactions first exceeded check transactions in early 2005.

21. We solve the system $4x + 3p = 59$, $5x - 6p = -14$. Solving the first equation for p, we find $p = -\frac{4}{3}x + \frac{59}{3}$. Substituting this value of p into the second equation, we have

$$5x - 6\left(-\frac{4}{3}x + \frac{59}{3}\right) = -14,\ 5x + 8x - 118 = -14,$$

$13x = 104$, and $x = 8$. Substituting this value of x into the equation $p = -\frac{4}{3}x + \frac{59}{3}$, we have

$p = -\frac{4}{3}(8) + \frac{59}{3} = \frac{27}{3} = 9$. Thus, the equilibrium quantity is 8000 units and the equilibrium price is $9.

23. We solve the system $p = -2x + 22$, $p = 3x + 12$. Substituting the first equation into the second, we find $-2x + 22 = 3x + 12$, so $5x = 10$ and $x = 2$. Substituting this value of x into the first equation, we obtain $p = -2(2) + 22 = 18$. Thus, the equilibrium quantity is 2000 units and the equilibrium price is $18.

25. Let x denote the number of DVD players produced per week, and p denote the price of each DVD player.

a. The slope of the demand curve is given by $\frac{\Delta p}{\Delta x} = -\frac{20}{250} = -\frac{2}{25}$. Using the point-slope form of the equation of a line with the point $(3000, 485)$, we have $p - 485 = -\frac{2}{25}(x - 3000)$, so $p = -\frac{2}{25}x + 240 + 485$ or $p = -0.08x + 725$.

b. From the given information, we know that the graph of the supply equation passes through the points $(0, 300)$ and $(2500, 525)$. Therefore, the slope of the supply curve is $m = \frac{525 - 300}{2500 - 0} = \frac{225}{2500} = 0.09$. Using the point-slope form of the equation of a line with the point $(0, 300)$, we find that $p - 300 = 0.09x$, so $p = 0.09x + 300$.

c. Equating the supply and demand equations, we have $-0.08x + 725 = 0.09x + 300$, so $0.17x = 425$ and $x = 2500$. Then $p = -0.08(2500) + 725 = 525$. We conclude that the equilibrium quantity is 2500 units and the equilibrium price is $525.

27. We solve the system $3x + p = 1500$, $2x - 3p = -1200$. Solving the first equation for p, we obtain $p = 1500 - 3x$. Substituting this value of p into the second equation, we obtain $2x - 3(1500 - 3x) = -1200$, so $11x = 3300$ and $x = 300$. Next, $p = 1500 - 3(300) = 600$. Thus, the equilibrium quantity is 300 and the equilibrium price is $600.

29. We solve the system of equations $p = 0.05x + 200$, $p = 0.025x + 50$, obtaining $0.025x + 50 = -0.05x + 200$, $0.075x = 150$, and so $x = 2000$. Thus, $p = -0.05(2000) + 200 = 100$, and so the equilibrium quantity is 2000 per month and the equilibrium price is $100 per unit.

31. a. We solve the system of equations $p = cx + d$, $p = ax + b$. Substituting the first into the second gives $cx + d = ax + b$, so $(c - a)x = b - d$ or $x = \frac{b - d}{c - a}$. Since $a < 0$ and $c > 0$, and $b > d > 0$, and $c - a \neq 0$, x is well-defined. Substituting this value of x into the second equation, we obtain

$$p = a\left(\frac{b - d}{c - a}\right) + b = \frac{ab - ad + bc - ab}{c - a} = \frac{bc - ad}{c - a}\ (1).$$ Therefore, the equilibrium quantity is $\frac{b - d}{c - a}$ and the equilibrium price is $\frac{bc - ad}{c - a}$.

b. If c is increased, the denominator in the expression for x increases and so x gets smaller. At the same time, the first term in equation (1) for p decreases (because a is negative) and so p gets larger. This analysis shows that if the unit price for producing the product is increased then the equilibrium quantity decreases while the equilibrium price increases.

c. If b is decreased, the numerator of the expression for x decreases while the denominator stays the same. Therefore x decreases. The expression for p also shows that p decreases. This analysis shows that if the (theoretical) upper bound for the unit price of a commodity is lowered, then both the equilibrium quantity and the equilibrium price drop.

33. True. $P(x) = R(x) - C(x) = sx - (cx + F) = (s - c)x - F$. Therefore, the firm is making a profit if $P(x) = (s - c)x - F > 0$; that is, if $x > \dfrac{F}{s - c}$ $(s \neq c)$.

35. Solving the two equations simultaneously to find the point(s) of intersection of L_1 and L_2, we obtain $m_1 x + b_1 = m_2 x + b_2$, so $(m_1 - m_2)x = b_2 - b_1$ (1).

a. If $m_1 = m_2$ and $b_2 \neq b_1$, then there is no solution for (1) and in this case L_1 and L_2 do not intersect.

b. If $m_1 \neq m_2$, then (1) can be solved (uniquely) for x, and this shows that L_1 and L_2 intersect at precisely one point.

c. If $m_1 = m_2$ and $b_1 = b_2$, then (1) is satisfied for all values of x, and this shows that L_1 and L_2 intersect at infinitely many points.

Technology Exercises   page 54

1. $(0.6, 6.2)$ **3.** $(3.8261, 0.1304)$ **5.** $(386.9091, 145.3939)$

7. a.

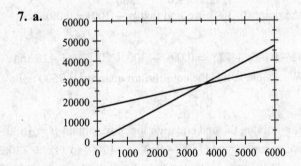

b.

From the graph, we see that the break-even point is approximately $(3548, 27997)$

c. The x-intercept is approximately 3548.

9. a. $C_1(x) = 34 + 0.18x$ and $C_2(x) = 28 + 0.22x$.

b.

c. $(150, 61)$

d. If the distance driven is less than or equal to 150 mi, rent from Acme Truck Leasing; if the distance driven is more than 150 mi, rent from Ace Truck Leasing.

11. a. $p = -\frac{1}{10}x + 284$; $p = \frac{1}{60}x + 60$

b.

The graphs intersect at roughly (1920, 92).

c. 1920/wk; $92/radio.

1.5   **The Method of Least Squares**

Problem-Solving Tips

You will find it helpful to organize data in least-squares problems using tables, as done in Examples 1–3 in the text.

Concept Questions   page 60

1. a. A scatter diagram is a graph showing the data points that describe the relationship between the two variables x and y.

b. The least squares line is the straight line that best fits a set of data points when the points are scattered about a straight line.

Exercises   page 60

1. a. We first summarize the data.

	x	y	x^2	xy
	1	4	1	4
	2	6	4	12
	3	8	9	24
	4	11	16	44
Sum	10	29	30	84

b.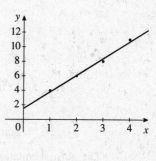

The normal equations are $4b + 10m = 29$ and $10b + 30m = 84$. Solving this system of equations, we obtain $m = 2.3$ and $b = 1.5$, so an equation is $y = 2.3x + 1.5$.

3. a. We first summarize the data.

x	y	x^2	xy
1	4.5	1	4.5
2	5	4	10
3	3	9	9
4	2	16	8
4	3.5	16	14
6	1	36	6
Sum 20	19	82	51.5

b.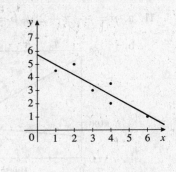

The normal equations are $6b + 20m = 19$ and $20b + 82m = 51.5$. The solutions are $m \approx -0.7717$ and $b \approx 5.7391$, so the required equation is $y = -0.772x + 5.739$.

5. a. We first summarize the data:

x	y	x^2	xy
1	3	1	3
2	5	4	10
3	5	9	15
4	7	16	28
5	8	25	40
Sum 15	28	55	96

b.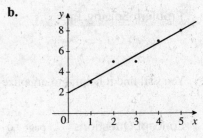

The normal equations are $55m + 15b = 96$ and $15m + 5b = 28$. Solving, we find $m = 1.2$ and $b = 2$, so the required equation is $y = 1.2x + 2$.

7. a. We first summarize the data:

x	y	x^2	xy
4	0.5	16	2
4.5	0.6	20.25	2.7
5	0.8	25	4
5.5	0.9	30.25	4.95
6	1.2	36	7.2
Sum 25	4	127.5	20.85

The normal equations are $5b + 25m = 4$ and $25b + 127.5m = 20.85$. The solutions are $m = 0.34$ and $b = -0.9$, so the required equation is $y = 0.34x - 0.9$.

b.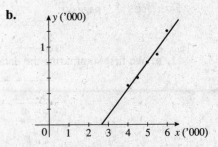

c. If $x = 6.4$, then $y = 0.34(6.4) - 0.9 = 1.276$, and so 1276 completed applications can be expected.

9. a. We first summarize the data:

. x	y	x^2	xy
1	436	1	436
2	438	4	876
3	428	9	1284
4	430	16	1720
5	426	25	2138
Sum 15	2158	55	6446

The normal equations are $5b + 15m = 2158$ and $15b + 55m = 6446$. Solving this system, we find $m = -2.8$ and $b = 440$. Thus, the equation of the least-squares line is $y = -2.8x + 440$.

b.

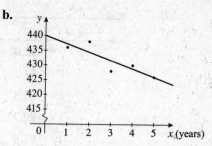

c. Two years from now, the average SAT verbal score in that area will be $y = -2.8\,(7) + 440 = 420.4$, or approximately 420.

11. a.

x	y	x^2	xy
0	154.5	0	0
1	381.8	1	381.8
2	654.5	4	1309
3	845	9	2535
Sum 6	2035.8	14	4225.8

The normal equations are $4b + 6m = 2035.8$ and $6b + 14m = 4225.8$. The solutions are $m = 234.42$ and $b = 157.32$, so the required equation is $y = 234.4x + 157.3$.

b. The projected number of Facebook users is $f\,(7) = 234.4\,(7) + 157.3 = 1798.1$, or approximately 1798.1 million.

13. a.

x	y	x^2	xy
1	20	1	20
2	24	4	48
3	26	9	78
4	28	16	112
5	32	25	160
Sum 15	130	55	418

The normal equations are $5b + 15m = 130$ and $15b + 55m = 418$. The solutions are $m = 2.8$ and $b = 17.6$, and so an equation of the line is $y = 2.8x + 17.6$.

b. When $x = 8$, $y = 2.8\,(8) + 17.6 = 40$. Hence, the state subsidy is expected to be $40 million for the eighth year.

15. a.

x	y	x^2	xy
1	26.1	1	26.1
2	27.2	4	54.4
3	28.9	9	86.7
4	31.1	16	124.4
5	32.6	25	163.0
Sum 15	145.9	55	454.6

The normal equations are $5b + 15m = 145.9$ and $15b + 55m = 454.6$. Solving this system, we find $m = 1.69$ and $b = 24.11$. Thus, the required equation is $y = f\,(x) = 1.69x + 24.11$.

b. The predicted global sales for 2014 are given by $f\,(8) = 1.69\,(8) + 24.11 = 37.63$, or 37.6 billion.

17.

x	y	x^2	xy
0	82.0	0	0
1	84.7	1	84.7
2	86.8	4	173.6
3	89.7	9	269.1
4	91.8	16	367.2
Sum 10	435	30	894.6

The normal equations are $5b + 10m = 435$ and $10b + 30m = 894.6$. The solutions are $m = 2.46$ and $b = 82.08$, so the required equation is $y = 2.46x + 82.1$.

b. The estimated number of credit union members in 2013 is $f(5) = 2.46(5) + 82.1 = 94.4$, or approximately 94.4 million.

19. a.

x	y	x^2	xy
0	29.4	0	0
1	32.2	1	32.2
2	34.8	4	69.6
3	37.7	9	113.1
4	40.4	16	161.6
Sum 10	174.5	30	376.5

The normal equations are $5b + 10m = 174.5$ and $10b + 30m = 376.5$. The solutions are $m = 2.75$ and $b = 29.4$, so $y = 2.75x + 29.4$.

b. The average rate of growth of the number of subscribers from 2006 through 2010 was 2.75 million per year.

21. a.

x	y	x^2	xy
0	6.4	0	0
1	6.8	1	6.8
2	7.1	4	14.2
3	7.4	9	22.2
4	7.6	16	30.4
Sum 10	35.3	30	73.6

The normal equations are $5b + 10m = 35.3$ and $10b + 30m = 73.6$. The solutions are $m = 0.3$ and $b = 6.46$, so the required equation is $y = 0.3x + 6.46$.

b. The rate of change is given by the slope of the least-squares line, that is, approximately \$0.3 billion/yr.

23. a. We summarize the data at right. The normal equations are $6b + 39m = 195.5$ and $39b + 271 = 1309$. The solutions are $b = 18.38$ and $m = 2.19$, so the required least-squares line is given by $y = 2.19x + 18.38$.

b. The average rate of increase is given by the slope of the least-squares line, namely \$2.19 billion/yr.

c. The revenue from overdraft fees in 2011 is $y = 2.19(11) + 18.38 = 42.47$, or approximately \$42.47 billion.

x	y	x^2	xy
4	27.5	16	110
5	29	25	145
6	31	36	186
7	34	49	238
8	36	64	288
9	38	81	342
Sum 39	195.5	271	1309

25. a.

x	y	x^2	xy
0	60	0	0
2	74	4	148
4	90	16	360
6	106	36	636
8	118	64	944
10	128	100	1280
12	150	144	1800
Sum	42	726	364	5168

The normal equations are $7b + 42m = 726$ and $42b + 364m = 5168$. The solutions are $m \approx 7.25$ and $b \approx 60.21$, so the required equation is $y = 7.25x + 60.21$.

b. $y = 7.25(11) + 60.21 = 139.96$, or $139.96 billion.

c. $7.25 billion/yr.

27. False. See Example 1 on page 56 of the text.

29. True.

Technology Exercises   page 67

1. $y = 2.3596x + 3.8639$

3. $y = -1.1948x + 3.5525$

5. a. $22.3x + 143.5$ **b.** $22.3 billion/yr   **c.** $366.5 billion

7. a. $y = 1.5857t + 6.6857$ **b.** $19.4 billion

9. a. $y = 1.751x + 7.9143$ **b.** $22 billion

CHAPTER 1   Concept Review Questions   page 68

1. ordered, abscissa (x-coordinate), ordinate (y-coordinate)

3. $\sqrt{(c-a)^2 + (d-b)^2}$

5. a. $\dfrac{y_2 - y_1}{x_2 - x_1}$ **b.** undefined   **c.** zero   **d.** positive

7. a. $y - y_1 = m(x - x_1)$, point-slope   **b.** $y = mx + b$; slope-intercept

9. $mx + b$

11. break-even

CHAPTER 1   Review Exercises   page 69

1. The distance is $d = \sqrt{(6-2)^2 + (4-1)^2} = \sqrt{4^2 + 3^2} = \sqrt{25} = 5$.

3. The distance is $d = \sqrt{[1-(-2)]^2 + [-7-(-3)]^2} = \sqrt{3^2 + (-4)^2} = \sqrt{9+16} = \sqrt{25} = 5$.

5. Substituting $x = -1$ and $y = -\frac{5}{4}$ into the left-hand side of the equation gives $6(-1) - 8\left(-\frac{5}{4}\right) - 16 = -12$. The equation is not satisfied, and so we conclude that the point $\left(-1, -\frac{5}{4}\right)$ does not lie on the line $6x - 8y - 16 = 0$.

7. An equation is $y = 4$.

9. The line passes through the points $(-2, 4)$ and $(3, 0)$, so its slope is $m = \dfrac{4 - 0}{-2 - 3} = -\dfrac{4}{5}$. An equation is $y - 0 = -\frac{4}{5}(x - 3)$, or $y = -\frac{4}{5}x + \frac{12}{5}$.

11. Writing the given equation in the form $y = -\frac{4}{3}x + 2$, we see that the slope of the given line is $-\frac{4}{3}$. Therefore, the slope of the required line is $\frac{3}{4}$ and an equation of the line is $y - 4 = \frac{3}{4}(x + 2)$, or $y = \frac{3}{4}x + \frac{11}{2}$.

13. Rewriting the given equation in slope-intercept form, we have $-5y = -3x + 6$, or $y = \frac{3}{5}x - \frac{6}{5}$. From this equation, we see that the slope of the line is $\frac{3}{5}$ and its y-intercept is $-\frac{6}{5}$.

15. The slope of the line joining the points $(-3, 4)$ and $(2, 1)$ is $m = \dfrac{1 - 4}{2 - (-3)} = -\dfrac{3}{5}$. Using the point-slope form of the equation of a line with the point $(-1, 3)$ and slope $-\frac{3}{5}$, we have $y - 3 = -\frac{3}{5}[x - (-1)]$, so $y = -\frac{3}{5}(x + 1) + 3 = -\frac{3}{5}x + \frac{12}{5}$.

17. Substituting $x = 2$ and $y = -4$ into the equation, we obtain $2(2) + k(-4) = -8$, so $-4k = -12$ and $k = 3$.

19. Setting $x = 0$ gives $y = -6$ as the y-intercept.
    Setting $y = 0$ gives $x = 8$ as the x-intercept.

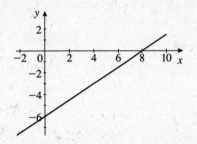

21. In 2015 (when $x = 5$), we have $S(5) = 6000(5) + 30{,}000 = 60{,}000$.

23. The slope of the line segment joining A and B is given by $m_1 = \dfrac{3 - 1}{5 - 1} = \dfrac{2}{4} = \dfrac{1}{2}$. The slope of the line segment joining B and C is $m_2 = \dfrac{5 - 3}{4 - 5} = \dfrac{2}{-1} = -2$. Since $m_1 = -1/m_2$, $\triangle ABC$ is a right triangle.

25. Let V denote the value of the building after t years.

    a. The rate of depreciation is $-\dfrac{\Delta V}{\Delta t} = \dfrac{6{,}000{,}000}{30} = 200{,}000$, or \$200,000/year.

    b. From part a, we know that the slope of the line is $-200{,}000$. Using the point-slope form of the equation of a line, we have $V - 0 = -200{,}000(t - 30)$, or $V = -200{,}000t + 6{,}000{,}000$. In the year 2020 (when $t = 10$), we have $V = -200{,}000(10) + 6{,}000{,}000 = 4{,}000{,}000$, or \$4,000,000.

27. Let x denote the number of units produced and sold.

   a. The cost function is $C(x) = 6x + 30{,}000$.

   b. The revenue function is $R(x) = 10x$.

   c. The profit function is $P(x) = R(x) - C(x) = 10x - (30{,}000 + 6x) = 4x - 30{,}000$.

   d. $P(6000) = 4(6000) - 30{,}000 = -6{,}000$, a loss of $6000; $P(8000) = 4(8000) - 30{,}000 = 2{,}000$, a profit of $2000; and $P(12{,}000) = 4(12{,}000) - 30{,}000 = 18{,}000$, a profit of $18{,}000.

29. a, b.

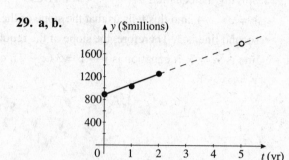

   c. The slope of L is $\dfrac{1251 - 887}{2 - 0} = 182$, so an equation of L is

$$y - 887 = 182(t - 0) \text{ or } y = 182t + 887.$$

   d. The amount consumers are projected to spend on Cyber Monday, 2014 ($t = 5$) is $182(5) + 887$, or $1.797 billion.

31. The slope of the supply curve is $\dfrac{\Delta p}{\Delta x} = \dfrac{100 - 50}{2000 - 200} = \dfrac{50}{1800} = \dfrac{1}{36}$. Using the point-slope form of the equation of a line with the point $(200, 50)$, we have $p - 50 = \frac{1}{36}(x - 200)$, so $p = \frac{1}{36}x - \frac{200}{36} + 50 = \frac{1}{36}x + \frac{1600}{36} = \frac{1}{36}x + \frac{400}{9}$, or $36p - x - 1600 = 0$.

33. a.

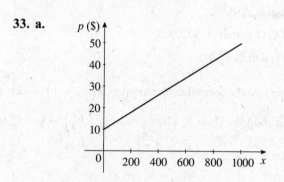

   b. The lowest price is $10 per unit.

   c. We solve the equation $0.04x + 10 = 20$, obtaining $x = 250$. Thus, the supplier will make 250 headphones available per week.

35. We solve the system $y = \frac{3}{4}x + 6$, $3x - 2y = -3$. Substituting the first equation into the second equation, we have $3x - 2\left(\frac{3}{4}x + 6\right) = -3$, $3x - \frac{3}{2}x - 12 = -3$, $\frac{3}{2}x = 9$, and $x = 6$. Substituting this value of x into the first equation, we have $y = \frac{3}{4}(6) + 6 = \frac{21}{2}$. Therefore, the point of intersection is $\left(6, \frac{21}{2}\right)$.

37. We solve the system $3x + p = 40$, $2x - p = -10$. Solving the first equation for p, we obtain $p = 40 - 3x$. Substituting this value of p into the second equation, we obtain $2x - (40 - 3x) = -10$, $5x - 40 = -10$, $5x = 30$, and $x = 6$. Next, $p = 40 - 3(6) = 40 - 18 = 22$. Thus, the equilibrium quantity is 6000 units and the equilibrium price is $22.

39. We solve the system of equations $2x + 7p - 1760 = 0$, $3x - 56p + 2680 = 0$. Solving the first equation for x yields $x = -\frac{7}{2}p + 880$, which when substituted into the second equation gives $3\left(-\frac{7}{2}p + 880\right) - 56p + 2680 = 0$, $-\frac{21}{2}p + 2640 - 56p = -2680$, $-21p + 5280 - 112p = -5360$, $-133p = -10{,}640$, and $p = \frac{10{,}640}{133} = 80$. Substituting this value of p into the expression for x, we find $x = -\frac{7}{2}(80) + 880 = 600$. Thus, the equilibrium quantity is 600 refrigerators and the equilibrium price is $80.

41. We solve the system $p = -0.02x + 40$, $p = 0.04x + 10$, obtaining $0.04x + 10 = -0.02x + 40$, $0.06x = 30$, $x = 500$, and $p = -0.02(500) + 40 = 30$. Thus, the equilibrium quantity is 500 units per week and the equilibrium price is \$30 per unit.

CHAPTER 1   **Before Moving On...**   page 71

1.

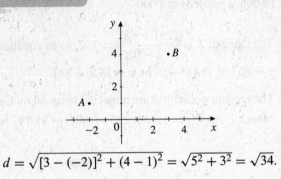

$$d = \sqrt{[3 - (-2)]^2 + (4 - 1)^2} = \sqrt{5^2 + 3^2} = \sqrt{34}.$$

2. Solving the equation $3x - y - 4 = 0$ gives $y = 3x - 4$, and this tells us that the slope of the second line is 3. Therefore, the slope of the required line is $m = 3$. It equation is $y - 1 = 3(x - 3)$, or $y = 3x - 8$.

3. The slope of the line passing through $(1, 2)$ and $(3, 5)$ is $m = \dfrac{5 - 2}{3 - 1} = \dfrac{3}{2}$. Solving $2x + 3y = 10$ gives $y = -\frac{2}{3}x + \frac{10}{3}$, and the slope of the line with this equation is $m_2 = -\frac{2}{3} = -\frac{1}{m_1}$. Thus, the two lines are perpendicular.

4. a. The unit cost is given by the coefficient of x in $C(x)$; that is, \$15.

   b. The monthly fixed cost is given by the constant term of $C(x)$; that is, \$22,000.

   c. The selling price is given by the coefficient of x in $R(x)$; that is, \$18.

5. Solving $2x - 3y = -2$ for x gives $x = \frac{3}{2}y - 1$. Substituting into the second equation gives $9\left(\frac{3}{2}y - 1\right) + 12y = 25$, so $\frac{27}{2}y - 9 + 12y = 25$, $27y - 18 + 24y = 50$, $51y = 68$, and $y = \frac{68}{51} = \frac{4}{3}$. Therefore, $x = \frac{3}{2}\left(\frac{4}{3}\right) - 1 = 1$, and so the point of intersection is $\left(1, \frac{4}{3}\right)$.

6. We solve the equation $S_1 = S_2$: $4.2 + 0.4t = 2.2 + 0.8t$, so $2 = 0.4t$ and $t = \frac{2}{0.4} = 5$. So Lowe's sales will surpass Best's in 5 years.

SYSTEMS OF LINEAR EQUATIONS AND MATRICES

2.1   Systems of Linear Equations: An Introduction

1. a. There may be no solution, a unique solution, or infinitely many solutions.

   b. There is no solution if the two lines represented by the given system of linear equations are parallel and distinct; there is a unique solution if the two lines intersect at precisely one point; there are infinitely many solutions if the two lines are parallel and coincident.

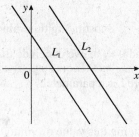
No solution

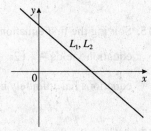
A unique solution   Infinitely many solutions

1. Solving the first equation for x, we find $x = 3y - 1$. Substituting this value of x into the second equation yields $4(3y - 1) + 3y = 11$, so $12y - 4 + 3y = 11$ and $y = 1$. Substituting this value of y into the first equation gives $x = 3(1) - 1 = 2$. Therefore, the unique solution of the system is $(2, 1)$.

3. Solving the first equation for x, we have $x = 7 - 4y$. Substituting this value of x into the second equation, we have $\frac{1}{2}(7 - 4y) + 2y = 5$, so $7 - 4y + 4y = 10$, and $7 = 10$. Clearly, this is impossible and we conclude that the system of equations has no solution.

5. Solving the first equation for x, we obtain $x = 7 - 2y$. Substituting this value of x into the second equation, we have $2(7 - 2y) - y = 4$, so $14 - 4y - y = 4$, $-5y = -10$, and $y = 2$. Then $x = 7 - 2(2) = 7 - 4 = 3$. We conclude that the solution to the system is $(3, 2)$.

7. Solving the first equation for x, we have $2x = 5y + 10$, so $x = \frac{5}{2}y + 5$. Substituting this value of x into the second equation, we have $6\left(\frac{5}{2}y + 5\right) - 15y = 30$, $15y + 30 - 15y = 30$, and $0 = 0$. This result tells us that the second equation is equivalent to the first. Thus, any ordered pair of numbers (x, y) satisfying the equation $2x - 5y = 10$ (or $6x - 15y = 30$) is a solution to the system. In particular, by assigning the value t to x, where t is any real number, we find that $y = -2 + \frac{2}{5}t$ so the ordered pair, $\left(t, \frac{2}{5}t - 2\right)$ is a solution to the system, and we conclude that the system has infinitely many solutions.

27

9. Solving the first equation for x, we obtain $4x - 5y = 14$, so $4x = 14 + 5y$, and $x = \frac{14}{4} + \frac{5}{4}y = \frac{7}{2} + \frac{5}{4}y$. Substituting this value of x into the second equation gives $2\left(\frac{7}{2} + \frac{5}{4}y\right) + 3y = -4$, so $7 + \frac{5}{2}y + 3y = -4$, $\frac{11}{2}y = -11$, and $y = -2$. Thus, $x = \frac{7}{2} + \frac{5}{4}(-2) = 1$. We conclude that the ordered pair $(1, -2)$ satisfies the given system of equations.

11. Solving the first equation for x, we obtain $2x = 3y + 6$, so $x = \frac{3}{2}y + 3$. Substituting this value of x into the second equation gives $6\left(\frac{3}{2}y + 3\right) - 9y = 12$, so $9y + 18 - 9y = 12$ and $18 = 12$. which is impossible. We conclude that the system of equations has no solution.

13. Solving the first equation for x, we obtain $-3x = -5y + 1$, so $x = \frac{5}{3}y - \frac{1}{3}$. Substituting this value of y into the second equation yields $2\left(\frac{5}{3}y - \frac{1}{3}\right) - 4y = -1$, $\frac{10}{3}y - \frac{2}{3} - 4y = -1$, $-\frac{2}{3}y = -\frac{1}{3}$, and $y = \frac{1}{2}$. Thus, $x = \frac{5}{3}\left(\frac{1}{2}\right) - \frac{1}{3} = \frac{1}{2}$, and the system has the unique solution $\left(\frac{1}{2}, \frac{1}{2}\right)$.

15. Solving the first equation for x, we obtain $3x = 6y + 2$, so $x = 2y + \frac{2}{3}$. Substituting this value of y into the second equation yields $-\frac{3}{2}\left(2y + \frac{2}{3}\right) + 3y = -1$, $-3y - 1 + 3y = -1$, and $0 = 0$. We conclude that the system of equations has infinitely many solutions of the form $\left(2t + \frac{2}{3}, t\right)$, where t is a parameter.

17. Solving the first equation for y, we obtain $y = -0.2x + 1.8$. Substituting this value of y into the second equation gives $0.4x + 0.3(-0.2x + 1.8) = 0.2$, $0.34x = -0.34$, and $x = -1$. Substituting this value of x into the first equation, we have $y = -0.2(-1) + 1.8 = 2$. Therefore, the solution is $(-1, 2)$.

19. Solving the first equation for y, we obtain $y = 2x - 3$. Substituting this value of y into the second equation yields $4x + k(2x - 3) = 4$, so $4x + 2xk - 3k = 4$, $2x(2 + k) = 4 + 3k$, and $x = \dfrac{4 + 3k}{2(2 + k)}$. Since x is not defined when the denominator of this last expression is zero, we conclude that the system has no solution when $k = -2$.

21. Solving the first equation for x in terms of y, we have $ax = by + c$ or $x = \dfrac{b}{a}y + \dfrac{c}{a}$ (provided $a \neq 0$). Substituting this value of x into the second equation gives $a\left(\dfrac{b}{a}y + \dfrac{c}{a}\right) + by = d$, $by + c + by = d$, $2by = d - c$, and $y = \dfrac{d - c}{2b}$ (provided $b \neq 0$). Substituting this into the expression for x gives $x = \dfrac{b}{a}\left(\dfrac{d - c}{2b}\right) + \dfrac{c}{a} = \dfrac{d - c}{2a} + \dfrac{c}{a} = \dfrac{c + d}{2a}$. Thus, the system has the unique solution $\left(\dfrac{c + d}{2a}, \dfrac{d - c}{2b}\right)$ if $a \neq 0$ and $b \neq 0$.

23. Let x and y denote the number of acres of corn and wheat planted, respectively. Then $x + y = 500$. Since the cost of cultivating corn is \$42/acre and that of wheat \$30/acre and Mr. Johnson has \$18,600 available for cultivation, we have $42x + 30y = 18,600$. Thus, the solution is found by solving the system of equations

$$
\begin{aligned}
x + y &= 500 \\
42x + 30y &= 18,600
\end{aligned}
$$

25. Let x denote the number of pounds of the \$8.00/lb coffee and y denote the number of pounds of the \$9/lb coffee. Then $x + y = 100$. Since the blended coffee sells for \$8.60/lb, we know that the blended mixture is worth $8.60\,(100) = \$860$. Therefore, $8x + 9y = 860$. Thus, the solution is found by solving the system of equations

$$\begin{aligned} x + y &= 1000 \\ 8x + 9y &= 860 \end{aligned}$$

27. Let x denote the number of children who ride the bus during the morning shift and y the number of adults who ride the bus during the morning shift. Then $x + y = 1000$. Since the total fare collected is \$1300, we have $0.5x + 1.5y = 1300$. Thus, the solution to the problem can be found by solving the system of equations

$$\begin{aligned} x + y &= 1000 \\ 0.5x + 1.5y &= 1300 \end{aligned}$$

29. Let x and y denote the costs of the ball and the bat, respectively. Then

$$\begin{aligned} x + y &= 110 \\ y - x &= 100 \end{aligned} \quad \text{or} \quad \begin{aligned} x + y &= 110 \\ -x + y &= 100 \end{aligned}$$

31. Let x be the amount of money invested at 6% in a savings account, y the amount of money invested at 8% in mutual funds, and z the amount of money invested at 12% in bonds. Since the total interest was \$21,600, we have $0.06x + 0.08y + 0.12z = 21{,}600$. Also, since the amount of Sid's investment in bonds is twice the amount of the investment in the savings account, we have $z = 2x$. Finally, the interest earned from his investment in bonds was equal to the interest earned on his money mutual funds, so $0.08y = 0.12z$. Thus, the solution to the problem can be found by solving the system of equations

$$\begin{aligned} 0.06x + 0.08y + 0.12z &= 21{,}600 \\ 2x - z &= 0 \\ 0.08y - 0.12z &= 0 \end{aligned}$$

33. The percentages must add up to 100%, so

$$\begin{aligned} x + y + z &= 100 \\ x + y \phantom{{}+ z} &= 67 \\ x \phantom{{}+ y} - z &= 17 \end{aligned}$$

35. Let x, y, and z denote the number of 100-lb. bags of grade A, grade B, and grade C fertilizers to be produced. The amount of nitrogen required is $18x + 20y + 24z$, and this must be equal to 26,400, so we have $18x + 20y + 24z = 26{,}400$. Similarly, the constraints on the use of phosphate and potassium lead to the equations $4x + 4y + 3z = 4900$ and $5x + 4y + 6z = 6200$, respectively. Thus we have the problem of finding the solution to the system

$$\begin{aligned} 18x + 20y + 24z &= 26{,}400 \quad &\text{(nitrogen)} \\ 4x + 4y + 3z &= 4900 \quad &\text{(phosphate)} \\ 5x + 4y + 6z &= 6200 \quad &\text{(potassium).} \end{aligned}$$

37. Let x, y, and z denote the number of compact, intermediate, and full-size cars to be purchased, respectively. The cost incurred in buying the specified number of cars is $18{,}000x + 27{,}000y + 36{,}000z$. Since the budget is \$2.25 million, we have the system

$$\begin{aligned} 18{,}000x + 27{,}000y + 36{,}000z &= 2{,}250{,}000 \\ x - 2y \phantom{{}+ z} &= 0 \\ x + y + z &= 100 \end{aligned}$$

39. Let x be the number of ounces of Food I used in the meal, y the number of ounces of Food II used in the meal, and z the number of ounces of Food III used in the meal. Since 100% of the daily requirement of proteins, carbohydrates, and iron is to be met by this meal, we have the system of linear equations

$$\begin{aligned} 10x + 6y + 8z &= 100 \\ 10x + 12y + 6z &= 100 \\ 5x + 4y + 12z &= 100 \end{aligned}$$

41. Let x, y, and z denote the numbers of front orchestra, rear orchestra, and front balcony seats sold for this performance, respectively. Then we have

$$\begin{aligned} x + y + z &= 1000 &&\text{(tickets sold total 1000)} \\ 80x + 60y + 50z &= 62{,}800 &&\text{(total revenue)} \\ x + y - 2z &= 400 &&\text{(relationship among different types of tickets)} \end{aligned}$$

43. Let x, y, and z denote the numbers of days spent in London, Paris, and Rome, respectively. Then we have

$$\begin{aligned} 280x + 330y + 260z &= 4060 &&\text{(hotel bills)} \\ 130x + 140y + 110z &= 1800 &&\text{(meals)} \\ x - y - z &= 0 &&\text{(since } x = y + z\text{)} \end{aligned}$$

45. True. In fact it has exactly one solution. Suppose the system is $y = m_1 x + b_1$, $y = m_2 x + b_2$, with $m_1 \neq m_2$. Then subtracting, we obtain $0 = (m_2 - m_1)x + (b_2 - b_1)$. Therefore, $x = \dfrac{b_1 - b_2}{m_2 - m_1}$ and

$$y = m_1 \left(\frac{b_1 - b_2}{m_2 - m_1} \right) + b_1 = \frac{m_1 b_1 - m_1 b_2 + m_2 b_1 - m_1 b_1}{m_2 - m_1} = \frac{m_2 b_1 - m_1 b_2}{m_2 - m_1}.$$

47. False. If all three lines are parallel and coincide, then the system has infinitely many solutions corresponding to all points on the (common) line.

2.2   Systems of Linear Equations: Unique Solutions

Problem-Solving Tips

When you come across new notation, make sure that you understand that notation. If you can't express the notation verbally, you haven't yet grasped its use. For example, in this section we introduced the notation $R_i \leftrightarrow R_j$. This notation tells us to interchange row i with row j.

1. Make sure you are familiar with the three row operations $R_i \leftrightarrow R_j$, $c R_i$, and $R_i + a R_j$.

2. Before writing the augmented matrix, make sure that the variables in all of the equations are on the left and the constants are on the right side of the equality symbol.

3. The last step of the Gauss-Jordan elimination method states that the matrix must be in row-reduced form. This means that the following must hold:

   a. Any row consisting entirely of zeros lies below any row with a nonzero entry.

   b. The first nonzero entry in each row is a 1.

   c. The leading 1 in any row lies to the right of any leading 1 in a row above that row.

   d. All columns containing a leading 1 are unit columns.

Concept Questions page 93

1. a. The two systems are equivalent to each other if they have precisely the same solutions.

  b. i. Interchange row i with row j.

   ii. Replace row i with c times row i.

   iii. Replace row i with the sum of row i and a times row j.

3. a. It lies below any other row having nonzero entries.

  b. It is a 1.

  c. The leading 1 in the lower row lies to the right of the leading 1 in the upper row.

  d. They are all 0.

Exercises page 93

1. $\begin{bmatrix} 2 & -3 & | & 7 \\ 3 & 1 & | & 4 \end{bmatrix}$

3. $\begin{bmatrix} 0 & -1 & 2 & | & 5 \\ 2 & 2 & -8 & | & 4 \\ 0 & 3 & 4 & | & 0 \end{bmatrix}$

5. $3x + 2y = -4$
  $x - y = 5$

7. $x + 3y + 2z = 4$
  $2x \qquad\quad = 5$
  $3x - 3y + 2z = 6$

9. Yes. Conditions 1–4 are satisfied (see page 86 of the text).

11. No. Condition 3 is violated. The first nonzero entry in the second row does not lie to the right of the first nonzero entry (1) in the first row.

13. Yes. Conditions 1–4 are satisfied.

15. No. Condition 2 and consequently condition 4 are not satisfied. The first nonzero entry in the last row is not a 1 and the column containing that entry does not have zeros elsewhere.

17. No. Condition 1 is violated. The first row consists entirely of zeros and it lies above row 2.

19. $\begin{bmatrix} ① & 3 & | & 4 \\ 2 & 4 & | & 6 \end{bmatrix} \xrightarrow{R_2 - 2R_1} \begin{bmatrix} 1 & 3 & | & 4 \\ 0 & -2 & | & -2 \end{bmatrix}$

21. $\begin{bmatrix} \boxed{-1} & 2 & | & 3 \\ 6 & 8 & | & 2 \end{bmatrix} \xrightarrow{-R_1} \begin{bmatrix} 1 & -2 & | & -3 \\ 6 & 8 & | & 2 \end{bmatrix} \xrightarrow{R_2 - 6R_1} \begin{bmatrix} 1 & -2 & | & -3 \\ 0 & 20 & | & 20 \end{bmatrix}$

23. $\begin{bmatrix} ② & 4 & 6 & | & 12 \\ 2 & 3 & 1 & | & 5 \\ 3 & -1 & 2 & | & 4 \end{bmatrix} \xrightarrow{\frac{1}{2}R_1} \begin{bmatrix} 1 & 2 & 3 & | & 6 \\ 2 & 3 & 1 & | & 5 \\ 3 & -1 & 2 & | & 4 \end{bmatrix} \xrightarrow[R_3 - 3R_1]{R_2 - 2R_1} \begin{bmatrix} 1 & 2 & 3 & | & 6 \\ 0 & -1 & -5 & | & -7 \\ 0 & -7 & -7 & | & -14 \end{bmatrix}$

25. $\begin{bmatrix} 0 & 1 & 3 & | & 4 \\ 2 & 4 & ① & | & 3 \\ 5 & 6 & 2 & | & -4 \end{bmatrix} \xrightarrow[R_3 - 2R_2]{R_1 - 3R_2} \begin{bmatrix} -6 & -11 & 0 & | & -5 \\ 2 & 4 & 1 & | & 3 \\ 1 & -2 & 0 & | & -10 \end{bmatrix}$

27. $\begin{bmatrix} 3 & 9 & | & 6 \\ 2 & 1 & | & 4 \end{bmatrix} \xrightarrow{\frac{1}{3}R_1} \begin{bmatrix} 1 & 3 & | & 2 \\ 2 & 1 & | & 4 \end{bmatrix} \xrightarrow{R_2 - 2R_1} \begin{bmatrix} 1 & 3 & | & 2 \\ 0 & -5 & | & 0 \end{bmatrix} \xrightarrow{-\frac{1}{5}R_2} \begin{bmatrix} 1 & 3 & | & 2 \\ 0 & 1 & | & 0 \end{bmatrix} \xrightarrow{R_1 - 3R_2} \begin{bmatrix} 1 & 0 & | & 2 \\ 0 & 1 & | & 0 \end{bmatrix}$

29. $\begin{bmatrix} 1 & 3 & 1 & | & 3 \\ 3 & 8 & 3 & | & 7 \\ 2 & -3 & 1 & | & -10 \end{bmatrix} \xrightarrow[R_3 - 2R_1]{R_2 - 3R_1} \begin{bmatrix} 1 & 3 & 1 & | & 3 \\ 0 & -1 & 0 & | & -2 \\ 0 & -9 & -1 & | & -16 \end{bmatrix} \xrightarrow{-R_2} \begin{bmatrix} 1 & 3 & 1 & | & 3 \\ 0 & 1 & 0 & | & 2 \\ 0 & -9 & -1 & | & -16 \end{bmatrix} \xrightarrow[R_3 + 9R_2]{R_1 - 3R_2}$

$\begin{bmatrix} 1 & 0 & 1 & | & -3 \\ 0 & 1 & 0 & | & 2 \\ 0 & 0 & -1 & | & 2 \end{bmatrix} \xrightarrow[-R_3]{R_1 + R_3} \begin{bmatrix} 1 & 0 & 0 & | & -1 \\ 0 & 1 & 0 & | & 2 \\ 0 & 0 & 1 & | & -2 \end{bmatrix}$

31. The augmented matrix is equivalent to the system of linear equations $3x + 9y = 6$, $2x + y = 4$. The ordered pair $(2, 0)$ is the solution to the system.

33. The augmented matrix is equivalent to the system of linear equations $x + 3y + z = 3$, $3x + 8y + 3z = 7$,

$2x - 3y + z = -10$. Reading off the solution from the last augmented matrix, $\begin{bmatrix} 1 & 0 & 0 & | & -1 \\ 0 & 1 & 0 & | & 2 \\ 0 & 0 & 1 & | & -2 \end{bmatrix}$, which is in

row-reduced form, we have $x = -1$, $y = 2$, and $z = -2$.

35. Using the Gauss-Jordan elimination method, we have

$\begin{bmatrix} 1 & 1 & | & 3 \\ 2 & -1 & | & 3 \end{bmatrix} \xrightarrow{R_2 - 2R_1} \begin{bmatrix} 1 & 1 & | & 3 \\ 0 & -3 & | & -3 \end{bmatrix} \xrightarrow{\frac{1}{3}R_2} \begin{bmatrix} 1 & 1 & | & 3 \\ 0 & 1 & | & 1 \end{bmatrix} \xrightarrow{R_1 - R_2} \begin{bmatrix} 1 & 0 & | & 2 \\ 0 & 1 & | & 1 \end{bmatrix}$. The solution is $(2, 1)$.

37. Using the Gauss-Jordan elimination method, we have

$\begin{bmatrix} 1 & -2 & | & 8 \\ 3 & 4 & | & 4 \end{bmatrix} \xrightarrow{R_2 - 3R_1} \begin{bmatrix} 1 & -2 & | & 8 \\ 0 & 10 & | & -20 \end{bmatrix} \xrightarrow{\frac{1}{10}R_2} \begin{bmatrix} 1 & -2 & | & 8 \\ 0 & 1 & | & -2 \end{bmatrix} \xrightarrow{R_1 + 2R_2} \begin{bmatrix} 1 & 0 & | & 4 \\ 0 & 1 & | & -2 \end{bmatrix}$.

The solution is $(4, -2)$.

39. Using the Gauss-Jordan elimination method, we have

$\begin{bmatrix} 2 & -3 & | & -8 \\ 4 & 1 & | & -2 \end{bmatrix} \xrightarrow{\frac{1}{2}R_1} \begin{bmatrix} 1 & -\frac{3}{2} & | & -4 \\ 4 & 1 & | & -2 \end{bmatrix} \xrightarrow{R_2 - 4R_1} \begin{bmatrix} 1 & -\frac{3}{2} & | & -4 \\ 0 & 7 & | & 14 \end{bmatrix} \xrightarrow{\frac{1}{7}R_2} \begin{bmatrix} 1 & -\frac{3}{2} & | & -4 \\ 0 & 1 & | & 2 \end{bmatrix} \xrightarrow{R_1 + \frac{3}{2}R_2} \begin{bmatrix} 1 & 0 & | & -1 \\ 0 & 1 & | & 2 \end{bmatrix}$.

The solution is $(-1, 2)$.

41. Using the Gauss-Jordan elimination method, we have

$\begin{bmatrix} 6 & 8 & | & 15 \\ 2 & -4 & | & -5 \end{bmatrix} \xrightarrow{\frac{1}{6}R_1} \begin{bmatrix} 1 & \frac{4}{3} & | & \frac{5}{2} \\ 2 & -4 & | & -5 \end{bmatrix} \xrightarrow{R_2 - 2R_1} \begin{bmatrix} 1 & \frac{4}{3} & | & \frac{5}{2} \\ 0 & -\frac{20}{3} & | & -10 \end{bmatrix} \xrightarrow{-\frac{3}{20}R_2} \begin{bmatrix} 1 & \frac{4}{3} & | & \frac{5}{2} \\ 0 & 1 & | & \frac{3}{2} \end{bmatrix} \xrightarrow{R_1 - \frac{4}{3}R_2} \begin{bmatrix} 1 & 0 & | & \frac{1}{2} \\ 0 & 1 & | & \frac{3}{2} \end{bmatrix}$.

The solution is $\left(\frac{1}{2}, \frac{3}{2}\right)$.

43. Using the Gauss-Jordan elimination method, we have

$\begin{bmatrix} 3 & -2 & | & 1 \\ 2 & 4 & | & 2 \end{bmatrix} \xrightarrow{\frac{1}{3}R_1} \begin{bmatrix} 1 & -\frac{2}{3} & | & \frac{1}{3} \\ 2 & 4 & | & 2 \end{bmatrix} \xrightarrow{R_2 - 2R_1} \begin{bmatrix} 1 & -\frac{2}{3} & | & \frac{1}{3} \\ 0 & \frac{16}{3} & | & \frac{4}{3} \end{bmatrix} \xrightarrow{\frac{3}{16}R_2} \begin{bmatrix} 1 & -\frac{2}{3} & | & \frac{1}{3} \\ 0 & 1 & | & \frac{1}{4} \end{bmatrix} \xrightarrow{R_1 + \frac{2}{3}R_2} \begin{bmatrix} 1 & 0 & | & \frac{1}{2} \\ 0 & 1 & | & \frac{1}{4} \end{bmatrix}$. The

solution is $\left(\frac{1}{2}, \frac{1}{4}\right)$.

45. $\begin{bmatrix} 2 & 1 & -2 & | & 4 \\ 1 & 3 & -1 & | & -3 \\ 3 & 4 & -1 & | & 7 \end{bmatrix} \xrightarrow{R_1 \leftrightarrow R_2} \begin{bmatrix} 1 & 3 & -1 & | & -3 \\ 2 & 1 & -2 & | & 4 \\ 3 & 4 & -1 & | & 7 \end{bmatrix} \xrightarrow[R_3 - 3R_1]{R_2 - 2R_1} \begin{bmatrix} 1 & 3 & -1 & | & -3 \\ 0 & -5 & 0 & | & 10 \\ 0 & -5 & 2 & | & 16 \end{bmatrix} \xrightarrow{-\frac{1}{5}R_2}$

$\begin{bmatrix} 1 & 3 & -1 & | & -3 \\ 0 & 1 & 0 & | & -2 \\ 0 & -5 & 2 & | & 16 \end{bmatrix} \xrightarrow[R_3 + 5R_2]{R_1 - 3R_2} \begin{bmatrix} 1 & 0 & -1 & | & 3 \\ 0 & 1 & 0 & | & -2 \\ 0 & 0 & 2 & | & 6 \end{bmatrix} \xrightarrow{\frac{1}{2}R_3} \begin{bmatrix} 1 & 0 & -1 & | & 3 \\ 0 & 1 & 0 & | & -2 \\ 0 & 0 & 1 & | & 3 \end{bmatrix} \xrightarrow{R_1 + R_3} \begin{bmatrix} 1 & 0 & 0 & | & 6 \\ 0 & 1 & 0 & | & -2 \\ 0 & 0 & 1 & | & 3 \end{bmatrix}.$

The solution is $(6, -2, 3)$.

47. $\begin{bmatrix} 2 & 2 & 1 & | & 9 \\ 1 & 0 & 1 & | & 4 \\ 0 & 4 & -3 & | & 17 \end{bmatrix} \xrightarrow{R_1 \leftrightarrow R_2} \begin{bmatrix} 1 & 0 & 1 & | & 4 \\ 2 & 2 & 1 & | & 9 \\ 0 & 4 & -3 & | & 17 \end{bmatrix} \xrightarrow{R_2 - 2R_1} \begin{bmatrix} 1 & 0 & 1 & | & 4 \\ 0 & 2 & -1 & | & 1 \\ 0 & 4 & -3 & | & 17 \end{bmatrix} \xrightarrow{\frac{1}{2}R_2} \begin{bmatrix} 1 & 0 & 1 & | & 4 \\ 0 & 1 & -\frac{1}{2} & | & \frac{1}{2} \\ 0 & 4 & -3 & | & 17 \end{bmatrix} \xrightarrow{R_3 - 4R_2}$

$\begin{bmatrix} 1 & 0 & 1 & | & 4 \\ 0 & 1 & -\frac{1}{2} & | & \frac{1}{2} \\ 0 & 0 & -1 & | & 15 \end{bmatrix} \xrightarrow{-R_3} \begin{bmatrix} 1 & 0 & 1 & | & 4 \\ 0 & 1 & -\frac{1}{2} & | & \frac{1}{2} \\ 0 & 0 & 1 & | & -15 \end{bmatrix} \xrightarrow[R_2 + \frac{1}{2}R_3]{R_1 - R_3} \begin{bmatrix} 1 & 0 & 0 & | & 19 \\ 0 & 1 & 0 & | & -7 \\ 0 & 0 & 1 & | & -15 \end{bmatrix}.$ The solution is $(19, -7, -15)$.

49. $\begin{bmatrix} 0 & -1 & 1 & | & 2 \\ 4 & -3 & 2 & | & 16 \\ 3 & 2 & 1 & | & 11 \end{bmatrix} \xrightarrow{R_1 \leftrightarrow R_2} \begin{bmatrix} 4 & -3 & 2 & | & 16 \\ 0 & -1 & 1 & | & 2 \\ 3 & 2 & 1 & | & 11 \end{bmatrix} \xrightarrow{R_1 - R_3} \begin{bmatrix} 1 & -5 & 1 & | & 5 \\ 0 & -1 & 1 & | & 2 \\ 3 & 2 & 1 & | & 11 \end{bmatrix} \xrightarrow[R_3 - 3R_1]{-R_2} \begin{bmatrix} 1 & -5 & 1 & | & 5 \\ 0 & 1 & -1 & | & -2 \\ 0 & 17 & -2 & | & -4 \end{bmatrix} \xrightarrow[R_3 - 17R_2]{R_1 + 5R_2}$

$\begin{bmatrix} 1 & 0 & -4 & | & -5 \\ 0 & 1 & -1 & | & -2 \\ 0 & 0 & 15 & | & 30 \end{bmatrix} \xrightarrow{\frac{1}{15}R_3} \begin{bmatrix} 1 & 0 & -4 & | & -5 \\ 0 & 1 & -1 & | & -2 \\ 0 & 0 & 1 & | & 2 \end{bmatrix} \xrightarrow[R_2 + R_3]{R_1 + 4R_3} \begin{bmatrix} 1 & 0 & 0 & | & 3 \\ 0 & 1 & 0 & | & 0 \\ 0 & 0 & 1 & | & 2 \end{bmatrix}.$ The solution is $(3, 0, 2)$.

51. Using the Gauss-Jordan elimination method, we have $\begin{bmatrix} 1 & -2 & 1 & | & 6 \\ 2 & 1 & -3 & | & -3 \\ 1 & -3 & 3 & | & 10 \end{bmatrix} \xrightarrow[R_3 - R_1]{R_2 - 2R_1} \begin{bmatrix} 1 & -2 & 1 & | & 6 \\ 0 & 5 & -5 & | & -15 \\ 0 & -1 & 2 & | & 4 \end{bmatrix} \xrightarrow{\frac{1}{5}R_2}$

$\begin{bmatrix} 1 & -2 & 1 & | & 6 \\ 0 & 1 & -1 & | & -3 \\ 0 & -1 & 2 & | & 4 \end{bmatrix} \xrightarrow[R_3 + R_2]{R_1 + 2R_2} \begin{bmatrix} 1 & 0 & -1 & | & 0 \\ 0 & 1 & -1 & | & -3 \\ 0 & 0 & 1 & | & 1 \end{bmatrix} \xrightarrow[R_2 + R_3]{R_1 + R_3} \begin{bmatrix} 1 & 0 & 0 & | & 1 \\ 0 & 1 & 0 & | & -2 \\ 0 & 0 & 1 & | & 1 \end{bmatrix}.$ The solution is $(1, -2, 1)$.

53. Using the Gauss-Jordan elimination method, we have

$\begin{bmatrix} 2 & 0 & 3 & | & -1 \\ 3 & -2 & 1 & | & 9 \\ 1 & 1 & 4 & | & 4 \end{bmatrix} \xrightarrow{R_1 \leftrightarrow R_3} \begin{bmatrix} 1 & 1 & 4 & | & 4 \\ 3 & -2 & 1 & | & 9 \\ 2 & 0 & 3 & | & -1 \end{bmatrix} \xrightarrow[R_3 - 2R_1]{R_2 - 3R_1} \begin{bmatrix} 1 & 1 & 4 & | & 4 \\ 0 & -5 & -11 & | & -3 \\ 0 & -2 & -5 & | & -9 \end{bmatrix} \xrightarrow{-\frac{1}{5}R_2}$

$\begin{bmatrix} 1 & 1 & 4 & | & 4 \\ 0 & 1 & \frac{11}{5} & | & \frac{3}{5} \\ 0 & -2 & -5 & | & -9 \end{bmatrix} \xrightarrow[R_3 + 2R_2]{R_1 - R_2} \begin{bmatrix} 1 & 0 & \frac{9}{5} & | & \frac{17}{5} \\ 0 & 1 & \frac{11}{5} & | & \frac{3}{5} \\ 0 & 0 & -\frac{3}{5} & | & -\frac{39}{5} \end{bmatrix} \xrightarrow{-\frac{5}{3}R_3} \begin{bmatrix} 1 & 0 & \frac{9}{5} & | & \frac{17}{5} \\ 0 & 1 & \frac{11}{5} & | & \frac{3}{5} \\ 0 & 0 & 1 & | & 13 \end{bmatrix} \xrightarrow[R_2 - \frac{11}{5}R_3]{R_1 - \frac{9}{5}R_3} \begin{bmatrix} 1 & 0 & 0 & | & -20 \\ 0 & 1 & 0 & | & -28 \\ 0 & 0 & 1 & | & 13 \end{bmatrix}.$

The solution is $(-20, -28, 13)$.

55. Using the Gauss-Jordan elimination method, we have

$$\begin{bmatrix} 1 & -1 & 3 & | & 14 \\ 1 & 1 & 1 & | & 6 \\ -2 & -1 & 1 & | & -4 \end{bmatrix} \xrightarrow[R_3+2R_1]{R_2-R_1} \begin{bmatrix} 1 & -1 & 3 & | & 14 \\ 0 & 2 & -2 & | & -8 \\ 0 & -3 & 7 & | & 24 \end{bmatrix} \xrightarrow{\frac{1}{2}R_2} \begin{bmatrix} 1 & -1 & 3 & | & 14 \\ 0 & 1 & -1 & | & -4 \\ 0 & -3 & 7 & | & 24 \end{bmatrix} \xrightarrow[R_3+3R_2]{R_1+R_2} \begin{bmatrix} 1 & 0 & 2 & | & 10 \\ 0 & 1 & -1 & | & -4 \\ 0 & 0 & 4 & | & 12 \end{bmatrix} \xrightarrow{\frac{1}{4}R_3}$$

$$\begin{bmatrix} 1 & 0 & 2 & | & 10 \\ 0 & 1 & -1 & | & -4 \\ 0 & 0 & 1 & | & 3 \end{bmatrix} \xrightarrow[R_2+R_3]{R_1-2R_3} \begin{bmatrix} 1 & 0 & 0 & | & 4 \\ 0 & 1 & 0 & | & -1 \\ 0 & 0 & 1 & | & 3 \end{bmatrix}. \text{ The solution is } (4, -1, 3).$$

57. Using the Gauss-Jordan elimination method, we have

$$\begin{bmatrix} 4 & 5 & | & 3 \\ 3 & k & | & 10 \end{bmatrix} \xrightarrow{R_1-R_2} \begin{bmatrix} 1 & 5-k & | & -7 \\ 3 & k & | & 10 \end{bmatrix} \xrightarrow{R_2-3R_1} \begin{bmatrix} 1 & 5-k & | & -7 \\ 0 & 4k-15 & | & 31 \end{bmatrix}.$$

In order for the system to have a unique solution we must have $4k - 15 \neq 0$, so $k \neq \frac{15}{4}$. With this condition, we have

$$\begin{bmatrix} 1 & 5-k & | & -7 \\ 0 & 4k-15 & | & 31 \end{bmatrix} \xrightarrow{\left(\frac{1}{4k-15}\right)R_2} \begin{bmatrix} 1 & 5-k & | & -7 \\ 0 & 4k-15 & | & \frac{31}{4k-15} \end{bmatrix} \xrightarrow{R_1-(5-k)R_2} \begin{bmatrix} 1 & 0 & | & \frac{3k-50}{4k-15} \\ 0 & 1 & | & \frac{31}{4k-15} \end{bmatrix}$$

Thus, the required solution is $x = \dfrac{3k - 50}{4k - 15}$, $y = \dfrac{31}{4k - 15}$.

59. We wish to solve the system of equations

$$x + y = 500$$
$$42x + 30y = 18{,}600$$

where x is the number of acres of corn planted and y is the number of acres of wheat planted. Using the Gauss-Jordan elimination method, we find

$$\begin{bmatrix} 1 & 1 & | & 500 \\ 42 & 30 & | & 18{,}600 \end{bmatrix} \xrightarrow{R_2-42R_1} \begin{bmatrix} 1 & 1 & | & 500 \\ 0 & -12 & | & -2400 \end{bmatrix} \xrightarrow{-\frac{1}{12}R_2} \begin{bmatrix} 1 & 1 & | & 500 \\ 0 & 1 & | & 200 \end{bmatrix} \xrightarrow{R_1-R_2} \begin{bmatrix} 1 & 0 & | & 300 \\ 0 & 1 & | & 200 \end{bmatrix}.$$

The solution to this system of equations is $x = 300$, $y = 200$. We conclude that Jacob should plant 300 acres of corn and 200 acres of wheat.

61. Let x denote the number of pounds of \$8/lb coffee and y the number of pounds of \$9/lb coffee. Then we wish to solve the system

$$x + y = 100$$
$$8x + 9y = 860$$

Using the Gauss-Jordan elimination method, we have $\begin{bmatrix} 1 & 1 & | & 100 \\ 8 & 9 & | & 860 \end{bmatrix} \xrightarrow{R_2-8R_1} \begin{bmatrix} 1 & 1 & | & 100 \\ 0 & 1 & | & 60 \end{bmatrix} \xrightarrow{R_1-R_2}$

$\begin{bmatrix} 1 & 0 & | & 40 \\ 0 & 1 & | & 60 \end{bmatrix}$. Therefore, 40 pounds of \$8/lb coffee and 60 pounds of \$9/lb coffee should be used in the 100-lb. mixture.

63. Let x and y denote the numbers of children and adults respectively who rode the bus during the morning shift. Then the solution to the problem can be found by solving the system of equations

$$x + y = 1000$$
$$0.5x + 1.5y = 1300$$

Using the Gauss-Jordan elimination method, we have $\begin{bmatrix} 1 & 1 & | & 1000 \\ 0.5 & 1.5 & | & 1300 \end{bmatrix} \xrightarrow{R_2 - 0.5R_1} \begin{bmatrix} 1 & 1 & | & 1000 \\ 0 & 1 & | & 800 \end{bmatrix} \xrightarrow{R_1 - R_2}$

$\begin{bmatrix} 1 & 0 & | & 200 \\ 0 & 1 & | & 800 \end{bmatrix}$. We conclude that 800 adults and 200 children rode the bus during the morning shift.

65. Let x and y denote the costs of the ball and the bat, respectively. Then $x + y = 110$ and $y - x = 100$. Using the Gauss-Jordan elimination method, we have

$$\begin{bmatrix} 1 & 1 & | & 110 \\ -1 & 1 & | & 100 \end{bmatrix} \xrightarrow{R_2 + R_1} \begin{bmatrix} 1 & 1 & | & 110 \\ 0 & 2 & | & 210 \end{bmatrix} \xrightarrow{\frac{1}{2}R_2} \begin{bmatrix} 1 & 1 & | & 110 \\ 0 & 1 & | & 105 \end{bmatrix} \xrightarrow{R_1 - R_2} \begin{bmatrix} 1 & 0 & | & 5 \\ 0 & 1 & | & 105 \end{bmatrix}.$$

Thus, $x = 5$ and $y = 105$, so the ball costs \$5 and the bat costs \$105.

67. Let x, y, and z, denote the amounts of money he should invest in a savings account, mutual funds, and bonds, respectively. Then we are required to solve the system

$$0.06x + 0.08y + 0.12z = 21,600$$
$$2x - z = 0$$
$$0.08y - 0.12z = 0$$

Using the Gauss-Jordan elimination method, we find

$$\begin{bmatrix} 0.06 & 0.08 & 0.12 & | & 21,600 \\ 2 & 0 & -1 & | & 0 \\ 0 & 0.08 & -0.12 & | & 0 \end{bmatrix} \xrightarrow[\frac{1}{0.08}R_3]{\frac{1}{0.06}R_1} \begin{bmatrix} 1 & \frac{4}{3} & 2 & | & 360,000 \\ 2 & 0 & -1 & | & 0 \\ 0 & 1 & -\frac{3}{2} & | & 0 \end{bmatrix} \xrightarrow{R_2 - 2R_1} \begin{bmatrix} 1 & \frac{4}{3} & 2 & | & 360,000 \\ 0 & -\frac{8}{3} & -5 & | & -720,000 \\ 0 & 1 & -\frac{3}{2} & | & 0 \end{bmatrix} \xrightarrow{-\frac{3}{8}R_2}$$

$$\begin{bmatrix} 1 & \frac{4}{3} & 2 & | & 360,000 \\ 0 & 1 & \frac{15}{8} & | & 270,000 \\ 0 & 1 & -\frac{3}{2} & | & 0 \end{bmatrix} \xrightarrow[R_3 - R_2]{R_1 - \frac{4}{3}R_2} \begin{bmatrix} 1 & 0 & -\frac{1}{2} & | & 0 \\ 0 & 1 & \frac{15}{8} & | & 270,000 \\ 0 & 0 & -\frac{27}{8} & | & -270,000 \end{bmatrix} \xrightarrow{-\frac{8}{27}R_3} \begin{bmatrix} 1 & 0 & -\frac{1}{2} & | & 0 \\ 0 & 1 & \frac{15}{8} & | & 270,000 \\ 0 & 0 & 1 & | & 80,000 \end{bmatrix} \xrightarrow[R_2 - \frac{15}{8}R_3]{R_1 + \frac{1}{2}R_3}$$

$$\begin{bmatrix} 1 & 0 & 0 & | & 40{,}000 \\ 0 & 1 & 0 & | & 120{,}000 \\ 0 & 0 & 1 & | & 80{,}000 \end{bmatrix}.$$

Therefore, Sid should invest \$40,000 in a savings account, \$120,000 in mutual funds, and \$80,000 in bonds.

69. We need to solve the system

$$\begin{aligned} x + y + z &= 100 \\ x + y \quad\;\; &= 67 \\ x \quad\;\; - z &= 17 \end{aligned}$$

Using the Gauss-Jordan elimination method, we have

$$\begin{bmatrix} 1 & 1 & 1 & | & 100 \\ 1 & 1 & 0 & | & 67 \\ 1 & 0 & -1 & | & 17 \end{bmatrix} \xrightarrow[R_3 - R_1]{R_2 - R_1} \begin{bmatrix} 1 & 1 & 1 & | & 100 \\ 0 & 0 & -1 & | & -33 \\ 0 & -1 & -2 & | & -83 \end{bmatrix} \xrightarrow{R_2 \leftrightarrow R_3} \begin{bmatrix} 1 & 1 & 1 & | & 100 \\ 0 & -1 & -2 & | & -83 \\ 0 & 0 & -1 & | & -33 \end{bmatrix} \begin{smallmatrix} -R_2 \\ \longrightarrow \\ -R_3 \end{smallmatrix}$$

$$\begin{bmatrix} 1 & 1 & 1 & | & 100 \\ 0 & 1 & 2 & | & 83 \\ 0 & 0 & 1 & | & 33 \end{bmatrix} \xrightarrow[R_2 - 2R_3]{R_1 - R_2} \begin{bmatrix} 1 & 0 & -1 & | & 17 \\ 0 & 1 & 0 & | & 17 \\ 0 & 0 & 1 & | & 33 \end{bmatrix} \xrightarrow{R_1 + R_3} \begin{bmatrix} 1 & 0 & 0 & | & 50 \\ 0 & 1 & 0 & | & 17 \\ 0 & 0 & 1 & | & 33 \end{bmatrix}.$$

Thus, $x = 50$, $y = 17$, and $z = 33$.

71. Refer to Exercise 2.1.35 on page 81 of the text. We obtain the following augmented matrices:

$$\begin{bmatrix} 18 & 20 & 24 & | & 26{,}400 \\ 4 & 4 & 3 & | & 4900 \\ 5 & 4 & 6 & | & 6200 \end{bmatrix} \xrightarrow{R_1 \leftrightarrow R_3} \begin{bmatrix} 5 & 4 & 6 & | & 6200 \\ 4 & 4 & 3 & | & 4900 \\ 18 & 20 & 24 & | & 26{,}400 \end{bmatrix} \xrightarrow{R_1 - R_2} \begin{bmatrix} 1 & 0 & 3 & | & 1300 \\ 4 & 4 & 3 & | & 4900 \\ 18 & 20 & 24 & | & 26{,}400 \end{bmatrix} \xrightarrow[R_3 - 18R_1]{R_2 - 4R_1}$$

$$\begin{bmatrix} 1 & 0 & 3 & | & 1300 \\ 0 & 4 & -9 & | & -300 \\ 0 & 20 & -30 & | & 3000 \end{bmatrix} \xrightarrow{\frac{1}{4}R_2} \begin{bmatrix} 1 & 0 & 3 & | & 1300 \\ 0 & 1 & -\frac{9}{4} & | & -75 \\ 0 & 20 & -30 & | & 3000 \end{bmatrix} \xrightarrow{R_3 - 20R_2} \begin{bmatrix} 1 & 0 & 3 & | & 1300 \\ 0 & 1 & -\frac{9}{4} & | & -75 \\ 0 & 0 & 15 & | & 4500 \end{bmatrix} \xrightarrow{\frac{1}{15}R_3}$$

$$\begin{bmatrix} 1 & 0 & 3 & | & 1300 \\ 0 & 1 & -\frac{9}{4} & | & -75 \\ 0 & 0 & 1 & | & 300 \end{bmatrix} \xrightarrow[R_2 + \frac{9}{4}R_3]{R_1 - 3R_3} \begin{bmatrix} 1 & 0 & 0 & | & 400 \\ 0 & 1 & 0 & | & 600 \\ 0 & 0 & 1 & | & 300 \end{bmatrix}.$$ We see that $x = 400$, $y = 600$, and $z = 300$. Therefore,

Lawnco should produce 400, 600, and 300 bags of grades A, B, and C fertilizer, respectively.

73. Let x, y, and z denote the numbers of compact, intermediate, and full-size cars, respectively, to be purchased. Then the problem can be solved by solving the system

$$\begin{aligned} 18{,}000x + 27{,}000y + 36{,}000z &= 2{,}250{,}000 \\ x - \quad\;\; 2y \quad\quad\quad\;\; &= 0 \\ x + \quad\;\; y + \quad\;\; z &= 100 \end{aligned}$$

Using the Gauss-Jordan elimination method, we have

$$\begin{bmatrix} 18{,}000 & 27{,}000 & 36{,}000 & | & 2{,}250{,}000 \\ 1 & -2 & 0 & | & 0 \\ 1 & 1 & -1 & | & 100 \end{bmatrix} \xrightarrow{R_1 \leftrightarrow R_3} \begin{bmatrix} 1 & 1 & 1 & | & 100 \\ 1 & -2 & 0 & | & 0 \\ 18{,}000 & 27{,}000 & 36{,}000 & | & 2{,}250{,}000 \end{bmatrix} \xrightarrow[R_3 - 18{,}000R_1]{R_2 - R_1}$$

$$\begin{bmatrix} 1 & 1 & 1 & | & 100 \\ 0 & -3 & -1 & | & -100 \\ 0 & 9000 & 18{,}000 & | & 450{,}000 \end{bmatrix} \xrightarrow{-\frac{1}{3}R_2} \begin{bmatrix} 1 & 1 & 1 & | & 100 \\ 0 & 1 & \frac{1}{3} & | & \frac{100}{3} \\ 0 & 9000 & 18{,}000 & | & 450{,}000 \end{bmatrix} \xrightarrow[R_3 - 9000R_2]{R_1 - R_2}$$

$$\begin{bmatrix} 1 & 0 & \frac{2}{3} & | & \frac{200}{3} \\ 0 & 1 & \frac{1}{3} & | & \frac{100}{3} \\ 0 & 0 & 15{,}000 & | & 150{,}000 \end{bmatrix} \xrightarrow{\frac{1}{15{,}000}R_3} \begin{bmatrix} 1 & 0 & \frac{2}{3} & | & \frac{200}{3} \\ 0 & 1 & \frac{1}{3} & | & \frac{100}{3} \\ 0 & 0 & 1 & | & 10 \end{bmatrix} \xrightarrow[R_2 - \frac{1}{3}R_3]{R_1 - \frac{2}{3}R_3} \begin{bmatrix} 1 & 0 & 0 & | & 60 \\ 0 & 1 & 0 & | & 30 \\ 0 & 0 & 1 & | & 10 \end{bmatrix}.$$

We conclude that 60 compact cars, 30 intermediate cars, and 10 full-size cars will be purchased.

75. Let x, y, and z, represent the numbers of ounces of Foods I, II, III used in the meal, respectively. Then the problem reduces to solving the following system of linear equations:

$$10x + 6y + 8z = 100$$
$$10x + 12y + 6z = 100$$
$$5x + 4y + 12z = 100$$

Using the Gauss-Jordan elimination method, we obtain

$$\begin{bmatrix} 10 & 6 & 8 & | & 100 \\ 10 & 12 & 6 & | & 100 \\ 5 & 4 & 12 & | & 100 \end{bmatrix} \xrightarrow{\frac{1}{10}R_1} \begin{bmatrix} 1 & \frac{3}{5} & \frac{4}{5} & | & 10 \\ 10 & 12 & 6 & | & 100 \\ 5 & 4 & 12 & | & 100 \end{bmatrix} \xrightarrow[R_3 - 5R_1]{R_2 - 10R_1} \begin{bmatrix} 1 & \frac{3}{5} & \frac{4}{5} & | & 10 \\ 0 & 6 & -2 & | & 0 \\ 0 & 1 & 8 & | & 50 \end{bmatrix} \xrightarrow{\frac{1}{6}R_2} \begin{bmatrix} 1 & \frac{3}{5} & \frac{4}{5} & | & 10 \\ 0 & 1 & -\frac{1}{3} & | & 0 \\ 0 & 1 & 8 & | & 50 \end{bmatrix} \xrightarrow[R_3 - R_2]{R_1 - \frac{3}{5}R_2}$$

$$\begin{bmatrix} 1 & 0 & 1 & | & 10 \\ 0 & 1 & -\frac{1}{3} & | & 0 \\ 0 & 0 & \frac{25}{3} & | & 50 \end{bmatrix} \xrightarrow{\frac{3}{25}R_3} \begin{bmatrix} 1 & 0 & 1 & | & 10 \\ 0 & 1 & -\frac{1}{3} & | & 0 \\ 0 & 0 & 1 & | & 6 \end{bmatrix} \xrightarrow[R_2 + \frac{1}{3}R_3]{R_1 - R_3} \begin{bmatrix} 1 & 0 & 0 & | & 4 \\ 0 & 1 & 0 & | & 2 \\ 0 & 0 & 1 & | & 6 \end{bmatrix}.$$

We conclude that 4 ounces of Food I, 2 ounces of Food II, and 6 ounces of Food III should be used to prepare the meal.

77. Let x, y, and z denote the numbers of front orchestra, rear orchestra, and front balcony seats sold for this performance. Then we are required to solve the system

$$x + y + z = 1000$$
$$80x + 60y + 50z = 62{,}800$$
$$x + y - 2z = 400$$

Using the Gauss-Jordan elimination method, we find

$$\begin{bmatrix} 1 & 1 & 1 & | & 1000 \\ 80 & 60 & 50 & | & 62{,}800 \\ 1 & 1 & -2 & | & 400 \end{bmatrix} \xrightarrow[R_3 - R_1]{R_2 - 80R_1} \begin{bmatrix} 1 & 1 & 1 & | & 1000 \\ 0 & -20 & -30 & | & -17{,}200 \\ 0 & 0 & -3 & | & -600 \end{bmatrix} \xrightarrow[-\frac{1}{3}R_3]{-\frac{1}{20}R_2} \begin{bmatrix} 1 & 1 & 1 & | & 1000 \\ 0 & 1 & \frac{3}{2} & | & 860 \\ 0 & 0 & 1 & | & 200 \end{bmatrix} \xrightarrow{R_1 - R_2}$$

$$\begin{bmatrix} 1 & 0 & -\frac{1}{2} & | & 140 \\ 0 & 1 & \frac{3}{2} & | & 860 \\ 0 & 0 & 1 & | & 200 \end{bmatrix} \xrightarrow[R_2 - \frac{3}{2}R_3]{R_1 + \frac{1}{2}R_3} \begin{bmatrix} 1 & 0 & 0 & | & 240 \\ 0 & 1 & 0 & | & 560 \\ 0 & 0 & 1 & | & 200 \end{bmatrix}.$$

We conclude that tickets for 240 front orchestra seats, 560 rear orchestra seats, and 200 front balcony seats were sold.

79. Let x, y, and z denote the number of days spent in London, Paris, and Rome, respectively. We have $280x + 330y + 260z = 4060$, $130x + 140y + 110z = 1800$, and $x - y - z = 0$ (since $x = y + z$). Using the Gauss-Jordan elimination method to solve the system, we have

$$\begin{bmatrix} 280 & 330 & 260 & | & 4060 \\ 130 & 140 & 110 & | & 1800 \\ 1 & -1 & -1 & | & 0 \end{bmatrix} \xrightarrow{R_1 \leftrightarrow R_3} \begin{bmatrix} 1 & -1 & -1 & | & 0 \\ 130 & 140 & 110 & | & 1800 \\ 280 & 330 & 260 & | & 4060 \end{bmatrix} \xrightarrow[R_3 - 280R_1]{R_2 - 130R_1} \begin{bmatrix} 1 & -1 & -1 & | & 0 \\ 0 & 270 & 240 & | & 1800 \\ 0 & 610 & 540 & | & 4060 \end{bmatrix} \xrightarrow{\frac{1}{270}R_2}$$

$$\begin{bmatrix} 1 & -1 & -1 & | & 0 \\ 0 & 1 & \frac{8}{9} & | & \frac{20}{3} \\ 0 & 610 & 540 & | & 4060 \end{bmatrix} \xrightarrow[R_3 - 610R_2]{R_1 + R_2} \begin{bmatrix} 1 & 0 & -\frac{1}{9} & | & \frac{20}{3} \\ 0 & 1 & \frac{8}{9} & | & \frac{20}{3} \\ 0 & 0 & -\frac{20}{9} & | & -\frac{20}{3} \end{bmatrix} \xrightarrow{-\frac{9}{20}R_3} \begin{bmatrix} 1 & 0 & -\frac{3}{23} & | & \frac{152}{23} \\ 0 & 1 & \frac{20}{23} & | & \frac{152}{23} \\ 0 & 0 & 1 & | & 3 \end{bmatrix} \xrightarrow[R_2 - \frac{20}{23}R_1]{R_1 + \frac{3}{23}R_3} \begin{bmatrix} 1 & 0 & 0 & | & 7 \\ 0 & 1 & 0 & | & 4 \\ 0 & 0 & 1 & | & 3 \end{bmatrix}.$$

The solution is $x = 7$, $y = 4$, and $z = 3$. Therefore, he spent 7 days in London, 4 days in Paris, and 3 days in Rome.

81. False. The constant cannot be zero. The system $\begin{cases} 2x + y = 1 \\ 3x - y = 2 \end{cases}$ is not equivalent to $\begin{cases} 2x + y = 1 \\ 0(3x - y) = 0 \ (2) \end{cases}$ or

$\begin{cases} 2x + y = 1 \\ 0 = 0 \end{cases}$

Technology Exercises   page 99

1. $(3, 1, -1, 2)$ **3.** $(5, 4, -3, -4)$ **5.** $(1, -1, 2, 0, 3)$

2.3   Systems of Linear Equations: Underdetermined and Overdetermined Systems

Problem-Solving Tips

After reading Theorem 1, try to express it in your own words. While you will not usually be required to prove these theorems in this course, you should understand their results. For example, Theorem 1 helps us decide before we solve a problem what the nature of the solution may be. Two cases are described: **a.** A system has an equal or greater number of equations as variables, and **b.** A system has fewer equations than variables. In case (a), the system may have no solution, one solution, or infinitely many solutions. In case (b) the system may have no solution or infinitely many solutions.

1. A system does not have a solution if any row of the augmented matrix representing the system has all zeros to the left of the vertical line and a nonzero entry to the right of the line.

2. If $(t, 1 - t, t)$, where t is a parameter, is a solution of a linear system of equations, then the system has infinitely many solutions, because t can be any real number. For example, if $t = 2$ then the solution is $(2, -1, 2)$, and if $t = -2$, then the solution is $(-2, 3, -2)$.

Concept Questions   page 106

1. There may be no solution, a unique solution, or infinitely many solutions.

3. No

Exercises   page 106

1. a. The system has one solution.

b. The solution is $(3, -1, 2)$.

3. a. The system has one solution.

b. The solution is $(2, 5)$.

5. a. The system has no solution. The last row contains all zeros to the left of the vertical line and a nonzero number (-1) to the right.

7. a. The system has infinitely many solutions.

b. Letting $x_3 = t$, we see that the solutions are given by $(4 - t, -2, t)$, where t is a parameter.

9. a. The system has no solution.

b. The last row contains all zeros to the left of the vertical line and a nonzero number (1) to its right.

11. a. The system has infinitely many solutions.

b. Letting $x_4 = t$, we see that the solutions are given by $(4, -1, 3 - t, t)$, where t is a parameter.

13. a. The system has infinitely many solutions.

b. Letting $x_3 = s$ and $x_4 = t$, the solutions are given by $(2 - 3s, 1 + s, s, t)$, where s and t are parameters.

15. Using the Gauss-Jordan elimination method, we have

$$
\begin{bmatrix} 2 & -1 & | & 3 \\ 1 & 2 & | & 4 \\ 2 & 3 & | & 7 \end{bmatrix}
\xrightarrow{R_1 \leftrightarrow R_2}
\begin{bmatrix} 1 & 2 & | & 4 \\ 2 & -1 & | & 3 \\ 2 & 3 & | & 7 \end{bmatrix}
\xrightarrow[R_3 - 2R_1]{R_2 - 2R_1}
\begin{bmatrix} 1 & 2 & | & 4 \\ 0 & -5 & | & -5 \\ 0 & -1 & | & -1 \end{bmatrix}
\xrightarrow{-\frac{1}{5}R_2}
\begin{bmatrix} 1 & 2 & | & 4 \\ 0 & 1 & | & 1 \\ 0 & -1 & | & -1 \end{bmatrix}
\xrightarrow[R_3 + R_2]{R_1 - 2R_2}
\begin{bmatrix} 1 & 0 & | & 2 \\ 0 & 1 & | & 1 \\ 0 & 0 & | & 0 \end{bmatrix}.
$$

The solution is $(2, 1)$.

17. Using the Gauss-Jordan elimination method, we have

$$
\begin{bmatrix} 3 & -2 & | & -3 \\ 2 & 1 & | & 3 \\ 1 & -2 & | & -5 \end{bmatrix}
\xrightarrow{R_1 \leftrightarrow R_3}
\begin{bmatrix} 1 & -2 & | & -5 \\ 2 & 1 & | & 3 \\ 3 & -2 & | & -3 \end{bmatrix}
\xrightarrow[R_3 - 3R_1]{R_2 - 2R_1}
\begin{bmatrix} 1 & -2 & | & -5 \\ 0 & 5 & | & 13 \\ 0 & 4 & | & 12 \end{bmatrix}
\xrightarrow{\frac{1}{5}R_2}
\begin{bmatrix} 1 & -2 & | & -5 \\ 0 & 1 & | & \frac{13}{5} \\ 0 & 4 & | & 12 \end{bmatrix}
\xrightarrow[R_3 - 4R_2]{R_1 + 2R_2}
\begin{bmatrix} 1 & 0 & | & \frac{1}{5} \\ 0 & 1 & | & \frac{13}{5} \\ 0 & 0 & | & \frac{8}{5} \end{bmatrix}.
$$

Since the last row implies the $0 = \frac{8}{5}$, we conclude that the system of equations is inconsistent and has no solution.

19.
$$
\begin{bmatrix} 3 & -2 & | & 5 \\ -1 & 3 & | & -4 \\ 2 & -4 & | & 6 \end{bmatrix}
\xrightarrow{R_1 \leftrightarrow R_2}
\begin{bmatrix} -1 & 3 & | & -4 \\ 3 & -2 & | & 5 \\ 2 & -4 & | & 6 \end{bmatrix}
\xrightarrow{-R_1}
\begin{bmatrix} 1 & -3 & | & 4 \\ 3 & -2 & | & 5 \\ 2 & -4 & | & 6 \end{bmatrix}
\xrightarrow[R_3 - 2R_1]{R_2 - 3R_1}
\begin{bmatrix} 1 & -3 & | & 4 \\ 0 & 7 & | & -7 \\ 0 & 2 & | & -2 \end{bmatrix}
\xrightarrow{\frac{1}{7}R_2}
$$

$$
\begin{bmatrix} 1 & -3 & | & 4 \\ 0 & 1 & | & -1 \\ 0 & 2 & | & -2 \end{bmatrix}
\xrightarrow[R_3 - 2R_2]{R_1 + 3R_2}
\begin{bmatrix} 1 & 0 & | & 1 \\ 0 & 1 & | & -1 \\ 0 & 0 & | & 0 \end{bmatrix}.
$$
We conclude that the solution is $(1, -1)$.

21.
$$
\begin{bmatrix} 1 & -2 & | & 2 \\ 7 & -14 & | & 14 \\ 3 & -6 & | & 6 \end{bmatrix}
\xrightarrow[R_3 - 3R_1]{R_2 - 7R_1}
\begin{bmatrix} 1 & -2 & | & 2 \\ 0 & 0 & | & 0 \\ 0 & 0 & | & 0 \end{bmatrix}.
$$
We conclude that the infinitely many solutions are given by $(2t + 2, t)$, where t is a parameter.

23. $\begin{bmatrix} 1 & 2 & 1 & | & -2 \\ -2 & -3 & -1 & | & 1 \\ 2 & 4 & 2 & | & -4 \end{bmatrix} \xrightarrow[R_3-2R_1]{R_2+2R_1} \begin{bmatrix} 1 & 2 & 1 & | & -2 \\ 0 & 1 & 1 & | & -3 \\ 0 & 0 & 0 & | & 0 \end{bmatrix} \xrightarrow{R_1-2R_2} \begin{bmatrix} 1 & 0 & -1 & | & 4 \\ 0 & 1 & 1 & | & -3 \\ 0 & 0 & 0 & | & 0 \end{bmatrix}$. Let $x_3 = t$ and we find that

$x_1 = 4 + t$ and $x_2 = -3 - t$. The infinitely many solutions are given by $(4+t, -3-t, t)$.

25. $\begin{bmatrix} 3 & 2 & | & 4 \\ -\frac{3}{2} & -1 & | & -2 \\ 6 & 4 & | & 8 \end{bmatrix} \xrightarrow{\frac{1}{3}R_1} \begin{bmatrix} 1 & \frac{2}{3} & | & \frac{4}{3} \\ -\frac{3}{2} & -1 & | & -2 \\ 6 & 4 & | & 8 \end{bmatrix} \xrightarrow[R_3-6R_1]{R_2+\frac{3}{2}R_1} \begin{bmatrix} 1 & \frac{2}{3} & | & \frac{4}{3} \\ 0 & 0 & | & 0 \\ 0 & 0 & | & 0 \end{bmatrix}$.

We conclude that the infinitely many solutions are given by $\left(\frac{4}{3} - \frac{2}{3}t, t\right)$, where t is a parameter.

27. $\begin{bmatrix} 1 & 1 & -2 & | & -3 \\ 2 & -1 & 3 & | & 7 \\ 1 & -2 & 5 & | & 0 \end{bmatrix} \xrightarrow[R_3-R_1]{R_2-2R_1} \begin{bmatrix} 1 & 1 & -2 & | & -3 \\ 0 & -3 & 7 & | & 13 \\ 0 & -3 & 7 & | & 3 \end{bmatrix} \xrightarrow{-\frac{1}{3}R_2} \begin{bmatrix} 1 & 1 & -2 & | & -3 \\ 0 & 1 & -\frac{7}{3} & | & -\frac{13}{3} \\ 0 & -3 & 7 & | & 3 \end{bmatrix} \xrightarrow[R_3+3R_2]{R_1-R_2} \begin{bmatrix} 1 & 0 & \frac{1}{3} & | & \frac{4}{3} \\ 0 & 1 & -\frac{7}{3} & | & -\frac{13}{3} \\ 0 & 0 & 0 & | & -10 \end{bmatrix}$.

The last row implies that $0 = -10$, which is impossible. We conclude that the system of equations is inconsistent and has no solution.

29. $\begin{bmatrix} 1 & -2 & 3 & | & 4 \\ 2 & 3 & -1 & | & 2 \\ 1 & 2 & -3 & | & -6 \end{bmatrix} \xrightarrow[R_3-R_1]{R_2-2R_1} \begin{bmatrix} 1 & -2 & 3 & | & 4 \\ 0 & 7 & -7 & | & -6 \\ 0 & 4 & -6 & | & -10 \end{bmatrix} \xrightarrow{\frac{1}{7}R_2} \begin{bmatrix} 1 & -2 & 3 & | & 4 \\ 0 & 1 & -1 & | & -\frac{6}{7} \\ 0 & 4 & -6 & | & -10 \end{bmatrix} \xrightarrow[R_3-4R_2]{R_1+2R_2}$

$\begin{bmatrix} 1 & 0 & 1 & | & \frac{16}{7} \\ 0 & 1 & -1 & | & -\frac{6}{7} \\ 0 & 0 & -2 & | & -\frac{46}{7} \end{bmatrix} \xrightarrow{-\frac{1}{2}R_3} \begin{bmatrix} 1 & 0 & 1 & | & \frac{16}{7} \\ 0 & 1 & -1 & | & -\frac{6}{7} \\ 0 & 0 & 1 & | & \frac{23}{7} \end{bmatrix} \xrightarrow[R_2+R_3]{R_1-R_3} \begin{bmatrix} 1 & 0 & 0 & | & -1 \\ 0 & 1 & 0 & | & \frac{17}{7} \\ 0 & 0 & 1 & | & \frac{23}{7} \end{bmatrix}$.

We conclude that the solution is $\left(-1, \frac{17}{7}, \frac{23}{7}\right)$.

31. $\begin{bmatrix} 4 & 1 & -1 & | & 4 \\ 8 & 2 & -2 & | & 8 \end{bmatrix} \xrightarrow{\frac{1}{4}R_1} \begin{bmatrix} 1 & \frac{1}{4} & -\frac{1}{4} & | & 1 \\ 8 & 2 & -2 & | & 8 \end{bmatrix} \xrightarrow{R_2-8R_1} \begin{bmatrix} 1 & \frac{1}{4} & -\frac{1}{4} & | & 1 \\ 0 & 0 & 0 & | & 0 \end{bmatrix}$.

We conclude that the infinitely many solutions are given by $\left(1 - \frac{1}{4}s + \frac{1}{4}t, s, t\right)$, where s and t are parameters.

33. $\begin{bmatrix} 2 & 1 & -3 & | & 1 \\ 1 & -1 & 2 & | & 1 \\ 5 & -2 & 3 & | & 6 \end{bmatrix} \xrightarrow{R_1 \leftrightarrow R_2} \begin{bmatrix} 1 & -1 & 2 & | & 1 \\ 2 & 1 & -3 & | & 1 \\ 5 & -2 & 3 & | & 6 \end{bmatrix} \xrightarrow[R_3-5R_1]{R_2-2R_1} \begin{bmatrix} 1 & -1 & 2 & | & 1 \\ 0 & 3 & -7 & | & -1 \\ 0 & 3 & -7 & | & 1 \end{bmatrix} \xrightarrow{\frac{1}{3}R_2}$

$\begin{bmatrix} 1 & -1 & 2 & | & 1 \\ 0 & 1 & -\frac{7}{3} & | & -\frac{1}{3} \\ 0 & 3 & -7 & | & 1 \end{bmatrix} \xrightarrow[R_3-3R_2]{R_1+R_2} \begin{bmatrix} 1 & 0 & -\frac{1}{3} & | & \frac{2}{3} \\ 0 & 1 & -\frac{7}{3} & | & -\frac{1}{3} \\ 0 & 0 & 0 & | & 2 \end{bmatrix}$.

The last row implies that $0 = 2$, which is impossible. We conclude that the system of equations is inconsistent and has no solution.

35.
$$\begin{bmatrix} 1 & 2 & -1 & -4 \\ 2 & 1 & 1 & 7 \\ 1 & 3 & 2 & 7 \\ 1 & -3 & 1 & 9 \end{bmatrix} \xrightarrow[\substack{R_2-2R_1 \\ R_3-R_1 \\ R_4-R_1}]{} \begin{bmatrix} 1 & 2 & -1 & -4 \\ 0 & -3 & 3 & 15 \\ 0 & 1 & 3 & 11 \\ 0 & -5 & 2 & 13 \end{bmatrix} \xrightarrow{-\frac{1}{3}R_2} \begin{bmatrix} 1 & 2 & -1 & -4 \\ 0 & 1 & -1 & -5 \\ 0 & 1 & 3 & 11 \\ 0 & -5 & 2 & 13 \end{bmatrix} \xrightarrow[\substack{R_1-2R_2 \\ R_3-R_2 \\ R_4+5R_2}]{}$$

$$\begin{bmatrix} 1 & 0 & 1 & 6 \\ 0 & 1 & -1 & -5 \\ 0 & 0 & 4 & 16 \\ 0 & 0 & -3 & -12 \end{bmatrix} \xrightarrow{\frac{1}{4}R_3} \begin{bmatrix} 1 & 0 & 1 & 6 \\ 0 & 1 & -1 & -5 \\ 0 & 0 & 1 & 4 \\ 0 & 0 & -3 & -12 \end{bmatrix} \xrightarrow[\substack{R_1-R_3 \\ R_2+R_3 \\ R_4+3R_3}]{} \begin{bmatrix} 1 & 0 & 0 & 2 \\ 0 & 1 & 0 & -1 \\ 0 & 0 & 1 & 4 \\ 0 & 0 & 0 & 0 \end{bmatrix}.$$

We conclude that the solution of the system is $(2, -1, 4)$.

37. Let x, y, and z represent the numbers of compact, mid-sized, and full-size cars, respectively, to be purchased. Then the problem can be solved by solving the system

$$x + y + z = 60$$
$$18,000x + 28,800y + 39,600z = 1,512,000$$

Using the Gauss-Jordan elimination method, we have

$$\begin{bmatrix} 1 & 1 & 1 & 60 \\ 18,000 & 28,800 & 39,600 & 1,512,000 \end{bmatrix} \xrightarrow{R_2-18,000R_1} \begin{bmatrix} 1 & 1 & 1 & 60 \\ 0 & 1080 & 21,600 & 432,000 \end{bmatrix} \xrightarrow{\frac{1}{1080}R_2}$$

$$\begin{bmatrix} 1 & 1 & 1 & 60 \\ 0 & 1 & 2 & 40 \end{bmatrix} \xrightarrow{R_1-R_2} \begin{bmatrix} 1 & 0 & -1 & 20 \\ 0 & 1 & 2 & 40 \end{bmatrix}.$$ We conclude that the solution is $(20 + z, 40 - 2z, z)$. Letting $z = 5$,

we see that one possible solution is $(25, 30, 5)$; that is Hartman should buy 25 compact, 30 mid-size, and 5 full-size cars. Letting $z = 10$, we see that another possible solution is $(30, 20, 10)$; that is, 30 compact, 20 mid-size, and 10 full-size cars.

39. Let x, y, and z denote the numbers of ounces of Foods I, II, and III, respectively, that the dietician includes in the meal. Then the problem can be solved by solving the system

$$400x + 1200y + 800z = 8800$$
$$110x + 570y + 340z = 2160$$
$$90x + 30y + 60z = 1020$$

Using the Gauss-Jordan elimination method, we have

$$\begin{bmatrix} 400 & 1200 & 800 & 8800 \\ 110 & 570 & 340 & 2160 \\ 90 & 30 & 60 & 1020 \end{bmatrix} \xrightarrow{\frac{1}{400}R_1} \begin{bmatrix} 1 & 3 & 2 & 22 \\ 110 & 570 & 340 & 2160 \\ 90 & 30 & 60 & 1020 \end{bmatrix} \xrightarrow[\substack{R_2-110R_1 \\ R_3-90R_1}]{} \begin{bmatrix} 1 & 3 & 2 & 22 \\ 0 & 240 & 120 & -260 \\ 0 & -240 & -120 & -960 \end{bmatrix} \xrightarrow{\frac{1}{240}R_2}$$

$$\begin{bmatrix} 1 & 3 & 2 & 22 \\ 0 & 1 & \frac{1}{2} & -\frac{13}{12} \\ 0 & -240 & -120 & -960 \end{bmatrix} \xrightarrow[\substack{R_1-3R_2 \\ R_3+240R_2}]{} \begin{bmatrix} 1 & 0 & \frac{1}{2} & \frac{101}{4} \\ 0 & 1 & \frac{1}{2} & -\frac{13}{12} \\ 0 & 0 & 0 & -1220 \end{bmatrix}.$$

The last row implies that $0 = -1220$, which is impossible. We conclude that the system of equations is inconsistent and has no solution—that is, the dietician cannot prepare a meal from these foods and meet the given requirements.

41. Let x, y, and z denote the amounts of money invested in stocks, bonds, and a money-market account, respectively. Then the problem can be solved by solving the system

$$\begin{aligned} x + y + z &= 100{,}000 \\ 6x + 4y + 2z &= 500{,}000 \\ x - y - 3z &= 0 \end{aligned}$$

Using the Gauss-Jordan elimination method, we have

$$\begin{bmatrix} 1 & 1 & 1 & | & 100{,}000 \\ 6 & 4 & 2 & | & 500{,}000 \\ 1 & -1 & -3 & | & 0 \end{bmatrix} \xrightarrow[R_3 - R_1]{R_2 - 6R_1} \begin{bmatrix} 1 & 1 & 1 & | & 100{,}000 \\ 0 & -2 & -4 & | & -100{,}000 \\ 0 & -2 & -4 & | & -100{,}000 \end{bmatrix} \xrightarrow{-\frac{1}{2}R_2} \begin{bmatrix} 1 & 1 & 1 & | & 100{,}000 \\ 0 & 1 & 2 & | & 50{,}000 \\ 0 & -2 & -4 & | & -100{,}000 \end{bmatrix} \xrightarrow[R_3 + 2R_2]{R_1 - R_2}$$

$$\begin{bmatrix} 1 & 0 & -1 & | & 50{,}000 \\ 0 & 1 & 2 & | & 50{,}000 \\ 0 & 0 & 0 & | & 0 \end{bmatrix}.$$

We conclude that the solution is $(50000 + z, 50000 - 2z, z)$. Therefore, one possible solution for the Garcias is to invest \$10,000 in a money-market account, \$60,000 in stocks, and \$30,000 in bonds. Another possible solution is for the Garcias to invest \$20,000 in a money-market account, \$70,000 in stocks, and \$10,000 in bonds.

43. a.

$$\begin{aligned} x_1 &&&&&& + x_6 && &= 1700 \\ x_1 &- x_2 &&&&&& + x_7 &= 700 \\ & x_2 &- x_3 &&&&&& &= 300 \\ && -x_3 &+ x_4 &&&&& &= 400 \\ &&& -x_4 &+ x_5 &&& + x_7 &= 700 \\ &&&& x_5 &+ x_6 &&& &= 1800 \end{aligned}$$

b.

$$\begin{bmatrix} 1 & 0 & 0 & 0 & 0 & 1 & 0 & | & 1700 \\ 1 & -1 & 0 & 0 & 0 & 0 & 1 & | & 700 \\ 0 & 1 & -1 & 0 & 0 & 0 & 0 & | & 300 \\ 0 & 0 & -1 & 1 & 0 & 0 & 0 & | & 400 \\ 0 & 0 & 0 & -1 & 1 & 0 & 1 & | & 700 \\ 0 & 0 & 0 & 0 & 1 & 1 & 0 & | & 1800 \end{bmatrix} \xrightarrow{R_2 - R_1} \begin{bmatrix} 1 & 0 & 0 & 0 & 0 & 1 & 0 & | & 1700 \\ 0 & -1 & 0 & 0 & 0 & -1 & 1 & | & -1000 \\ 0 & 1 & -1 & 0 & 0 & 0 & 0 & | & 300 \\ 0 & 0 & -1 & 1 & 0 & 0 & 0 & | & 400 \\ 0 & 0 & 0 & -1 & 1 & 0 & 1 & | & 700 \\ 0 & 0 & 0 & 0 & 1 & 1 & 0 & | & 1800 \end{bmatrix} \xrightarrow{-R_2}$$

$$\begin{bmatrix} 1 & 0 & 0 & 0 & 0 & 1 & 0 & | & 1700 \\ 0 & 1 & 0 & 0 & 0 & 1 & -1 & | & 1000 \\ 0 & 1 & -1 & 0 & 0 & 0 & 0 & | & 300 \\ 0 & 0 & -1 & 1 & 0 & 0 & 0 & | & 400 \\ 0 & 0 & 0 & -1 & 1 & 0 & 1 & | & 700 \\ 0 & 0 & 0 & 0 & 1 & 1 & 0 & | & 1800 \end{bmatrix} \xrightarrow{R_3 - R_2} \begin{bmatrix} 1 & 0 & 0 & 0 & 0 & 1 & 0 & | & 1700 \\ 0 & 1 & 0 & 0 & 0 & 1 & -1 & | & 1000 \\ 0 & 0 & -1 & 0 & 0 & -1 & 1 & | & -700 \\ 0 & 0 & -1 & 1 & 0 & 0 & 0 & | & 400 \\ 0 & 0 & 0 & -1 & 1 & 0 & 1 & | & 700 \\ 0 & 0 & 0 & 0 & 1 & 1 & 0 & | & 1800 \end{bmatrix} \xrightarrow{-R_3}$$

$$\begin{bmatrix} 1 & 0 & 0 & 0 & 0 & 1 & 0 & | & 1700 \\ 0 & 1 & 0 & 0 & 0 & 1 & -1 & | & 1000 \\ 0 & 0 & 1 & 0 & 0 & 1 & -1 & | & 700 \\ 0 & 0 & -1 & 1 & 0 & 0 & 0 & | & 400 \\ 0 & 0 & 0 & -1 & 1 & 0 & 1 & | & 700 \\ 0 & 0 & 0 & 0 & 1 & 1 & 0 & | & 1800 \end{bmatrix} \xrightarrow{R_4 + R_3} \begin{bmatrix} 1 & 0 & 0 & 0 & 0 & 1 & 0 & | & 1700 \\ 0 & 1 & 0 & 0 & 0 & 1 & -1 & | & 1000 \\ 0 & 0 & 1 & 0 & 0 & 1 & -1 & | & 700 \\ 0 & 0 & 0 & 1 & 0 & 1 & -1 & | & 1100 \\ 0 & 0 & 0 & -1 & 1 & 0 & 1 & | & 700 \\ 0 & 0 & 0 & 0 & 1 & 1 & 0 & | & 1800 \end{bmatrix} \xrightarrow{R_5 + R_4}$$

$$\begin{bmatrix} 1 & 0 & 0 & 0 & 0 & 1 & 0 & | & 1700 \\ 0 & 1 & 0 & 0 & 0 & 1 & -1 & | & 1000 \\ 0 & 0 & 1 & 0 & 0 & 1 & -1 & | & 700 \\ 0 & 0 & 0 & 1 & 0 & 1 & -1 & | & 1100 \\ 0 & 0 & 0 & 0 & 1 & 1 & 0 & | & 1800 \\ 0 & 0 & 0 & 0 & 1 & 1 & 0 & | & 1800 \end{bmatrix} \xrightarrow{R_6 - R_5} \begin{bmatrix} 1 & 0 & 0 & 0 & 0 & 1 & 0 & | & 1700 \\ 0 & 1 & 0 & 0 & 0 & 1 & -1 & | & 1000 \\ 0 & 0 & 1 & 0 & 0 & 1 & -1 & | & 700 \\ 0 & 0 & 0 & 1 & 0 & 1 & -1 & | & 1100 \\ 0 & 0 & 0 & 0 & 1 & 1 & 0 & | & 1800 \\ 0 & 0 & 0 & 0 & 0 & 0 & 0 & | & 0 \end{bmatrix}.$$

We conclude that the solution of the system is

$(1700 - s, 1000 - s + t, 700 - s + t, 1100 - s + t, 1800 - s, s, t)$. Two possible traffic patterns are

$(900, 1000, 700, 1100, 1000, 800, 800)$ and $(1000, 1100, 800, 1200, 1100, 700, 800)$.

c. x_6 must have at least 300 cars/hour.

45. We solve the given system using the Gauss-Jordan elimination method. We have

$$\begin{bmatrix} 2 & 3 & | & 2 \\ 1 & 4 & | & 6 \\ 5 & k & | & 2 \end{bmatrix} \xrightarrow{R_1 \leftrightarrow R_2} \begin{bmatrix} 1 & 4 & | & 6 \\ 2 & 3 & | & 2 \\ 5 & k & | & 2 \end{bmatrix} \xrightarrow[R_3 - 5R_1]{R_2 - 2R_1} \begin{bmatrix} 1 & 4 & | & 6 \\ 0 & -5 & | & -10 \\ 0 & k-20 & | & -28 \end{bmatrix} \xrightarrow{-\frac{1}{5}R_2} \begin{bmatrix} 1 & 4 & | & 6 \\ 0 & 1 & | & 2 \\ 0 & k-20 & | & -28 \end{bmatrix} \xrightarrow[R_3 - (k-20)R_2]{R_1 - 4R_2}$$

$$\begin{bmatrix} 1 & 0 & | & -2 \\ 0 & 1 & | & 2 \\ 0 & 0 & | & 12-2k \end{bmatrix}.$$

From the last matrix, we see that the system has a solution if and only if $x = -2$, $y = 2$, $12 - 2k = 0$, or $k = 6$. (All the entries in the last row of the matrix must be equal to zero.)

47. Using the Gauss-Jordan elimination method, we have

$$\begin{bmatrix} 3 & 2 & -1 & | & 8 \\ 2 & 0 & -2 & | & 4 \\ 0 & 1 & k & | & 1 \end{bmatrix} \xrightarrow{R_1 - R_2} \begin{bmatrix} 1 & 2 & 1 & | & 4 \\ 2 & 0 & -2 & | & 4 \\ 0 & 1 & k & | & 1 \end{bmatrix} \xrightarrow{R_2 - 2R_1} \begin{bmatrix} 1 & 2 & 1 & | & 4 \\ 0 & -4 & -4 & | & -4 \\ 0 & 1 & k & | & 1 \end{bmatrix} \xrightarrow{-\frac{1}{4}R_2} \begin{bmatrix} 1 & 2 & 1 & | & 4 \\ 0 & 1 & 1 & | & 1 \\ 0 & 1 & k & | & 1 \end{bmatrix} \xrightarrow[R_3 - R_2]{R_1 - 2R_2}$$

$$\begin{bmatrix} 1 & 0 & -1 & | & 2 \\ 0 & 1 & 1 & | & 1 \\ 0 & 0 & k-1 & | & 0 \end{bmatrix}.$$

In order for the system to have infinitely many solutions, we must have $k - 1 = 0$, so $k = 1$. In this case, we

have the final augmented matrix $\begin{bmatrix} 1 & 0 & -1 & | & 2 \\ 0 & 1 & 1 & | & 1 \\ 0 & 0 & 0 & | & 0 \end{bmatrix}$, from which we conclude that the infinitely many solutions are

$x = 2 + a$, $y = 1 - a$, $z - a$; that is, $(2 + a, 1 - a, a)$, where a is an arbitrary parameter.

49. Using the Gauss-Jordan elimination method, we have

$$\begin{bmatrix} 2 & -3 & 4 & | & 12 \\ 6 & -9 & k & | & 36 \end{bmatrix} \xrightarrow{\frac{1}{2}R_1} \begin{bmatrix} 1 & -\frac{3}{2} & 2 & | & 6 \\ 6 & -9 & k & | & 36 \end{bmatrix} \xrightarrow{R_2 - 6R_1} \begin{bmatrix} 1 & -\frac{3}{2} & 2 & | & 6 \\ 0 & 0 & k-12 & | & 0 \end{bmatrix}.$$ The system has a solution provided

$k - 12 = 0$; that is, $k = 12$.

51. Put $u = 1/x$ and $v = 1/y$. Then the system can be written as

$$3u - 4v = -15$$
$$5u + 6v = 13.$$

Using the Gauss-Jordan elimination method, we have

$$\begin{bmatrix} 3 & -4 & | & -15 \\ 5 & 6 & | & 13 \end{bmatrix} \xrightarrow{2R_1 - R_2} \begin{bmatrix} 1 & -14 & | & -43 \\ 5 & 6 & | & 13 \end{bmatrix} \xrightarrow{R_2 - 5R_1} \begin{bmatrix} 1 & -14 & | & -43 \\ 0 & 76 & | & 228 \end{bmatrix} \xrightarrow{\frac{1}{76}R_2} \begin{bmatrix} 1 & -14 & | & -43 \\ 0 & 1 & | & 3 \end{bmatrix} \xrightarrow{R_1 + 14R_2}$$

$$\begin{bmatrix} 1 & 0 & | & -1 \\ 0 & 1 & | & 3 \end{bmatrix}.$$ Thus, $u = 1/x = -1$ and $v = 1/y = 3$, so $x = -1$ and $y = \frac{1}{3}$.

53. False.

Technology Exercises   page 110

1. $(1 + t, 2 + t, t)$, where t is a parameter   **3.** $\left(-\frac{17}{7} + \frac{6}{7}t, 3 - t, -\frac{18}{7} + \frac{1}{7}t, t\right)$, where t is a parameter

5. No solution

2.4   Matrices

Problem-Solving Tips

1. If a matrix has size $m \times n$, then it has m rows and n columns. For example, a 4×3 matrix has 4 rows and 3 columns.

2. The sum and difference of two matrices A and B are defined only if A and B have the same size. To find the sum (difference) of A and B, we add (subtract) the corresponding entries in the two matrices.

3. To find the scalar product of a real number c and a matrix A, multiply each entry in A by c.

Concept Questions   page 117

1. a. A matrix is an ordered rectangular array of real numbers.

  b. A matrix has size (or dimension) $m \times n$ if it has m rows and n columns.

  c. A row matrix has size $1 \times n$.   **d.** A column matrix has size $m \times 1$.   **e.** A square matrix has size $n \times n$.

3. a. Matrices A and B must have the same size (numbers of rows and columns).

  b. The matrix A can have any size.

Exercises   page 118

1. The size of A is 4×4; the size of B is 4×3; the size of C is 1×5, and the size of D is 4×1.

3. These are entries of the matrix B. The entry b_{13} refers to the entry in the first row and third column and is equal to 2. Similarly, $b_{31} = 3$, and $b_{43} = 8$.

5. The column matrix is the matrix D. The transpose of the matrix D is $D^T = \begin{bmatrix} 1 & 3 & -2 & 0 \end{bmatrix}$.

7. A has size 3×2; B has size 3×2; C and D have size 3×3.

9. $A + B = \begin{bmatrix} -1 & 2 \\ 3 & -2 \\ 4 & 0 \end{bmatrix} + \begin{bmatrix} 2 & 4 \\ 3 & 1 \\ -2 & 2 \end{bmatrix} = \begin{bmatrix} 1 & 6 \\ 6 & -1 \\ 2 & 2 \end{bmatrix}$.

11. $C - D = \begin{bmatrix} 3 & -1 & 0 \\ 2 & -2 & 3 \\ 4 & 6 & 2 \end{bmatrix} - \begin{bmatrix} 2 & -2 & 4 \\ 3 & 6 & 2 \\ -2 & 3 & 1 \end{bmatrix} = \begin{bmatrix} 1 & 1 & -4 \\ -1 & -8 & 1 \\ 6 & 3 & 1 \end{bmatrix}$.

13. $\begin{bmatrix} 2 & -1 & 3 \\ 9 & 2 & 1 \end{bmatrix} + \begin{bmatrix} -1 & 2 & -1 \\ -6 & 4 & 2 \end{bmatrix} = \begin{bmatrix} 1 & 1 & 2 \\ 3 & 6 & 3 \end{bmatrix}$

15. $\begin{bmatrix} 2 & -3 & 4 & -1 \\ 3 & 1 & 0 & 0 \end{bmatrix} + \begin{bmatrix} 4 & 3 & -2 & -4 \\ 6 & 2 & 0 & -3 \end{bmatrix} = \begin{bmatrix} 6 & 0 & 2 & -5 \\ 9 & 3 & 0 & -3 \end{bmatrix}$.

17. $\begin{bmatrix} 1.2 & 4.5 & -4.2 \\ 8.2 & 6.3 & -3.2 \end{bmatrix} - \begin{bmatrix} 3.1 & 1.5 & -3.6 \\ 2.2 & -3.3 & -4.4 \end{bmatrix} = \begin{bmatrix} -1.9 & 3.0 & -0.6 \\ 6.0 & 9.6 & 1.2 \end{bmatrix}$.

19. $3 \begin{bmatrix} 1 & 1 & -3 \\ 3 & 2 & 3 \\ 7 & -1 & 6 \end{bmatrix} + 4 \begin{bmatrix} -2 & -1 & 8 \\ 4 & 2 & 2 \\ 3 & 6 & 3 \end{bmatrix} = \begin{bmatrix} 3 & 3 & -9 \\ 9 & 6 & 9 \\ 21 & -3 & 18 \end{bmatrix} + \begin{bmatrix} -8 & -4 & 32 \\ 16 & 8 & 8 \\ 12 & 24 & 12 \end{bmatrix} = \begin{bmatrix} -5 & -1 & 23 \\ 25 & 14 & 17 \\ 33 & 21 & 30 \end{bmatrix}$.

21. $\frac{1}{2} \begin{bmatrix} 1 & 0 & 0 & -4 \\ 3 & 0 & -1 & 6 \\ -2 & 1 & -4 & 2 \end{bmatrix} + \frac{4}{3} \begin{bmatrix} 3 & 0 & -1 & 4 \\ -2 & 1 & -6 & 2 \\ 8 & 2 & 0 & -2 \end{bmatrix} - \frac{1}{3} \begin{bmatrix} 3 & -9 & -1 & 0 \\ 6 & 2 & 0 & -6 \\ 0 & 1 & -3 & 1 \end{bmatrix} = \begin{bmatrix} \frac{7}{2} & 3 & -1 & \frac{10}{3} \\ -\frac{19}{6} & \frac{2}{3} & -\frac{17}{2} & \frac{23}{3} \\ \frac{29}{3} & \frac{17}{6} & -1 & -2 \end{bmatrix}$.

23. $\begin{bmatrix} 2x - 2 & 3 & 2 \\ 2 & 4 & y - 2 \\ 2z & -3 & 2 \end{bmatrix} = \begin{bmatrix} 3 & u & 2 \\ 2 & 4 & 5 \\ 4 & -3 & 2 \end{bmatrix}$.

Now by the definition of equality of matrices, $u = 3$, $2x - 2 = 3$, and $2x = 5$ (so $x = \frac{5}{2}$), $y - 2 = 5$ (so $y = 7$), and $2z = 4$ (so $z = 2$).

25. $\begin{bmatrix} 1 & x \\ 2y & -3 \end{bmatrix} - 4 \begin{bmatrix} 2 & -2 \\ 0 & 3 \end{bmatrix} = \begin{bmatrix} 3z & 10 \\ 4 & -u \end{bmatrix}$; $\begin{bmatrix} -7 & x + 8 \\ 2y & -15 \end{bmatrix} = \begin{bmatrix} 3z & 10 \\ 4 & -u \end{bmatrix}$.

Now by the definition of equality of matrices, $-u = -15$ (so $u = 15$), $x + 8 = 10$ (so $x = 2$), $2y = 4$ (so $y = 2$), and $3z = -7$ (so $z = -\frac{7}{3}$).

27. $2A + X = 3B$ implies that $X = 3B - 2A$; that is,

$$X = 3 \begin{bmatrix} 2 & -3 \\ 1 & -2 \end{bmatrix} - 2 \begin{bmatrix} -2 & 1 \\ 0 & 3 \end{bmatrix} = \begin{bmatrix} 6 & -9 \\ 3 & -6 \end{bmatrix} - \begin{bmatrix} -4 & 2 \\ 0 & 6 \end{bmatrix} = \begin{bmatrix} 10 & -11 \\ 3 & -12 \end{bmatrix}.$$

29. To verify the Commutative Law for matrix addition, let us show that $A + B = B + A$.

Now $A + B = \begin{bmatrix} 2 & -4 & 3 \\ 4 & 2 & 1 \end{bmatrix} + \begin{bmatrix} 4 & 3 & 2 \\ 1 & 0 & 4 \end{bmatrix} = \begin{bmatrix} 6 & -1 & 5 \\ 5 & 2 & 5 \end{bmatrix} = \begin{bmatrix} 4 & 3 & 2 \\ 1 & 0 & 4 \end{bmatrix} + \begin{bmatrix} 2 & -4 & 3 \\ 4 & 2 & 1 \end{bmatrix} = B + A.$

31. $(3 + 5) A = 8A = 8 \begin{bmatrix} 3 & 1 \\ 2 & 4 \\ -4 & 0 \end{bmatrix} = \begin{bmatrix} 24 & 8 \\ 16 & 32 \\ -32 & 0 \end{bmatrix} = 3 \begin{bmatrix} 3 & 1 \\ 2 & 4 \\ -4 & 0 \end{bmatrix} + 5 \begin{bmatrix} 3 & 1 \\ 2 & 4 \\ -4 & 0 \end{bmatrix} = 3A + 5A.$

33. $4(A + B) = 4 \left(\begin{bmatrix} 3 & 1 \\ 2 & 4 \\ -4 & 0 \end{bmatrix} + \begin{bmatrix} 1 & 2 \\ -1 & 0 \\ 3 & 2 \end{bmatrix} \right) = 4 \begin{bmatrix} 4 & 3 \\ 1 & 4 \\ -1 & 2 \end{bmatrix} = \begin{bmatrix} 16 & 12 \\ 4 & 16 \\ -4 & 8 \end{bmatrix}$ and

$4A + 4B = 4 \begin{bmatrix} 3 & 1 \\ 2 & 4 \\ -4 & 0 \end{bmatrix} + 4 \begin{bmatrix} 1 & 2 \\ -1 & 0 \\ 3 & 2 \end{bmatrix} = \begin{bmatrix} 16 & 12 \\ 4 & 16 \\ -4 & 8 \end{bmatrix}.$

35. $\begin{bmatrix} 3 & 2 & -1 & 5 \end{bmatrix}^T = \begin{bmatrix} 3 \\ 2 \\ -1 \\ 5 \end{bmatrix}.$

37. $\begin{bmatrix} 1 & -1 & 2 \\ 3 & 4 & 2 \\ 0 & 1 & 0 \end{bmatrix}^T = \begin{bmatrix} 1 & 3 & 0 \\ -1 & 4 & 1 \\ 2 & 2 & 0 \end{bmatrix}.$

39.

	1	2	3	4
Mr. Cross	220	215	210	205
Mr. Jones	220	210	200	195
Mr. Smith	215	205	195	190

41. $A = $

	White	Black	Hispanic
Women	82.6	80.5	91.2
Men	78.0	73.9	84.8

and $B = $

	Women	Men
White	82.6	78.0
Black	80.5	73.9
Hispanic	91.2	84.8

43. a.

	6-month	1-year	$2\frac{1}{2}$-year	5-year
Current week	0.17	0.27	0.41	0.87
Previous week	0.17	0.27	0.42	0.88
One year ago	0.22	0.34	0.52	1.15

b. $a_{12} = 0.27$, $a_{22} = 0.27$. The average yield for 1-year CDs was the same the week of January 14 as it was the previous week.

c. $a_{13} = 0.41$, $a_{23} = 0.42$. The average yield for $2\frac{1}{2}$-year CDs was 0.01% lower the week of January 14 than it was the previous week.

d. $a_{33} = 0.52$, $a_{34} = 1.15$. A year ago, the average yield for $2\frac{1}{2}$-year CDs was 0.52%, much lower than the 1.15% average yield for 5-year CDs at that time.

45. a. $A =$

	Textbooks	Fiction	Nonfiction	Reference
Hardcover	5280	1680	2320	1890
Paperback	1940	2810	1490	2070

b. $B =$

	Textbooks	Fiction	Nonfiction	Reference
Hardcover	6340	2220	1790	1980
Paperback	2050	3100	1720	2710

c. $C =$

	Textbooks	Fiction	Nonfiction	Reference
Hardcover	11,620	3900	4110	3870
Paperback	3990	5910	3210	4780

47. a.

	Large-cap	Small-cap	International	Bonds	Cash
Conservative	15	0	5	50	30
Moderately conservative	25	5	10	50	10
Moderate	35	10	15	35	5
Moderately aggressive	45	15	20	15	5
Aggressive	50	20	25	0	5

b. $a_{12} = 0$, so in the conservative portfolio, there is no investment in small-cap.

c. $a_{13} = 5$, $a_{23} = 10$, $a_{33} = 15$, $a_{43} = 20$, and $a_{53} = 25$. The more aggressive the portfolio, the larger the proportion of international investment.

d. Each row sums to 100, as expected: the investments in each portfolio must add up to 100%.

49. a.

	NY	NY Co-ops	NJ	Conn	
$A =$	3.83	3.67	3.78	3.79	30-year fixed
	3.16	2.98	3.03	3.03	15-year fixed
	2.99	2.96	2.97	2.45	Adjustable

	NY	NY Co-ops	NJ	Conn	
$B =$	3.84	3.75	3.81	3.80	30-year fixed
	3.15	2.97	3.04	3.02	15-year fixed
	2.99	2.95	2.96	2.44	Adjustable

b. $a_{12} = 3.67$ and $b_{12} = 3.75$. These give the 30-year fixed rates in the New York Co-ops for the weeks ending February 15 and 22, 2013.

c. $a_{33} = 2.97$ and $b_{33} = 2.96$. These give the adjustable rates in NJ for the weeks ending February 15 and 22, 2013.

d. $\frac{1}{4}(a_{11} + a_{12} + a_{13} + a_{14}) = \frac{1}{4}(3.83 + 3.67 + 3.78 + 3.79) = 3.7675$. Thus, the average 30-year fixed rate in New York, the New York Co-Ops, New Jersey, and Connecticut for the week ending February 15, 2013 was 3.7675%.

e. $\frac{1}{2}(a_{34} + b_{34}) = \frac{1}{2}(2.45 + 2.44) = 2.445$, so the average adjustable rate in New Jersey for the weeks ending February 15 and 22, 2013 was 2.445%.

51. $B = (1.03)\,A = 1.03 \begin{bmatrix} 340 & 360 & 380 \\ 410 & 430 & 440 \\ 620 & 660 & 700 \end{bmatrix} \begin{matrix} \text{I} \\ \text{II} \\ \text{III} \end{matrix} = \begin{matrix} & M_1 & M_2 & M_3 \\ \text{I} \\ \text{II} \\ \text{III} \end{matrix} \begin{bmatrix} 350.2 & 370.8 & 391.4 \\ 422.3 & 442.9 & 453.2 \\ 638.6 & 679.8 & 721 \end{bmatrix}$

53. True. Each element in $A + B$ is obtained by adding together the corresponding elements in A and B. Therefore, the matrix $c(A + B)$ is obtained by multiplying each element in $A + B$ by c. On the other hand, cA is obtained by multiplying each element in A by c and cB is obtained by multiplying each element in B by c, and $cA + cB$ is obtained by adding the corresponding elements in cA and cB. Thus $c(A + B) = cA + cB$.

55. False. Take $\begin{bmatrix} 1 & 2 \\ 3 & 4 \end{bmatrix}$ and $c = 2$. Then $cA = 2\begin{bmatrix} 1 & 2 \\ 3 & 4 \end{bmatrix} = \begin{bmatrix} 2 & 4 \\ 6 & 8 \end{bmatrix}$ and $(cA)^T = \begin{bmatrix} 2 & 6 \\ 4 & 8 \end{bmatrix}$. On the other hand,

$$\frac{1}{c}A^T = \frac{1}{2}\begin{bmatrix} 1 & 3 \\ 2 & 4 \end{bmatrix} = \begin{bmatrix} \frac{1}{2} & \frac{3}{2} \\ 1 & 2 \end{bmatrix} \neq (cA)^T.$$

Technology Exercises   page 125

1. $\begin{bmatrix} 15 & 38.75 & -67.5 & 33.75 \\ 51.25 & 40 & 52.5 & -38.75 \\ 21.25 & 35 & -65 & 105 \end{bmatrix}$

3. $\begin{bmatrix} -5 & 6.3 & -6.8 & 3.9 \\ 1 & 0.5 & 5.4 & -4.8 \\ 0.5 & 4.2 & -3.5 & 5.6 \end{bmatrix}$

5. $\begin{bmatrix} 16.44 & -3.65 & -3.66 & 0.63 \\ 12.77 & 10.64 & 2.58 & 0.05 \\ 5.09 & 0.28 & -10.84 & 17.64 \end{bmatrix}$

7. $\begin{bmatrix} 22.2 & -0.3 & -12 & 4.5 \\ 21.6 & 17.7 & 9 & -4.2 \\ 8.7 & 4.2 & -20.7 & 33.6 \end{bmatrix}$

2.5   Multiplication of Matrices

Problem-Solving Tips

1. The **matrix product** of two matrices A and B is defined only if the number of columns in A is equal to the number of rows in B. The **scalar product** of a real number c and a matrix A is always defined.

2. We can write a system of equations in the matrix form $AX = B$, where A is the coefficient matrix, X is the column matrix of unknowns, and B is the column matrix of constants.

Concept Questions   page 132

1. Scalar multiplication involves multiplying a matrix A by a scalar c (result: cA); whereas matrix multiplication involves the product of two matrices. Example: $3 \begin{bmatrix} 1 & 2 \\ 2 & 3 \end{bmatrix} = \begin{bmatrix} 3 & 6 \\ 6 & 9 \end{bmatrix}$ and $\begin{bmatrix} 2 & 1 \\ 3 & 0 \end{bmatrix} \begin{bmatrix} 1 & 3 & 2 \\ 1 & 4 & 3 \end{bmatrix} = \begin{bmatrix} 3 & 10 & 7 \\ 3 & 9 & 6 \end{bmatrix}$.

Exercises   page 132

1.

Size of A $\overset{\overbrace{\qquad\text{Same}\qquad}}{2 \times 3 \qquad\qquad 3 \times 5}$ Size of B

$\underbrace{\qquad (2 \times 5) \qquad}$
Size of AB

Size of B $\overset{\overbrace{\text{Not the same}}}{3 \times 5 \qquad\qquad 2 \times 3}$ Size of A

BA is undefined

3.

Size of A $\overset{\overbrace{\qquad\text{Same}\qquad}}{1 \times 7 \qquad\qquad 7 \times 1}$ Size of B

$\underbrace{\qquad (1 \times 1) \qquad}$
Size of AB

Size of B $\overset{\overbrace{\qquad\text{Same}\qquad}}{7 \times 1 \qquad\qquad 1 \times 7}$ Size of A

$\underbrace{\qquad (7 \times 7) \qquad}$
Size of BA

5. If AB and BA are defined, then $n = s$ and $m = t$.

7. $\begin{bmatrix} 1 & 2 \\ 3 & 0 \end{bmatrix} \begin{bmatrix} 1 \\ -1 \end{bmatrix} = \begin{bmatrix} -1 \\ 3 \end{bmatrix}$.

9. $\begin{bmatrix} 4 & 1 & 2 \\ -1 & 2 & 4 \end{bmatrix} \begin{bmatrix} 4 \\ 1 \\ -2 \end{bmatrix} = \begin{bmatrix} 13 \\ -10 \end{bmatrix}$.

11. $\begin{bmatrix} -1 & 2 \\ 3 & 1 \end{bmatrix} \begin{bmatrix} 2 & 4 \\ 3 & 1 \end{bmatrix} = \begin{bmatrix} 4 & -2 \\ 9 & 13 \end{bmatrix}$.

13. $\begin{bmatrix} 2 & 1 & 2 \\ 3 & 2 & 4 \end{bmatrix} \begin{bmatrix} -1 & 2 \\ 4 & 3 \\ 0 & 1 \end{bmatrix} = \begin{bmatrix} 2 & 9 \\ 5 & 16 \end{bmatrix}$.

15. $\begin{bmatrix} 0.1 & 0.9 \\ 0.2 & 0.8 \end{bmatrix} \begin{bmatrix} 1.2 & 0.4 \\ 0.5 & 2.1 \end{bmatrix} = \begin{bmatrix} 0.1\,(1.2) + 0.9\,(0.5) & 0.1\,(0.4) + 0.9\,(2.1) \\ 0.2\,(1.2) + 0.8\,(0.5) & 0.2\,(0.4) + 0.8\,(2.1) \end{bmatrix} = \begin{bmatrix} 0.57 & 1.93 \\ 0.64 & 1.76 \end{bmatrix}$.

17. $\begin{bmatrix} 6 & -3 & 0 \\ -2 & 1 & -8 \\ 4 & -4 & 9 \end{bmatrix} \begin{bmatrix} 1 & 0 & 0 \\ 0 & 1 & 0 \\ 0 & 0 & 1 \end{bmatrix} = \begin{bmatrix} 6 & -3 & 0 \\ -2 & 1 & -8 \\ 4 & -4 & 9 \end{bmatrix}$.

19. $\begin{bmatrix} 3 & 0 & -2 & 1 \\ 1 & 2 & 0 & -1 \end{bmatrix} \begin{bmatrix} 2 & 1 & -2 \\ -1 & 2 & 0 \\ 0 & 0 & 1 \\ -1 & -2 & 2 \end{bmatrix} = \begin{bmatrix} 5 & 1 & -6 \\ 1 & 7 & -4 \end{bmatrix}$.

21. $4 \begin{bmatrix} 1 & -2 & 0 \\ 2 & -1 & 1 \\ 3 & 0 & -1 \end{bmatrix} \begin{bmatrix} 1 & 3 & 1 \\ 1 & 4 & 0 \\ 0 & 1 & -2 \end{bmatrix} = \begin{bmatrix} -4 & -20 & 4 \\ 4 & 12 & 0 \\ 12 & 32 & 20 \end{bmatrix}$.

23. $\begin{bmatrix} 1 & 0 \\ 0 & 1 \end{bmatrix} \begin{bmatrix} 4 & -3 & 2 \\ 7 & 1 & -5 \end{bmatrix} \begin{bmatrix} 1 & 0 & 0 \\ 0 & 1 & 0 \\ 0 & 0 & 1 \end{bmatrix} = \begin{bmatrix} 1 & 0 \\ 0 & 1 \end{bmatrix} \begin{bmatrix} 4 & -3 & 2 \\ 7 & 1 & -5 \end{bmatrix} = \begin{bmatrix} 4 & -3 & 2 \\ 7 & 1 & -5 \end{bmatrix}$.

25. To verify the associative law for matrix multiplication, we will show that $(AB)C = A(BC)$:

$$AB = \begin{bmatrix} 1 & 0 & -2 \\ 1 & -3 & 2 \\ -2 & 1 & 1 \end{bmatrix} \begin{bmatrix} 3 & 1 & 0 \\ 2 & 2 & 0 \\ 1 & -3 & -1 \end{bmatrix} = \begin{bmatrix} 1 & 7 & 2 \\ -1 & -11 & -2 \\ -3 & -3 & -1 \end{bmatrix}, \text{ so}$$

$$(AB)C = \begin{bmatrix} 1 & 7 & 2 \\ -1 & -11 & -2 \\ -3 & -3 & -1 \end{bmatrix} \begin{bmatrix} 2 & -1 & 0 \\ 1 & -1 & 2 \\ 3 & -2 & 1 \end{bmatrix} = \begin{bmatrix} 15 & -12 & 16 \\ -19 & 16 & -24 \\ -12 & 8 & -7 \end{bmatrix}.$$

On the other hand, $BC = \begin{bmatrix} 3 & 1 & 0 \\ 2 & 2 & 0 \\ 1 & -3 & -1 \end{bmatrix} \begin{bmatrix} 2 & -1 & 0 \\ 1 & -1 & 2 \\ 3 & -2 & 1 \end{bmatrix} = \begin{bmatrix} 7 & -4 & 2 \\ 6 & -4 & 4 \\ -4 & 4 & -7 \end{bmatrix}$, so

$$A(BC) = \begin{bmatrix} 1 & 0 & -2 \\ 1 & -3 & 2 \\ -2 & 1 & 1 \end{bmatrix} \begin{bmatrix} 7 & -4 & 2 \\ 6 & -4 & 4 \\ -4 & 4 & -7 \end{bmatrix} = \begin{bmatrix} 15 & -12 & 16 \\ -19 & 16 & -24 \\ -12 & 8 & -7 \end{bmatrix}.$$

27. $AB = \begin{bmatrix} 1 & 2 \\ 3 & 4 \end{bmatrix} \begin{bmatrix} 2 & 1 \\ 4 & 3 \end{bmatrix} = \begin{bmatrix} 10 & 7 \\ 22 & 15 \end{bmatrix}$ and $BA = \begin{bmatrix} 2 & 1 \\ 4 & 3 \end{bmatrix} \begin{bmatrix} 1 & 2 \\ 3 & 4 \end{bmatrix} = \begin{bmatrix} 5 & 8 \\ 13 & 20 \end{bmatrix}$.

Therefore, $AB \neq BA$ and matrix multiplication is not commutative.

29. $AB = \begin{bmatrix} 3 & 0 \\ 8 & 0 \end{bmatrix} \begin{bmatrix} 0 & 0 \\ 4 & 5 \end{bmatrix} = \begin{bmatrix} 0 & 0 \\ 0 & 0 \end{bmatrix}$. Thus, $AB = 0$, but neither A nor B is the zero matrix. Therefore, $AB = 0$,

does not imply that A or B is the zero matrix.

31. $\begin{bmatrix} a & b \\ c & d \end{bmatrix} \begin{bmatrix} 1 & 0 \\ -1 & 3 \end{bmatrix} = \begin{bmatrix} a-b & 3b \\ c-d & 3d \end{bmatrix} = \begin{bmatrix} -1 & -3 \\ 3 & 6 \end{bmatrix}$. Thus, $3b = -3$ (so $b = -1$), $3d = 6$ (so $d = 2$), $a - b = -1$

(so $a = b - 1 = -2$), and $c - d = 3$ (so $c = d + 3 = 5$). Therefore, $A = \begin{bmatrix} -2 & -1 \\ 5 & 2 \end{bmatrix}$.

33. Let $B = \begin{bmatrix} a & b \\ c & d \end{bmatrix}$. Then $AB = I$ gives $\begin{bmatrix} 2 & 1 \\ -2 & 2 \end{bmatrix} \begin{bmatrix} a & b \\ c & d \end{bmatrix} = \begin{bmatrix} 1 & 0 \\ 0 & 1 \end{bmatrix}$, so $\begin{bmatrix} 2a+c & 2b+d \\ -2a+2c & -2b+2d \end{bmatrix} = \begin{bmatrix} 1 & 0 \\ 0 & 1 \end{bmatrix}$.

By the equality of matrices, we have

$$\begin{array}{ccc} 2a + & c = 1 \\ -2a + 2c = 0 \end{array} \quad \text{and} \quad \begin{array}{ccccc} 2b & + & d & = & 0 \\ -2b & + & 2d & = & 1 \end{array}$$

The first system of equations gives $a = \frac{1}{3}$ and $c = \frac{1}{3}$, and the second gives $b = -\frac{1}{6}$ and $d = \frac{1}{3}$. Thus,

$$B = \begin{bmatrix} \frac{1}{3} & -\frac{1}{6} \\ \frac{1}{3} & \frac{1}{3} \end{bmatrix}.$$

35. Let $A = \begin{bmatrix} a_1 & b_1 \\ 0 & d_1 \end{bmatrix}$ and $B = \begin{bmatrix} a_2 & b_2 \\ 0 & d_2 \end{bmatrix}$ be two upper triangular matrices.

a. $A + B = \begin{bmatrix} a_1 & b_1 \\ 0 & d_1 \end{bmatrix} + \begin{bmatrix} a_2 & b_2 \\ 0 & d_2 \end{bmatrix} = \begin{bmatrix} a_1 + a_2 & b_1 + b_2 \\ 0 & d_1 + d_2 \end{bmatrix}$, which is a 2×2 upper triangular matrix. Also,

$$AB = \begin{bmatrix} a_1 & b_1 \\ 0 & d_1 \end{bmatrix} \begin{bmatrix} a_2 & b_2 \\ 0 & d_2 \end{bmatrix} = \begin{bmatrix} a_1 a_2 & a_1 b_2 + b_1 d_2 \\ 0 & d_1 d_2 \end{bmatrix} \text{ is a } 2 \times 2 \text{ upper triangular matrix.}$$

b. We compute $BA = \begin{bmatrix} a_2 & b_2 \\ 0 & d_2 \end{bmatrix} \begin{bmatrix} a_1 & b_1 \\ 0 & d_1 \end{bmatrix} = \begin{bmatrix} a_1 a_2 & a_2 b_1 + b_2 d_1 \\ 0 & d_1 d_2 \end{bmatrix} \neq AB$. From the result of part (a), we see that $AB \neq BA$ in general.

37. a. $A^T = \begin{bmatrix} 2 & 5 \\ 4 & -6 \end{bmatrix}$ and $\left(A^T \right)^T = \begin{bmatrix} 2 & 4 \\ 5 & -6 \end{bmatrix} = A.$

b. $(A + B)^T = \begin{bmatrix} 6 & 12 \\ -2 & -3 \end{bmatrix}^T = \begin{bmatrix} 6 & -2 \\ 12 & -3 \end{bmatrix}$ and $A^T + B^T = \begin{bmatrix} 2 & 5 \\ 4 & -6 \end{bmatrix} + \begin{bmatrix} 4 & -7 \\ 8 & 3 \end{bmatrix} = \begin{bmatrix} 6 & -2 \\ 12 & -3 \end{bmatrix}.$

c. $AB = \begin{bmatrix} 2 & 4 \\ 5 & -6 \end{bmatrix} \begin{bmatrix} 4 & 8 \\ -7 & 3 \end{bmatrix} = \begin{bmatrix} -20 & 28 \\ 62 & 22 \end{bmatrix}$, so $(AB)^T = \begin{bmatrix} -20 & 62 \\ 28 & 22 \end{bmatrix}$, and

$$B^T A^T = \begin{bmatrix} 4 & -7 \\ 8 & 3 \end{bmatrix} \begin{bmatrix} 2 & 5 \\ 4 & -6 \end{bmatrix} = \begin{bmatrix} -20 & 62 \\ 28 & 22 \end{bmatrix} = (AB)^T.$$

39. The given system of linear equations can be represented by the matrix equation $AX = B$, where

$$A = \begin{bmatrix} 2 & -3 \\ 3 & -4 \end{bmatrix}, X = \begin{bmatrix} x \\ y \end{bmatrix}, \text{ and } B = \begin{bmatrix} 7 \\ 8 \end{bmatrix}.$$

41. The given system of linear equations can be represented by the matrix equation $AX = B$, where

$$A = \begin{bmatrix} 2 & -3 & 4 \\ 0 & 2 & -3 \\ 1 & -1 & 2 \end{bmatrix}, X = \begin{bmatrix} x \\ y \\ z \end{bmatrix}, \text{ and } B = \begin{bmatrix} 6 \\ 7 \\ 4 \end{bmatrix}.$$

43. The given system of linear equations can be represented by the matrix equation $AX = B$, where

$$A = \begin{bmatrix} -1 & 1 & 1 \\ 2 & -1 & -1 \\ -3 & 2 & 4 \end{bmatrix}, X = \begin{bmatrix} x_1 \\ x_2 \\ x_3 \end{bmatrix}, \text{ and } B = \begin{bmatrix} 0 \\ 2 \\ 4 \end{bmatrix}.$$

45. a. $AB = \begin{bmatrix} 200 & 300 & 100 & 200 \\ 100 & 200 & 400 & 0 \end{bmatrix} \begin{bmatrix} 27 \\ 24 \\ 63 \\ 56 \end{bmatrix} = \begin{bmatrix} 30{,}100 \\ 32{,}700 \end{bmatrix}.$

b. The first entry shows that Olivia's total stockholdings are \$30,100, while the second entry shows that Isabella's stockholdings are \$32,700.

47. a. $AB = \begin{bmatrix} 12 & 14 & 20 & 10 \end{bmatrix} \begin{bmatrix} 20 & 30 & 40 \\ 20 & 20 & 55 \\ 15 & 35 & 45 \\ 25 & 30 & 40 \end{bmatrix} = \begin{matrix} A & B & C \\ \begin{bmatrix} 1070 & 1640 & 2550 \end{bmatrix} \end{matrix}.$

The matrix AB gives the number of cabins in each category; that is, 1070 category A cabins, 1640 category B cabins, and 2550 category C cabins on each cruise.

b. $ABC = \begin{bmatrix} 12 & 14 & 20 & 10 \end{bmatrix} \begin{bmatrix} 20 & 30 & 40 \\ 20 & 20 & 55 \\ 15 & 35 & 45 \\ 25 & 30 & 40 \end{bmatrix} \begin{bmatrix} 8000 \\ 10,000 \\ 7000 \end{bmatrix} = \begin{bmatrix} 1070 & 1640 & 2550 \end{bmatrix} \begin{bmatrix} 8000 \\ 10,000 \\ 7000 \end{bmatrix} = 42,810,000.$

The company's total revenue for 2015 was $42,810,000.

49. $C = \begin{bmatrix} 200 & 300 & 240 & 120 \end{bmatrix} \begin{bmatrix} 821.50 & 838.60 & 831.38 \\ 55.48 & 55.26 & 53.37 \\ 714.01 & 718.41 & 718.90 \\ 181.21 & 181.73 & 182.94 \end{bmatrix} = \begin{bmatrix} 374,051.60 & 378,524.00 & 376,775.80 \end{bmatrix}.$

51. a.
$A = \begin{matrix} & \text{N. Kroner} & \text{S. Kronor} & \text{D. Kroner} & \text{Rubles} \\ \text{Ava} & 82 & 68 & 62 & 1200 \\ \text{Ella} & 64 & 74 & 44 & 1600 \end{matrix}$

b.
$B = \begin{matrix} \text{N. Kroner} & 0.1751 \\ \text{S. Kronor} & 0.1560 \\ \text{D. Kroner} & 0.1747 \\ \text{Rubles} & 0.0325 \end{matrix}$

c. $AB = \begin{bmatrix} 82 & 68 & 62 & 1200 \\ 64 & 74 & 44 & 1600 \end{bmatrix} \begin{bmatrix} 0.1751 \\ 0.1560 \\ 0.1747 \\ 0.0325 \end{bmatrix} = \begin{bmatrix} 74.7976 \\ 82.4372 \end{bmatrix}$, so Ava will have $74.80 and Ella $82.44.

53. a. $BA = \begin{bmatrix} 1 & 1 & 1 \end{bmatrix} \begin{bmatrix} 60 & 80 & 120 & 40 \\ 20 & 30 & 60 & 10 \\ 10 & 15 & 30 & 5 \end{bmatrix} = \begin{bmatrix} 90 & 125 & 210 & 55 \end{bmatrix}$. The entries give the total numbers of model I, II, III, and IV houses built in the three states.

b. $AC^T = \begin{bmatrix} 60 & 80 & 120 & 40 \\ 20 & 30 & 60 & 10 \\ 10 & 15 & 30 & 5 \end{bmatrix} \begin{bmatrix} 1 \\ 1 \\ 1 \\ 1 \end{bmatrix} = \begin{bmatrix} 300 \\ 120 \\ 60 \end{bmatrix}$. The entries give the total numbers of houses built in each of the three states.

55. The column vector that represents the admission prices is $B = \begin{bmatrix} 4 \\ 6 \\ 8 \end{bmatrix}$. The column vector that gives the gross

receipts for each theater is $AB = \begin{bmatrix} 225 & 110 & 50 \\ 75 & 180 & 225 \\ 280 & 85 & 110 \\ 0 & 250 & 225 \end{bmatrix} \begin{bmatrix} 4 \\ 6 \\ 8 \end{bmatrix} = \begin{bmatrix} 1960 \\ 3180 \\ 2510 \\ 3300 \end{bmatrix}$.

The total revenue collected is given by $1960 + 3180 + 2510 + 3300$, or \$10,950.

57. $BA = \begin{bmatrix} 30{,}000 & 40{,}000 & 20{,}000 \end{bmatrix} \begin{matrix} \text{Dem.} \quad \text{Rep.} \quad \text{Ind.} \\ \begin{bmatrix} 0.50 & 0.30 & 0.20 \\ 0.45 & 0.40 & 0.15 \\ 0.40 & 0.50 & 0.10 \end{bmatrix} \end{matrix} = \begin{matrix} \text{Dem.} \quad \text{Rep.} \quad \text{Ind.} \\ \begin{bmatrix} 41{,}000 & 35{,}000 & 14{,}000 \end{bmatrix} \end{matrix}$.

59. $AB = \begin{bmatrix} 2700 & 3000 \\ 800 & 700 \\ 500 & 300 \end{bmatrix} \begin{bmatrix} 0.25 & 0.20 & 0.30 & 0.25 \\ 0.30 & 0.35 & 0.25 & 0.10 \end{bmatrix} = \begin{bmatrix} 1575 & 1590 & 1560 & 975 \\ 410 & 405 & 415 & 270 \\ 215 & 205 & 225 & 155 \end{bmatrix}$.

61. $AC = \begin{bmatrix} 80 & 60 & 40 \end{bmatrix} \begin{bmatrix} 0.17 \\ 0.21 \\ 0.24 \end{bmatrix} = \begin{bmatrix} 35.8 \end{bmatrix}$ and $BD = \begin{bmatrix} 300 & 150 & 250 \end{bmatrix} \begin{bmatrix} 0.12 \\ 0.15 \\ 0.17 \end{bmatrix} = \begin{bmatrix} 101 \end{bmatrix}$, so

$AC + BD = \begin{bmatrix} 136.8 \end{bmatrix}$, or \$136.80. This represents Cindy's long distance bill for phone calls to those three cities.

63. a. $MA^T = \begin{bmatrix} 400 & 1200 & 800 \\ 110 & 570 & 340 \\ 90 & 30 & 60 \end{bmatrix} \begin{bmatrix} 7 \\ 1 \\ 6 \end{bmatrix} = \begin{bmatrix} 8800 \\ 3380 \\ 1020 \end{bmatrix}$. The amounts of vitamin A, vitamin C, and calcium taken by a

girl in the first meal are 8800, 3380, and 1020 units, respectively.

b. $MB^T = \begin{bmatrix} 400 & 1200 & 800 \\ 110 & 570 & 340 \\ 90 & 30 & 60 \end{bmatrix} \begin{bmatrix} 9 \\ 3 \\ 2 \end{bmatrix} = \begin{bmatrix} 8800 \\ 3380 \\ 1020 \end{bmatrix}$. The amounts of vitamin A, vitamin C, and calcium taken by

a girl in the second meal are 8800, 3380, and 1020 units, respectively.

c. $M(A+B)^T = \begin{bmatrix} 400 & 1200 & 800 \\ 110 & 570 & 340 \\ 90 & 30 & 60 \end{bmatrix} \begin{bmatrix} 16 \\ 4 \\ 8 \end{bmatrix} = \begin{bmatrix} 17{,}600 \\ 6760 \\ 2040 \end{bmatrix}$. The amounts of vitamin A, vitamin C, and calcium

taken by a girl in the two meals are 17,600, 6760, and 2040 units, respectively.

65. False. Let A be a matrix of order 2×3 and let B be a matrix of order 3×2. Then AB and BA are both defined, although neither A nor B is a square matrix.

67. True. In order for the sum $B + C$ to be defined, B and C must have the same size, and in order for the product of A and $B + C$ to be defined, the number of columns of A must be equal to the number of rows of $B + C$.

Technology Exercises   page 140

1. $\begin{bmatrix} 18.66 & 15.2 & -12 \\ 24.48 & 41.88 & 89.82 \\ 15.39 & 7.16 & -1.25 \end{bmatrix}$

3. $\begin{bmatrix} 20.09 & 20.61 & -1.3 \\ 44.42 & 71.6 & 64.89 \\ 20.97 & 7.17 & -60.65 \end{bmatrix}$

5. $\begin{bmatrix} 32.89 & 13.63 & -57.17 \\ -12.85 & -8.37 & 256.92 \\ 13.48 & 14.29 & 181.64 \end{bmatrix}$

7. $\begin{bmatrix} 128.59 & 123.08 & -32.50 \\ 246.73 & 403.12 & 481.52 \\ 125.06 & 47.01 & -264.81 \end{bmatrix}$

9. $AB = \begin{bmatrix} 87 & 68 & 110 & 82 \\ 119 & 176 & 221 & 143 \\ 51 & 128 & 142 & 94 \\ 28 & 174 & 174 & 112 \end{bmatrix}$ and $BA = \begin{bmatrix} 113 & 117 & 72 & 101 & 90 \\ 72 & 85 & 36 & 72 & 76 \\ 81 & 69 & 76 & 87 & 30 \\ 133 & 157 & 56 & 121 & 146 \\ 154 & 157 & 94 & 127 & 122 \end{bmatrix}$.

11. $AC + AD = \begin{bmatrix} 170 & 18.1 & 133.1 & -106.3 & 341.3 \\ 349 & 226.5 & 324.1 & 164 & 506.4 \\ 245.2 & 157.7 & 231.5 & 125.5 & 312.9 \\ 310 & 245.2 & 291 & 274.3 & 354.2 \end{bmatrix}$.

2.6   The Inverse of a Square Matrix

Problem-Solving Tips

The problem-solving skills that you learned in earlier sections are building blocks for the rest of the course. You can't skip a section or a concept and hope to understand the material in a new section—it just won't work. For example, in this section we discussed the process for finding the inverse of a matrix. You need to use the Gauss-Jordan elimination method of elimination to find the inverse of a matrix, so if you don't know how to use that method to solve a system of equations, you won't be able to find the inverse of a matrix. If you are having difficulty, you may need to go back and review the earlier section before you go on.

1. Not every square matrix has an inverse. If there is a row to the left of the vertical line in the augmented matrix containing all zeros, then the matrix does not have an inverse.

2. You can use the formula $A^{-1} = \dfrac{1}{D}\begin{bmatrix} a & b \\ c & d \end{bmatrix}$, where $D = ad - bc \neq 0$, to find the inverse of a 2×2 matrix. Note that if $D = 0$, the matrix does not have an inverse.

3. The inverse of a matrix can be used to find the solution of a system of n equations in n unknowns.

Concept Questions   page 148

1. The inverse of a square matrix A is the matrix A^{-1} satisfying the conditions

$$AA^{-1} = A^{-1}A = I$$

3. The formula for finding the inverse of a 2×2 matrix are given on page 145 of the text.

Exercises   page 148

1. $\begin{bmatrix} 1 & -3 \\ 1 & -2 \end{bmatrix} \begin{bmatrix} -2 & 3 \\ -1 & 1 \end{bmatrix} = \begin{bmatrix} 1 & 0 \\ 0 & 1 \end{bmatrix}$ and $\begin{bmatrix} -2 & 3 \\ -1 & 1 \end{bmatrix} \begin{bmatrix} 1 & -3 \\ 1 & -2 \end{bmatrix} = \begin{bmatrix} 1 & 0 \\ 0 & 1 \end{bmatrix}$

3. $\begin{bmatrix} 3 & 2 & 3 \\ 2 & 2 & 1 \\ 2 & 1 & 1 \end{bmatrix} \begin{bmatrix} -\frac{1}{3} & -\frac{1}{3} & \frac{4}{3} \\ 0 & 1 & -1 \\ \frac{2}{3} & -\frac{1}{3} & -\frac{2}{3} \end{bmatrix} = \begin{bmatrix} 1 & 0 & 0 \\ 0 & 1 & 0 \\ 0 & 0 & 1 \end{bmatrix}$ and $\begin{bmatrix} -\frac{1}{3} & -\frac{1}{3} & \frac{4}{3} \\ 0 & 1 & -1 \\ \frac{2}{3} & -\frac{1}{3} & -\frac{2}{3} \end{bmatrix} \begin{bmatrix} 3 & 2 & 3 \\ 2 & 2 & 1 \\ 2 & 1 & 1 \end{bmatrix} = \begin{bmatrix} 1 & 0 & 0 \\ 0 & 1 & 0 \\ 0 & 0 & 1 \end{bmatrix}$.

5. Using Formula (13), we find $A^{-1} = \dfrac{1}{(2)(3) - (1)(5)} \begin{bmatrix} 3 & -5 \\ -1 & 2 \end{bmatrix} = \begin{bmatrix} 3 & -5 \\ -1 & 2 \end{bmatrix}$.

7. Since $ad - bc = (3)(2) - (-2)(-3) = 6 - 6 = 0$, the inverse does not exist.

9. $\left[\begin{array}{ccc|ccc} 2 & -3 & -4 & 1 & 0 & 0 \\ 0 & 0 & -1 & 0 & 1 & 0 \\ 1 & -2 & 1 & 0 & 0 & 1 \end{array} \right] \xrightarrow{R_1 \leftrightarrow R_3} \left[\begin{array}{ccc|ccc} 1 & -2 & 1 & 0 & 0 & 1 \\ 0 & 0 & -1 & 0 & 1 & 0 \\ 2 & -3 & -4 & 1 & 0 & 0 \end{array} \right] \xrightarrow{R_3 - 2R_1} \left[\begin{array}{ccc|ccc} 1 & -2 & 1 & 0 & 0 & 1 \\ 0 & 0 & -1 & 0 & 1 & 0 \\ 0 & 1 & -6 & 1 & 0 & -2 \end{array} \right] \xrightarrow{R_2 \leftrightarrow R_3}$

$\left[\begin{array}{ccc|ccc} 1 & -2 & 1 & 0 & 0 & 1 \\ 0 & 1 & -6 & 1 & 0 & -2 \\ 0 & 0 & -1 & 0 & 1 & 0 \end{array} \right] \xrightarrow[{-R_3}]{R_1 + 2R_2} \left[\begin{array}{ccc|ccc} 1 & 0 & -11 & 2 & 0 & -3 \\ 0 & 1 & -6 & 1 & 0 & -2 \\ 0 & 0 & 1 & 0 & -1 & 0 \end{array} \right] \xrightarrow[{R_2 + 6R_3}]{R_1 + 11R_3} \left[\begin{array}{ccc|ccc} 1 & 0 & 0 & 2 & -11 & -3 \\ 0 & 1 & 0 & 1 & -6 & -2 \\ 0 & 0 & 1 & 0 & -1 & 0 \end{array} \right]$.

Therefore, the required inverse is $\begin{bmatrix} 2 & -11 & -3 \\ 1 & -6 & -2 \\ 0 & -1 & 0 \end{bmatrix}$.

11. $\left[\begin{array}{ccc|ccc} 4 & 2 & 2 & 1 & 0 & 0 \\ -1 & -3 & 4 & 0 & 1 & 0 \\ 3 & -1 & 6 & 0 & 0 & 1 \end{array} \right] \xrightarrow{R_1 - R_3} \left[\begin{array}{ccc|ccc} 1 & 3 & -4 & 1 & 0 & -1 \\ -1 & -3 & 4 & 0 & 1 & 0 \\ 3 & -1 & 6 & 0 & 0 & 1 \end{array} \right] \xrightarrow{R_2 + R_1} \left[\begin{array}{ccc|ccc} 1 & 3 & -4 & 1 & 0 & -1 \\ 0 & 0 & 0 & 1 & 1 & -1 \\ 3 & -1 & 6 & 0 & 0 & 1 \end{array} \right]$.

Because there is a row of zeros to the left of the vertical line, we see that the inverse does not exist.

13.
$$\begin{bmatrix} 1 & 4 & -1 & | & 1 & 0 & 0 \\ 2 & 3 & -2 & | & 0 & 1 & 0 \\ -1 & 2 & 3 & | & 0 & 0 & 1 \end{bmatrix} \xrightarrow[R_3 + R_1]{R_2 - 2R_1} \begin{bmatrix} 1 & 4 & -1 & | & 1 & 0 & 0 \\ 0 & -5 & 0 & | & -2 & 1 & 0 \\ 0 & 6 & 2 & | & 1 & 0 & 1 \end{bmatrix} \xrightarrow{R_2 + R_3} \begin{bmatrix} 1 & 4 & -1 & | & 1 & 0 & 0 \\ 0 & 1 & 2 & | & -1 & 1 & 1 \\ 0 & 6 & 2 & | & 1 & 0 & 1 \end{bmatrix} \xrightarrow[R_3 - 6R_2]{R_1 - 4R_2}$$

$$\begin{bmatrix} 1 & 0 & -9 & | & 5 & -4 & -4 \\ 0 & 1 & 2 & | & -1 & 1 & 1 \\ 0 & 0 & -10 & | & 7 & -6 & -5 \end{bmatrix} \xrightarrow{-\frac{1}{10}R_3} \begin{bmatrix} 1 & 0 & -9 & | & 5 & -4 & -4 \\ 0 & 1 & 2 & | & -1 & 1 & 1 \\ 0 & 0 & 1 & | & -\frac{7}{10} & \frac{3}{5} & \frac{1}{2} \end{bmatrix} \xrightarrow[R_2 - 2R_3]{R_1 + 9R_3} \begin{bmatrix} 1 & 0 & 0 & | & -\frac{13}{10} & \frac{7}{5} & \frac{1}{2} \\ 0 & 1 & 0 & | & \frac{2}{5} & -\frac{1}{5} & 0 \\ 0 & 0 & 1 & | & -\frac{7}{10} & \frac{3}{5} & \frac{1}{2} \end{bmatrix}, \text{ so}$$

$$A^{-1} = \begin{bmatrix} -\frac{13}{10} & \frac{7}{5} & \frac{1}{2} \\ \frac{2}{5} & -\frac{1}{5} & 0 \\ -\frac{7}{10} & \frac{3}{5} & \frac{1}{2} \end{bmatrix}.$$

15.
$$\begin{bmatrix} 1 & 1 & -1 & 1 & | & 1 & 0 & 0 & 0 \\ 2 & 1 & 1 & 0 & | & 0 & 1 & 0 & 0 \\ 2 & 1 & 0 & 1 & | & 0 & 0 & 1 & 0 \\ 2 & -1 & -1 & 3 & | & 0 & 0 & 0 & 1 \end{bmatrix} \xrightarrow[R_4 - 2R_1]{\substack{R_2 - 2R_1 \\ R_3 - 2R_1}} \begin{bmatrix} 1 & 1 & -1 & 1 & | & 1 & 0 & 0 & 0 \\ 0 & -1 & 3 & -2 & | & -2 & 1 & 0 & 0 \\ 0 & -1 & 2 & -1 & | & -2 & 0 & 1 & 0 \\ 0 & -3 & 1 & 1 & | & -2 & 0 & 0 & 1 \end{bmatrix} \xrightarrow{-R_2}$$

$$\begin{bmatrix} 1 & 1 & -1 & 1 & | & 1 & 0 & 0 & 0 \\ 0 & 1 & -3 & 2 & | & 2 & -1 & 0 & 0 \\ 0 & -1 & 2 & -1 & | & -2 & 0 & 1 & 0 \\ 0 & -3 & 1 & 1 & | & -2 & 0 & 0 & 1 \end{bmatrix} \xrightarrow[R_4 + 3R_2]{\substack{R_1 - R_2 \\ R_3 + R_2}} \begin{bmatrix} 1 & 0 & 2 & -1 & | & -1 & 1 & 0 & 0 \\ 0 & 1 & -3 & 2 & | & 2 & -1 & 0 & 0 \\ 0 & 0 & -1 & 1 & | & 0 & -1 & 1 & 0 \\ 0 & 0 & -8 & 7 & | & 4 & -3 & 0 & 1 \end{bmatrix} \xrightarrow{-R_3}$$

$$\begin{bmatrix} 1 & 0 & 2 & -1 & | & -1 & 1 & 0 & 0 \\ 0 & 1 & -3 & 2 & | & 2 & -1 & 0 & 0 \\ 0 & 0 & 1 & -1 & | & 0 & 1 & -1 & 0 \\ 0 & 0 & -8 & 7 & | & 4 & -3 & 0 & 1 \end{bmatrix} \xrightarrow[R_4 + 8R_3]{\substack{R_1 - 2R_3 \\ R_2 + 3R_3}} \begin{bmatrix} 1 & 0 & 0 & 1 & | & -1 & -1 & 2 & 0 \\ 0 & 1 & 0 & -1 & | & 2 & 2 & -3 & 0 \\ 0 & 0 & 1 & -1 & | & 0 & 1 & -1 & 0 \\ 0 & 0 & 0 & -1 & | & 4 & 5 & -8 & 1 \end{bmatrix} \xrightarrow[\substack{R_3 - R_4 \\ -R_4}]{\substack{R_1 + R_4 \\ R_2 - R_4}}$$

$$\begin{bmatrix} 1 & 0 & 0 & 0 & | & 3 & 4 & -6 & 1 \\ 0 & 1 & 0 & 0 & | & -2 & -3 & 5 & -1 \\ 0 & 0 & 1 & 0 & | & -4 & -4 & 7 & -1 \\ 0 & 0 & 0 & 1 & | & -4 & -5 & 8 & -1 \end{bmatrix}. \text{ Thus, the required inverse is } A^{-1} = \begin{bmatrix} 3 & 4 & -6 & 1 \\ -2 & -3 & 5 & -1 \\ -4 & -4 & 7 & -1 \\ -4 & -5 & 8 & -1 \end{bmatrix}.$$

We can verify our result by showing that $A^{-1}A = A$: $\begin{bmatrix} 3 & 4 & -6 & 1 \\ -2 & -3 & 5 & -1 \\ -4 & -4 & 7 & -1 \\ -4 & -5 & 8 & -1 \end{bmatrix} \begin{bmatrix} 1 & 1 & -1 & 1 \\ 2 & 1 & 1 & 0 \\ 2 & 1 & 0 & 1 \\ 2 & -1 & -1 & 3 \end{bmatrix} = \begin{bmatrix} 1 & 0 & 0 & 0 \\ 0 & 1 & 0 & 0 \\ 0 & 0 & 1 & 0 \\ 0 & 0 & 0 & 1 \end{bmatrix}.$

17. a. $AX = B$, where $A = \begin{bmatrix} 2 & 5 \\ 1 & 3 \end{bmatrix}, X = \begin{bmatrix} x \\ y \end{bmatrix},$ **b.** $X = A^{-1}B = \begin{bmatrix} 3 & -5 \\ -1 & 2 \end{bmatrix}\begin{bmatrix} 3 \\ 2 \end{bmatrix} = \begin{bmatrix} -1 \\ 1 \end{bmatrix}.$

$B = \begin{bmatrix} 3 \\ 2 \end{bmatrix}.$

19. a. $AX = B$, where $A = \begin{bmatrix} 2 & -3 & -4 \\ 0 & 0 & -1 \\ 1 & -2 & 1 \end{bmatrix}$, $X = \begin{bmatrix} x \\ y \\ z \end{bmatrix}$, $B = \begin{bmatrix} 4 \\ 3 \\ -8 \end{bmatrix}$.

b. $X = A^{-1}B = \begin{bmatrix} 2 & -11 & -3 \\ 1 & -6 & -2 \\ 0 & -1 & 0 \end{bmatrix} \begin{bmatrix} 4 \\ 3 \\ -8 \end{bmatrix} = \begin{bmatrix} -1 \\ 2 \\ -3 \end{bmatrix}$.

21. a. $AX = B$, where $A = \begin{bmatrix} 1 & 4 & -1 \\ 2 & 3 & -2 \\ -1 & 2 & 3 \end{bmatrix}$, $X = \begin{bmatrix} x \\ y \\ z \end{bmatrix}$, $B = \begin{bmatrix} 3 \\ 1 \\ 7 \end{bmatrix}$.

b. $X = A^{-1}B = \begin{bmatrix} -\frac{13}{10} & \frac{7}{5} & \frac{1}{2} \\ \frac{2}{5} & -\frac{1}{5} & 0 \\ -\frac{7}{10} & \frac{3}{5} & \frac{1}{2} \end{bmatrix} \begin{bmatrix} 3 \\ 1 \\ 7 \end{bmatrix} = \begin{bmatrix} 1 \\ 1 \\ 2 \end{bmatrix}$.

23. a. $AX = B$, where $A = \begin{bmatrix} 1 & 1 & -1 & 1 \\ 2 & 1 & 1 & 0 \\ 2 & 1 & 0 & 1 \\ 2 & -1 & -1 & 3 \end{bmatrix}$, $X = \begin{bmatrix} x_1 \\ x_2 \\ x_3 \\ x_4 \end{bmatrix}$, $B = \begin{bmatrix} 6 \\ 4 \\ 7 \\ 9 \end{bmatrix}$.

b. $X = A^{-1}B = \begin{bmatrix} 3 & 4 & -6 & 1 \\ -2 & -3 & 5 & -1 \\ -4 & -4 & 7 & -1 \\ -4 & -5 & 8 & -1 \end{bmatrix} \begin{bmatrix} 6 \\ 4 \\ 7 \\ 9 \end{bmatrix} = \begin{bmatrix} 1 \\ 2 \\ 0 \\ 3 \end{bmatrix}$.

25. a. $A = \begin{bmatrix} 1 & 2 \\ 2 & -1 \end{bmatrix}$, $X = \begin{bmatrix} x \\ y \end{bmatrix}$, $B = \begin{bmatrix} b_1 \\ b_2 \end{bmatrix}$.

b. i. $X = A^{-1}B = \begin{bmatrix} 0.2 & 0.4 \\ 0.4 & -0.2 \end{bmatrix} \begin{bmatrix} 14 \\ 5 \end{bmatrix} = \begin{bmatrix} 4.8 \\ 4.6 \end{bmatrix}$, and we conclude that $x = 4.8$ and $y = 4.6$.

ii. $X = A^{-1}B = \begin{bmatrix} 0.2 & 0.4 \\ 0.4 & -0.2 \end{bmatrix} \begin{bmatrix} 4 \\ -1 \end{bmatrix} = \begin{bmatrix} 0.4 \\ 1.8 \end{bmatrix}$, and we conclude that $x = 0.4$ and $y = 1.8$.

27. a. First we find A^{-1}:
$$\left[\begin{array}{ccc|ccc} 1 & 2 & 1 & 1 & 0 & 0 \\ 1 & 1 & 1 & 0 & 1 & 0 \\ 3 & 1 & 1 & 0 & 0 & 1 \end{array}\right] \xrightarrow[R_3-3R_1]{R_2-R_1} \left[\begin{array}{ccc|ccc} 1 & 2 & 1 & 1 & 0 & 0 \\ 0 & -1 & 0 & -1 & 1 & 0 \\ 0 & -5 & -2 & -3 & 0 & 1 \end{array}\right] \xrightarrow{-R_2}$$

$$\left[\begin{array}{ccc|ccc} 1 & 2 & 1 & 1 & 0 & 0 \\ 0 & 1 & 0 & 1 & -1 & 0 \\ 0 & -5 & -2 & -3 & 0 & 1 \end{array}\right] \xrightarrow[R_3+5R_2]{R_1-2R_2} \left[\begin{array}{ccc|ccc} 1 & 0 & 1 & -1 & 2 & 0 \\ 0 & 1 & 0 & 1 & -1 & 0 \\ 0 & 0 & -2 & 2 & -5 & 1 \end{array}\right] \xrightarrow{-\frac{1}{2}R_3} \left[\begin{array}{ccc|ccc} 1 & 0 & 1 & -1 & 2 & 0 \\ 0 & 1 & 0 & 1 & -1 & 0 \\ 0 & 0 & 1 & -1 & \frac{5}{2} & -\frac{1}{2} \end{array}\right] \xrightarrow{R_1-R_3}$$

$$\left[\begin{array}{ccc|ccc} 1 & 0 & 0 & 0 & -\frac{1}{2} & \frac{1}{2} \\ 0 & 1 & 0 & 1 & -1 & 0 \\ 0 & 0 & 1 & -1 & \frac{5}{2} & -\frac{1}{2} \end{array}\right]. \text{ Thus, } \left[\begin{array}{ccc} 1 & 2 & 1 \\ 1 & 1 & 1 \\ 3 & 1 & 1 \end{array}\right]\left[\begin{array}{c} x \\ y \\ z \end{array}\right] = \left[\begin{array}{c} b_1 \\ b_2 \\ b_3 \end{array}\right].$$

b. i.
$$\left[\begin{array}{c} x \\ y \\ z \end{array}\right] = \left[\begin{array}{ccc} 0 & -\frac{1}{2} & \frac{1}{2} \\ 1 & -1 & 0 \\ -1 & \frac{5}{2} & -\frac{1}{2} \end{array}\right]\left[\begin{array}{c} 7 \\ 4 \\ 2 \end{array}\right] = \left[\begin{array}{c} -1 \\ 3 \\ 2 \end{array}\right], \text{ and we conclude that } x=-1, y=3, \text{ and } z=2.$$

ii.
$$\left[\begin{array}{c} x \\ y \\ z \end{array}\right] = \left[\begin{array}{ccc} 0 & -\frac{1}{2} & \frac{1}{2} \\ 1 & -1 & 0 \\ -1 & \frac{5}{2} & -\frac{1}{2} \end{array}\right]\left[\begin{array}{c} 5 \\ -3 \\ -1 \end{array}\right] = \left[\begin{array}{c} 1 \\ 8 \\ -12 \end{array}\right], \text{ and we conclude that } x=1, y=8, \text{ and } z=-12.$$

29. a.
$$\left[\begin{array}{ccc|ccc} 3 & 2 & -1 & 1 & 0 & 0 \\ 2 & -3 & 1 & 0 & 1 & 0 \\ 1 & -1 & -1 & 0 & 0 & 1 \end{array}\right] \xrightarrow{R_1 \leftrightarrow R_3} \left[\begin{array}{ccc|ccc} 1 & -1 & -1 & 0 & 0 & 1 \\ 2 & -3 & 1 & 0 & 1 & 0 \\ 3 & 2 & -1 & 1 & 0 & 0 \end{array}\right] \xrightarrow[R_3-3R_1]{R_2-2R_1} \left[\begin{array}{ccc|ccc} 1 & -1 & -1 & 0 & 0 & 1 \\ 0 & -1 & 3 & 0 & 1 & -2 \\ 0 & 5 & 2 & 1 & 0 & -3 \end{array}\right] \xrightarrow{-R_2}$$

$$\left[\begin{array}{ccc|ccc} 1 & -1 & -1 & 0 & 0 & 1 \\ 0 & 1 & -3 & 0 & -1 & 2 \\ 0 & 5 & 2 & 1 & 0 & -3 \end{array}\right] \xrightarrow[R_3-5R_2]{R_1+R_2} \left[\begin{array}{ccc|ccc} 1 & 0 & -4 & 0 & -1 & 3 \\ 0 & 1 & -3 & 0 & -1 & 2 \\ 0 & 0 & 17 & 1 & 5 & -13 \end{array}\right] \xrightarrow{\frac{1}{17}R_3} \left[\begin{array}{ccc|ccc} 1 & 0 & -4 & 0 & -1 & 3 \\ 0 & 1 & -3 & 0 & -1 & 2 \\ 0 & 0 & 1 & \frac{1}{17} & \frac{5}{17} & -\frac{13}{17} \end{array}\right] \xrightarrow[R_2+3R_3]{R_1+4R_3}$$

$$\left[\begin{array}{ccc|ccc} 1 & 0 & 0 & \frac{4}{17} & \frac{3}{17} & -\frac{1}{17} \\ 0 & 1 & 0 & \frac{3}{17} & -\frac{2}{17} & -\frac{5}{17} \\ 0 & 0 & 1 & \frac{1}{17} & \frac{5}{17} & -\frac{13}{17} \end{array}\right]. \text{ Therefore, } A^{-1} = \left[\begin{array}{ccc} \frac{4}{17} & \frac{3}{17} & -\frac{1}{17} \\ \frac{3}{17} & -\frac{2}{17} & -\frac{5}{17} \\ \frac{1}{17} & \frac{5}{17} & -\frac{13}{17} \end{array}\right] \text{ and } \left[\begin{array}{ccc} 3 & 2 & -1 \\ 2 & -3 & 1 \\ 1 & -1 & -1 \end{array}\right]\left[\begin{array}{c} x \\ y \\ z \end{array}\right] = \left[\begin{array}{c} b_1 \\ b_2 \\ b_3 \end{array}\right].$$

b. i.
$$\left[\begin{array}{c} x \\ y \\ z \end{array}\right] = \left[\begin{array}{ccc} \frac{4}{17} & \frac{3}{17} & -\frac{1}{17} \\ \frac{3}{17} & -\frac{2}{17} & -\frac{5}{17} \\ \frac{1}{17} & \frac{5}{17} & -\frac{13}{17} \end{array}\right]\left[\begin{array}{c} 2 \\ -2 \\ 4 \end{array}\right] = \left[\begin{array}{c} -\frac{2}{17} \\ -\frac{10}{17} \\ -\frac{60}{17} \end{array}\right]. \text{ We conclude that } x=-\frac{2}{17}, y=-\frac{10}{17}, \text{ and } z=-\frac{60}{17}.$$

ii.
$$\left[\begin{array}{c} x \\ y \\ z \end{array}\right] = \left[\begin{array}{ccc} \frac{4}{17} & \frac{3}{17} & -\frac{1}{17} \\ \frac{3}{17} & -\frac{2}{17} & -\frac{5}{17} \\ \frac{1}{17} & \frac{5}{17} & -\frac{13}{17} \end{array}\right]\left[\begin{array}{c} 8 \\ -3 \\ 6 \end{array}\right] = \left[\begin{array}{c} 1 \\ 0 \\ -5 \end{array}\right]. \text{ We conclude that } x=1, y=0, \text{ and } z=-5.$$

31. a. $AX = B_1$ and $AX = B_2$, where $A = \begin{bmatrix} 1 & 1 & 1 & 1 \\ 1 & -1 & -1 & 1 \\ 0 & 1 & 2 & 2 \\ 1 & 2 & 1 & -2 \end{bmatrix}$, $X = \begin{bmatrix} x_1 \\ x_2 \\ x_3 \\ x_4 \end{bmatrix}$, $B_1 = \begin{bmatrix} 1 \\ -1 \\ 4 \\ 0 \end{bmatrix}$, and $B_2 = \begin{bmatrix} 2 \\ 8 \\ 4 \\ -1 \end{bmatrix}$.

We first find A^{-1}: $\left[\begin{array}{cccc|cccc} 1 & 1 & 1 & 1 & 1 & 0 & 0 & 0 \\ 1 & -1 & -1 & 1 & 0 & 1 & 0 & 0 \\ 0 & 1 & 2 & 2 & 0 & 0 & 1 & 0 \\ 1 & 2 & 1 & -2 & 0 & 0 & 0 & 1 \end{array}\right]$ $\xrightarrow[R_4 - R_1]{R_2 - R_1}$ $\left[\begin{array}{cccc|cccc} 1 & 1 & 1 & 1 & 1 & 0 & 0 & 0 \\ 0 & -2 & -2 & 0 & -1 & 1 & 0 & 0 \\ 0 & 1 & 2 & 2 & 0 & 0 & 1 & 0 \\ 0 & 1 & 0 & -3 & -1 & 0 & 0 & 1 \end{array}\right]$ $\xrightarrow{R_2 \leftrightarrow R_3}$

$\left[\begin{array}{cccc|cccc} 1 & 1 & 1 & 1 & 1 & 0 & 0 & 0 \\ 0 & 1 & 2 & 2 & 0 & 0 & 1 & 0 \\ 0 & -2 & -2 & 0 & -1 & 1 & 0 & 0 \\ 0 & 1 & 0 & -3 & -1 & 0 & 0 & 1 \end{array}\right]$ $\xrightarrow[R_4 - R_2]{\overset{R_1 - R_2}{R_3 + 2R_2}}$ $\left[\begin{array}{cccc|cccc} 1 & 0 & -1 & -1 & 1 & 0 & -1 & 0 \\ 0 & 1 & 2 & 2 & 0 & 0 & 1 & 0 \\ 0 & 0 & 2 & 4 & -1 & 1 & 2 & 0 \\ 0 & 0 & -2 & -5 & -1 & 0 & -1 & 1 \end{array}\right]$ $\xrightarrow{\frac{1}{2}R_3}$

$\left[\begin{array}{cccc|cccc} 1 & 0 & -1 & -1 & 1 & 0 & -1 & 0 \\ 0 & 1 & 2 & 2 & 0 & 0 & 1 & 0 \\ 0 & 0 & 1 & 2 & -\frac{1}{2} & \frac{1}{2} & 1 & 0 \\ 0 & 0 & -2 & -5 & -1 & 0 & -1 & 1 \end{array}\right]$ $\xrightarrow[R_4 + 2R_3]{\overset{R_1 + R_3}{R_2 - 2R_3}}$ $\left[\begin{array}{cccc|cccc} 1 & 0 & 0 & 1 & \frac{1}{2} & \frac{1}{2} & 0 & 0 \\ 0 & 1 & 0 & -2 & 1 & -1 & -1 & 0 \\ 0 & 0 & 1 & 2 & -\frac{1}{2} & \frac{1}{2} & 1 & 0 \\ 0 & 0 & 0 & -1 & -2 & 1 & 1 & 1 \end{array}\right]$ $\xrightarrow[-R_4]{\overset{R_1 + R_4}{\overset{R_2 - 2R_4}{R_3 + 2R_4}}}$

$\left[\begin{array}{cccc|cccc} 1 & 0 & 0 & 0 & -\frac{3}{2} & \frac{3}{2} & 1 & 1 \\ 0 & 1 & 0 & 0 & 5 & -3 & -3 & -2 \\ 0 & 0 & 1 & 0 & -\frac{9}{2} & \frac{5}{2} & 3 & 2 \\ 0 & 0 & 0 & 1 & 2 & -1 & -1 & -1 \end{array}\right]$. Therefore, $A^{-1} = \begin{bmatrix} -\frac{3}{2} & \frac{3}{2} & 1 & 1 \\ 5 & -3 & -3 & -2 \\ -\frac{9}{2} & \frac{5}{2} & 3 & 2 \\ 2 & -1 & -1 & -1 \end{bmatrix}$.

b. i. $\begin{bmatrix} x_1 \\ x_2 \\ x_3 \\ x_4 \end{bmatrix} = \begin{bmatrix} -\frac{3}{2} & \frac{3}{2} & 1 & 1 \\ 5 & -3 & -3 & -2 \\ -\frac{9}{2} & \frac{5}{2} & 3 & 2 \\ 2 & -1 & -1 & -1 \end{bmatrix} \begin{bmatrix} 1 \\ -1 \\ 4 \\ 0 \end{bmatrix} = \begin{bmatrix} 1 \\ -4 \\ 5 \\ -1 \end{bmatrix}$, so $x_1 = 1$, $x_2 = -4$, $x_3 = 5$, and $x_4 = -1$.

ii. $\begin{bmatrix} x_1 \\ x_2 \\ x_3 \\ x_4 \end{bmatrix} = \begin{bmatrix} -\frac{3}{2} & \frac{3}{2} & 1 & 1 \\ 5 & -3 & -3 & -2 \\ -\frac{9}{2} & \frac{5}{2} & 3 & 2 \\ 2 & -1 & -1 & -1 \end{bmatrix} \begin{bmatrix} 2 \\ 8 \\ 4 \\ -1 \end{bmatrix} = \begin{bmatrix} 12 \\ -24 \\ 21 \\ -7 \end{bmatrix}$, so $x_1 = 12$, $x_2 = -24$, $x_3 = 21$, and $x_4 = -7$.

33. a. Using Formula (13), we find $A^{-1} = \dfrac{1}{(2)(-5) - (-4)(3)} \begin{bmatrix} -5 & -3 \\ 4 & 2 \end{bmatrix} = \begin{bmatrix} -\frac{5}{2} & -\frac{3}{2} \\ 2 & 1 \end{bmatrix}$.

b. Using Formula (13) once again, we find $\left(A^{-1}\right)^{-1} = \dfrac{1}{\left(-\frac{5}{2}\right)(1) - 2\left(-\frac{3}{2}\right)} \begin{bmatrix} 1 & \frac{3}{2} \\ -2 & -\frac{5}{2} \end{bmatrix} = \begin{bmatrix} 2 & 3 \\ -4 & -5 \end{bmatrix} = A.$

35. a. $ABC = \begin{bmatrix} 2 & -5 \\ 1 & -3 \end{bmatrix} \begin{bmatrix} 4 & 3 \\ 1 & 1 \end{bmatrix} \begin{bmatrix} 2 & 3 \\ -2 & 1 \end{bmatrix} = \begin{bmatrix} 2 & -5 \\ 1 & -3 \end{bmatrix} \begin{bmatrix} 2 & 15 \\ 0 & 4 \end{bmatrix} = \begin{bmatrix} 4 & 10 \\ 2 & 3 \end{bmatrix}$. Using the formula for the inverse

of a 2×2 matrix, we find $A^{-1} = \begin{bmatrix} 3 & -5 \\ 1 & -2 \end{bmatrix}$, $B^{-1} = \begin{bmatrix} 1 & -3 \\ -1 & 4 \end{bmatrix}$, and $C^{-1} = \begin{bmatrix} \frac{1}{8} & -\frac{3}{8} \\ \frac{1}{4} & \frac{1}{4} \end{bmatrix}$.

b. Using the formula for the inverse of a 2×2 matrix, we find $(ABC)^{-1} = \begin{bmatrix} -\frac{3}{8} & \frac{5}{4} \\ \frac{1}{4} & -\frac{1}{2} \end{bmatrix}$, while

$C^{-1}B^{-1}A^{-1} = \begin{bmatrix} \frac{1}{8} & -\frac{3}{8} \\ \frac{1}{4} & \frac{1}{4} \end{bmatrix} \begin{bmatrix} 1 & -3 \\ -1 & 4 \end{bmatrix} \begin{bmatrix} 3 & -5 \\ 1 & -2 \end{bmatrix} = \begin{bmatrix} \frac{1}{8} & -\frac{3}{8} \\ \frac{1}{4} & \frac{1}{4} \end{bmatrix} \begin{bmatrix} 0 & 1 \\ 1 & -3 \end{bmatrix} = \begin{bmatrix} -\frac{3}{8} & \frac{5}{4} \\ \frac{1}{4} & -\frac{1}{2} \end{bmatrix}$.

Therefore, $(ABC)^{-1} = C^{-1}B^{-1}A^{-1}$.

37. Multiplying both sides of the equation on the right by $\begin{bmatrix} 1 & 2 \\ 3 & -1 \end{bmatrix}^{-1}$, we obtain

$A \begin{bmatrix} 1 & 2 \\ 3 & -1 \end{bmatrix} \begin{bmatrix} 1 & 2 \\ 3 & -1 \end{bmatrix}^{-1} = \begin{bmatrix} 2 & 1 \\ 3 & -2 \end{bmatrix} \begin{bmatrix} 1 & 2 \\ 3 & -1 \end{bmatrix}^{-1}$, so $A = \begin{bmatrix} 2 & 1 \\ 3 & -2 \end{bmatrix} \begin{bmatrix} \frac{1}{7} & \frac{2}{7} \\ \frac{3}{7} & -\frac{1}{7} \end{bmatrix} = \begin{bmatrix} \frac{5}{7} & \frac{3}{7} \\ -\frac{3}{7} & \frac{8}{7} \end{bmatrix}$.

39. Let x denote the number of copies of the deluxe edition and y the number of copies of the standard edition demanded per month when the unit prices are p and q dollars, respectively. Then the three systems of linear equations

$$\begin{array}{lll} 5x + y = 20{,}000 & \qquad 5x + y = 25{,}000 & \qquad 5x + y = 25{,}000 \\ x + 3y = 15{,}000 & \qquad x + 3y = 15{,}000 & \qquad x + 3y = 20{,}000 \end{array}$$

give the quantity demanded of each edition at the stated price. These systems may be written in the form

$AX = B_1$, $AX = B_2$, and $AX = B_3$, where $A = \begin{bmatrix} 5 & 1 \\ 1 & 3 \end{bmatrix}$, $B_1 = \begin{bmatrix} 20{,}000 \\ 15{,}000 \end{bmatrix}$, $B_2 = \begin{bmatrix} 25{,}000 \\ 15{,}000 \end{bmatrix}$, and

$B_3 = \begin{bmatrix} 25{,}000 \\ 20{,}000 \end{bmatrix}$. Using the formula for the inverse of a 2×2 matrix, with $a = 5$, $b = 1$, $c = 1$, $d = 3$, and

$D = ad - bc = (5)(3) - (1)(1) = 14$, we find that $A^{-1} = \begin{bmatrix} \frac{3}{14} & -\frac{1}{14} \\ -\frac{1}{14} & \frac{5}{14} \end{bmatrix}$.

a. $\begin{bmatrix} x \\ y \end{bmatrix} = \begin{bmatrix} \frac{3}{14} & -\frac{1}{14} \\ -\frac{1}{14} & \frac{5}{14} \end{bmatrix} \begin{bmatrix} 20{,}000 \\ 15{,}000 \end{bmatrix} = \begin{bmatrix} 3214 \\ 3929 \end{bmatrix}$. **b.** $\begin{bmatrix} x \\ y \end{bmatrix} = \begin{bmatrix} \frac{3}{14} & -\frac{1}{14} \\ -\frac{1}{14} & \frac{5}{14} \end{bmatrix} \begin{bmatrix} 25{,}000 \\ 15{,}000 \end{bmatrix} = \begin{bmatrix} 4286 \\ 3571 \end{bmatrix}$.

c. $\begin{bmatrix} x \\ y \end{bmatrix} = \begin{bmatrix} \frac{3}{14} & -\frac{1}{14} \\ -\frac{1}{14} & \frac{5}{14} \end{bmatrix} \begin{bmatrix} 25{,}000 \\ 20{,}000 \end{bmatrix} = \begin{bmatrix} 3929 \\ 5357 \end{bmatrix}$.

41. Let x, y, and z denote the number of acres of soybeans, corn, and wheat to be cultivated, respectively. Furthermore, let a, b, and c denote the amount of land available, the amount of labor available, and the amount of money available for seeds, respectively. Then we have the system

$$
\begin{aligned}
x + y + z &= a && \text{(land)} \\
2x + 6y + 6z &= b && \text{(labor)} \\
12x + 20y + 8z &= c && \text{(seeds)}
\end{aligned}
$$

The system can be written in the form $AX = B$, where $A = \begin{bmatrix} 1 & 1 & 1 \\ 2 & 6 & 6 \\ 12 & 20 & 8 \end{bmatrix}$, $X = \begin{bmatrix} x \\ y \\ z \end{bmatrix}$, and $B = \begin{bmatrix} a \\ b \\ c \end{bmatrix}$.

Using the technique for finding A^{-1} developed in this section, we find $A^{-1} = \begin{bmatrix} \frac{3}{2} & -\frac{1}{4} & 0 \\ -\frac{7}{6} & \frac{1}{12} & \frac{1}{12} \\ \frac{2}{3} & \frac{1}{6} & -\frac{1}{12} \end{bmatrix}$.

a. Here $a = 1000$, $b = 4400$, and $c = 13{,}200$. Therefore

$$
X = A^{-1}B = \begin{bmatrix} \frac{3}{2} & -\frac{1}{4} & 0 \\ -\frac{7}{6} & \frac{1}{12} & \frac{1}{12} \\ \frac{2}{3} & \frac{1}{6} & -\frac{1}{12} \end{bmatrix} \begin{bmatrix} 1000 \\ 4400 \\ 13{,}200 \end{bmatrix} = \begin{bmatrix} 400 \\ 300 \\ 300 \end{bmatrix}
$$

so Jackson Farms should cultivate 400, 300, and 300 acres of soybeans, corn, and wheat, respectively.

b. Here $a = 1200$, $b = 5200$, and $c = 16{,}400$. Therefore

$$
X = A^{-1}B = \begin{bmatrix} \frac{3}{2} & -\frac{1}{4} & 0 \\ -\frac{7}{6} & \frac{1}{12} & \frac{1}{12} \\ \frac{2}{3} & \frac{1}{6} & -\frac{1}{12} \end{bmatrix} \begin{bmatrix} 1200 \\ 5200 \\ 16{,}400 \end{bmatrix} = \begin{bmatrix} 500 \\ 400 \\ 300 \end{bmatrix}
$$

so Jackson Farms should cultivate 500, 400, and 300 acres of soybeans, corn, and wheat, respectively.

43. Let x, y, and z denote the amount to be invested in high-, medium-, and low-risk stocks, respectively. Next, let a denote the amount to be invested and let c denote the return on the investments. Then we have the system

$$
\begin{aligned}
x + y + z &= a \\
x + y - z &= 0 && \text{(because } z = x + y) \\
0.15x + 0.1y + 0.06z &= c
\end{aligned}
$$

The system is equivalent to the matrix equation $AX = B$, where

$A = \begin{bmatrix} 1 & 1 & 1 \\ 1 & 1 & -1 \\ 0.15 & 0.10 & 0.06 \end{bmatrix}$, $X = \begin{bmatrix} x \\ y \\ z \end{bmatrix}$, and $B = \begin{bmatrix} a \\ 0 \\ c \end{bmatrix}$. We calculate $A^{-1} = \begin{bmatrix} -1.6 & -0.4 & 20 \\ 2.1 & 0.9 & -20 \\ 0.5 & -0.5 & 0 \end{bmatrix}$.

a. Here $a = 200{,}000$ and $c = 20{,}000$, so $X = A^{-1}B = \begin{bmatrix} -1.6 & -0.4 & 20 \\ 2.1 & 0.9 & -20 \\ 0.5 & -0.5 & 0 \end{bmatrix} \begin{bmatrix} 200{,}000 \\ 0 \\ 20{,}000 \end{bmatrix} = \begin{bmatrix} 80{,}000 \\ 20{,}000 \\ 100{,}000 \end{bmatrix}$, so

the club should invest \$80,000 in high-risk, \$20,000 in medium-risk, and \$100,000 in low risk stocks.

b. Here $a = 220{,}000$ and $c = 22{,}000$. The solution is $x = 88{,}000$, $y = 22{,}000$, and $z = 110{,}000$; that is, the club should invest \$88,000 in high-risk, \$22,000 in medium-risk, and \$110,000 in low-risk stocks.

c. Here $a = 240{,}000$ and $c = 22{,}000$. The result is \$56,000 in high-risk stocks, \$64,000 in medium-risk stocks, and \$120,000 in low-risk stocks.

45. In order for the inverse of A to exist, $D = ad - bc \neq 0$. Here $a = 1$, $b = 2$, $c = k$, and $d = 3$, so we must have $(1)(3) - (2)(k) \neq 0$, or $k \neq \frac{3}{2}$. Therefore, A^{-1} has an inverse provided $k \neq \frac{3}{2}$. Using Formula (13), we have

$$A^{-1} = \frac{1}{3 - 2k} \begin{bmatrix} 3 & -2 \\ -k & 1 \end{bmatrix}.$$

47. From the computation $\begin{bmatrix} a & 0 & | & 1 & 0 \\ 0 & d & | & 0 & 0 \end{bmatrix} \begin{array}{c} \frac{1}{a} R_1 \\ \xrightarrow{} \\ \frac{1}{d} R_2 \end{array} \begin{bmatrix} 1 & 0 & | & \frac{1}{a} & 0 \\ 0 & 1 & | & 0 & \frac{1}{d} \end{bmatrix}$, we see that A^{-1} exists provided $ad \neq 0$. Also, we can see that an $n \times n$ diagonal matrix A has an inverse provided all main diagonal entries are nonzero. In fact, the inverse is an $n \times n$ matrix whose main diagonal entries are the reciprocals of the corresponding entries of A.

49. True. Multiplying both sides of the equation by cA yields

$$I = (cA)(cA)^{-1} = (cA)\left[\frac{1}{c}\left(A^{-1}\right)\right] = c\left(\frac{1}{c}\right)AA^{-1} = I.$$

51. True. $AX = B$ can have a unique solution only if A^{-1} exists, in which case the solution is found as follows: $A^{-1}(AX) = A^{-1}B$, $(A^{-1}A)X = A^{-1}B$, $IX = A^{-1}B$, and $X = A^{-1}B$.

Technology Exercises   page 155

1. $\begin{bmatrix} 0.36 & 0.04 & -0.36 \\ 0.06 & 0.05 & 0.20 \\ -0.19 & 0.10 & 0.09 \end{bmatrix}$

3. $\begin{bmatrix} 0.01 & -0.09 & 0.31 & -0.11 \\ -0.25 & 0.58 & -0.15 & -0.02 \\ 0.86 & -0.42 & 0.07 & -0.37 \\ -0.27 & 0.01 & -0.05 & 0.31 \end{bmatrix}$

5. $\begin{bmatrix} 0.30 & 0.85 & -0.10 & -0.77 & -0.11 \\ -0.21 & 0.10 & 0.01 & -0.26 & 0.21 \\ 0.03 & -0.16 & 0.12 & -0.01 & 0.03 \\ -0.14 & -0.46 & 0.13 & 0.71 & -0.05 \\ 0.10 & -0.05 & -0.10 & -0.03 & 0.11 \end{bmatrix}$

7. $x = 1.2$, $y = 3.6$, and $z = 2.7$.

9. $x_1 = 2.50$, $x_2 = -0.88$, $x_3 = 0.70$, and $x_4 = 0.51$.

2.7   Leontief Input-Output Model

Concept Questions   page 160

1. X represents total output, AX represents internal consumption, and D represents consumer demand.

Exercises   page 160

1. a. The amount of agricultural products consumed in the production of \$100 million worth of manufactured goods is given by $100 \cdot 0.10$, or \$10 million.

b. The amount of manufactured goods required to produce $200 million of all goods in the economy is given by $200 (0.1 + 0.4 + 0.3) = 160$, or $160 million.

c. From the input-output matrix, we see that the agricultural sector consumes the greatest amount of agricultural products, 0.4 units, in the production of each unit of goods in that sector. The manufacturing and transportation sectors consume the least, at 0.1 units each.

3. Multiplying both sides of the given equation on the left by $(I - A)^{-1}$, we see that $X = (I - A)^{-1} D$.

Now $I - A = \begin{bmatrix} 1 & 0 \\ 0 & 1 \end{bmatrix} - \begin{bmatrix} 0.4 & 0.2 \\ 0.3 & 0.1 \end{bmatrix} = \begin{bmatrix} 0.6 & -0.2 \\ -0.3 & 0.9 \end{bmatrix}$. Using the formula for the inverse of a 2×2 matrix, we

find $(I - A)^{-1} = \begin{bmatrix} 1.875 & 0.417 \\ 0.625 & 1.25 \end{bmatrix}$. Thus, $(I - A)^{-1} X = \begin{bmatrix} 1.875 & 0.417 \\ 0.625 & 1.25 \end{bmatrix} \begin{bmatrix} 10 \\ 12 \end{bmatrix} = \begin{bmatrix} 23.754 \\ 21.25 \end{bmatrix}$.

5. We first compute $I - A = \begin{bmatrix} 1 & 0 \\ 0 & 1 \end{bmatrix} - \begin{bmatrix} 0.5 & 0.2 \\ 0.2 & 0.5 \end{bmatrix} = \begin{bmatrix} 0.5 & -0.2 \\ -0.2 & 0.5 \end{bmatrix}$. Using the formula for the inverse of a

2×2 matrix, we find $(I - A)^{-1} = \begin{bmatrix} 2.381 & 0.952 \\ 0.952 & 2.381 \end{bmatrix}$, so $\begin{bmatrix} x \\ y \end{bmatrix} = \begin{bmatrix} 2.381 & 0.952 \\ 0.952 & 2.381 \end{bmatrix} \begin{bmatrix} 10 \\ 20 \end{bmatrix} = \begin{bmatrix} 42.85 \\ 57.14 \end{bmatrix}$.

7. We verify $(I - A)(I - A)^{-1} = \begin{bmatrix} 0.92 & -0.60 & -0.30 \\ -0.04 & 0.98 & -0.01 \\ -0.02 & 0 & 0.94 \end{bmatrix} \begin{bmatrix} 1.12 & 0.69 & 0.37 \\ 0.05 & 1.05 & 0.03 \\ 0.02 & 0.01 & 1.07 \end{bmatrix} = \begin{bmatrix} 1 & 0 & 0 \\ 0 & 1 & 0 \\ 0 & 0 & 1 \end{bmatrix}$.

9. a. $A = \begin{bmatrix} 0.2 & 0.4 \\ 0.3 & 0.3 \end{bmatrix}$ and $(I - A) = \begin{bmatrix} 1 & 0 \\ 0 & 1 \end{bmatrix} - \begin{bmatrix} 0.2 & 0.4 \\ 0.3 & 0.3 \end{bmatrix} = \begin{bmatrix} 0.8 & -0.4 \\ -0.3 & 0.7 \end{bmatrix}$.

Using the formula for the inverse of a 2×2 matrix, we find $(I - A)^{-1} = \begin{bmatrix} 1.591 & 0.909 \\ 0.682 & 1.818 \end{bmatrix}$.

Thus, $\begin{bmatrix} x \\ y \end{bmatrix} = \begin{bmatrix} 1.591 & 0.909 \\ 0.682 & 1.818 \end{bmatrix} \begin{bmatrix} 120 \\ 140 \end{bmatrix} = \begin{bmatrix} 318.18 \\ 336.36 \end{bmatrix}$. To fulfill consumer demand, $318.2 million worth of

agricultural products and $336.4 million worth of manufactured goods should be produced.

b. The net value of goods consumed in the internal process of production is

$AX = X - D = \begin{bmatrix} 318.18 \\ 336.36 \end{bmatrix} - \begin{bmatrix} 120 \\ 140 \end{bmatrix} = \begin{bmatrix} 198.18 \\ 196.36 \end{bmatrix}$, or $198.2 million of agricultural products and

$196.4 million worth of manufactured goods.

11. a. $I - A = \begin{bmatrix} 1 & 0 & 0 \\ 0 & 1 & 0 \\ 0 & 0 & 1 \end{bmatrix} - \begin{bmatrix} 0.4 & 0.1 & 0.1 \\ 0.1 & 0.4 & 0.3 \\ 0.2 & 0.2 & 0.2 \end{bmatrix} = \begin{bmatrix} 0.6 & -0.1 & -0.1 \\ -0.1 & 0.6 & -0.3 \\ -0.2 & -0.2 & 0.8 \end{bmatrix}$.

Using the methods of Section 2.6, we next compute the inverse $(I - A)^{-1}$ and calculate

$$X = (1 - A)^{-1} D = \begin{bmatrix} 1.875 & 0.446 & 0.402 \\ 0.625 & 2.054 & 0.848 \\ 0.625 & 0.625 & 1.563 \end{bmatrix} \begin{bmatrix} 200 \\ 100 \\ 60 \end{bmatrix} = \begin{bmatrix} 443.72 \\ 381.28 \\ 281.28 \end{bmatrix}.$$

Therefore, to fulfill demand, approximately \$443.7 million worth of agricultural products, \$381.3 million worth of manufactured goods, and \$281.3 million worth of transportation should be produced.

b. To meet the gross output, the value of goods and transportation consumed in the internal process of production is

$$AX = X - D = \begin{bmatrix} 443.72 \\ 381.28 \\ 281.28 \end{bmatrix} - \begin{bmatrix} 200 \\ 100 \\ 60 \end{bmatrix} = \begin{bmatrix} 243.72 \\ 281.28 \\ 221.28 \end{bmatrix},$$ or approximately \$243.7 million worth of agricultural

products, \$281.3 million worth of manufactured goods, and \$221.3 million worth of transportation.

13. We want to solve the equation $(I - A) X = D$ for X, the total output matrix. First, we compute

$I - A = \begin{bmatrix} 1 & 0 \\ 0 & 1 \end{bmatrix} - \begin{bmatrix} 0.4 & 0.2 \\ 0.3 & 0.5 \end{bmatrix} = \begin{bmatrix} 0.6 & -0.2 \\ -0.3 & 0.5 \end{bmatrix}$. Using the formula for the inverse of a 2×2 matrix, we find

$(I - A)^{-1} = \begin{bmatrix} 2.08 & 0.833 \\ 1.25 & 2.5 \end{bmatrix}$. Therefore, $X = (I - A)^{-1} D = \begin{bmatrix} 2.08 & 0.833 \\ 1.25 & 2.5 \end{bmatrix} \begin{bmatrix} 12 \\ 24 \end{bmatrix} = \begin{bmatrix} 45 \\ 75 \end{bmatrix}$.

We conclude that \$45 million worth of goods of one industry and \$75 million worth of goods of the other industry must be produced.

15. First, we compute $I - A = \begin{bmatrix} 1 & 0 & 0 \\ 0 & 1 & 0 \\ 0 & 0 & 1 \end{bmatrix} - \begin{bmatrix} 0.2 & 0.4 & 0.2 \\ 0.5 & 0 & 0.5 \\ 0 & 0.2 & 0 \end{bmatrix} = \begin{bmatrix} 0.8 & -0.4 & -0.2 \\ -0.5 & 1 & -0.5 \\ 0 & -0.2 & 1 \end{bmatrix}$.

Next, using the Gauss-Jordan elimination method, we find $(I - A)^{-1} = \begin{bmatrix} 1.8 & 0.88 & 0.80 \\ 1 & 1.6 & 1 \\ 0.2 & 0.32 & 1.20 \end{bmatrix}$.

Then $\begin{bmatrix} x \\ y \\ z \end{bmatrix} = \begin{bmatrix} 1.8 & 0.88 & 0.80 \\ 1 & 1.6 & 1 \\ 0.2 & 0.32 & 1.20 \end{bmatrix} \begin{bmatrix} 10 \\ 5 \\ 15 \end{bmatrix} = \begin{bmatrix} 34.4 \\ 33 \\ 21.6 \end{bmatrix}$. We conclude that \$34.4 million worth of goods of one

industry, \$33 million worth of a second industry, and \$21.6 million worth of a third industry should be produced.

Technology Exercises   page 164

1. The final outputs of the first, second, third, and fourth industries are 602.62, 502.30, 572.57, and 523.46 million dollars, respectively.

3. The final outputs of the first, second, third, and fourth industries are 143.06, 132.98, 188.59, and 125.53 million dollars, respectively.

CHAPTER 2	Concept Review Questions	page 165

1. a. one, many, no   **b.** one, many, no   **3.** $R_i \leftrightarrow R_j, cR_i, R_i + aR_j$, solution

5. size, entries   **7.** $m \times n, n \times m, a_{ji}$

9. a. Columns, rows   **b.** $m \times p$ **11.** $A^{-1}A, AA^{-1}$, singular

CHAPTER 2	Review Exercises	page 166

1. $\begin{bmatrix} 1 & 2 \\ -1 & 3 \\ 2 & 1 \end{bmatrix} + \begin{bmatrix} 1 & 0 \\ 0 & 1 \\ 1 & 2 \end{bmatrix} = \begin{bmatrix} 2 & 2 \\ -1 & 4 \\ 3 & 3 \end{bmatrix}.$

3. $\begin{bmatrix} -3 & 2 & 1 \end{bmatrix} \begin{bmatrix} 2 & 1 \\ -1 & 0 \\ 2 & 1 \end{bmatrix} = \begin{bmatrix} -6 & -2 \end{bmatrix}.$

5. By the equality of matrices, $x = 2, z = 1, y = 3$, and $w = 3$.

7. By the equality of matrices, $a + 3 = 6$ (so $a = 3$), $-1 = e + 2$ (so $e = -3$), $b = 4, c + 1 = -1$ (so $c = -2$), and $d = 2$.

9. $2A + 3B = 2 \begin{bmatrix} 1 & 3 & 1 \\ -2 & 1 & 3 \\ 4 & 0 & 2 \end{bmatrix} + 3 \begin{bmatrix} 2 & 1 & 3 \\ -2 & -1 & -1 \\ 1 & 4 & 2 \end{bmatrix} = \begin{bmatrix} 2 & 6 & 2 \\ -4 & 2 & 6 \\ 8 & 0 & 4 \end{bmatrix} + \begin{bmatrix} 6 & 3 & 9 \\ -6 & -3 & -3 \\ 3 & 12 & 6 \end{bmatrix} = \begin{bmatrix} 8 & 9 & 11 \\ -10 & -1 & 3 \\ 11 & 12 & 10 \end{bmatrix}.$

11. $3A = 3 \begin{bmatrix} 1 & 3 & 1 \\ -2 & 1 & 3 \\ 4 & 0 & 2 \end{bmatrix} = \begin{bmatrix} 3 & 9 & 3 \\ -6 & 3 & 9 \\ 12 & 0 & 6 \end{bmatrix}$ and $2(3A) = 2 \begin{bmatrix} 3 & 9 & 3 \\ -6 & 3 & 9 \\ 12 & 0 & 6 \end{bmatrix} = \begin{bmatrix} 6 & 18 & 6 \\ -12 & 6 & 18 \\ 24 & 0 & 12 \end{bmatrix}.$

13. $B - C = \begin{bmatrix} 2 & 1 & 3 \\ -2 & -1 & -1 \\ 1 & 4 & 2 \end{bmatrix} - \begin{bmatrix} 3 & -1 & 2 \\ 1 & 6 & 4 \\ 2 & 1 & 3 \end{bmatrix} = \begin{bmatrix} -1 & 2 & 1 \\ -3 & -7 & -5 \\ -1 & 3 & -1 \end{bmatrix}$ and so

$A(B - C) = \begin{bmatrix} 1 & 3 & 1 \\ -2 & 1 & 3 \\ 4 & 0 & 2 \end{bmatrix} \begin{bmatrix} -1 & 2 & 1 \\ -3 & -7 & -5 \\ -1 & 3 & -1 \end{bmatrix} = \begin{bmatrix} -11 & -16 & -15 \\ -4 & -2 & -10 \\ -6 & 14 & 2 \end{bmatrix}.$

15. $BC = \begin{bmatrix} 2 & 1 & 3 \\ -2 & -1 & -1 \\ 1 & 4 & 2 \end{bmatrix} \begin{bmatrix} 3 & -1 & 2 \\ 1 & 6 & 4 \\ 2 & 1 & 3 \end{bmatrix} = \begin{bmatrix} 13 & 7 & 17 \\ -9 & -5 & -11 \\ 11 & 25 & 24 \end{bmatrix}$, so

$ABC = \begin{bmatrix} 1 & 3 & 1 \\ -2 & 1 & 3 \\ 4 & 0 & 2 \end{bmatrix} \begin{bmatrix} 13 & 7 & 17 \\ -9 & -5 & -11 \\ 11 & 25 & 24 \end{bmatrix} = \begin{bmatrix} -3 & 17 & 8 \\ -2 & 56 & 27 \\ 74 & 78 & 116 \end{bmatrix}.$

17. Using the Gauss-Jordan elimination method, we find

$$\begin{bmatrix} 2 & -3 & | & 5 \\ 3 & 4 & | & -1 \end{bmatrix} \xrightarrow{\frac{1}{2}R_1} \begin{bmatrix} 1 & -\frac{3}{2} & | & \frac{5}{2} \\ 3 & 4 & | & -1 \end{bmatrix} \xrightarrow{R_2 - 3R_1} \begin{bmatrix} 1 & -\frac{3}{2} & | & \frac{5}{2} \\ 0 & \frac{17}{2} & | & -\frac{17}{2} \end{bmatrix} \xrightarrow{\frac{2}{17}R_2} \begin{bmatrix} 1 & -\frac{3}{2} & | & \frac{5}{2} \\ 0 & 1 & | & -1 \end{bmatrix} \xrightarrow{R_1 + \frac{3}{2}R_2} \begin{bmatrix} 1 & 0 & | & 1 \\ 0 & 1 & | & -1 \end{bmatrix}.$$

We conclude that $x = 1$ and $y = -1$.

19. $\begin{bmatrix} 1 & -1 & 2 & | & 5 \\ 3 & 2 & 1 & | & 10 \\ 2 & -3 & -2 & | & -10 \end{bmatrix} \xrightarrow[R_3 - 2R_1]{R_2 - 3R_1} \begin{bmatrix} 1 & -1 & 2 & | & 5 \\ 0 & 5 & -5 & | & -5 \\ 0 & -1 & -6 & | & -20 \end{bmatrix} \xrightarrow{\frac{1}{5}R_2} \begin{bmatrix} 1 & -1 & 2 & | & 5 \\ 0 & 1 & -1 & | & -1 \\ 0 & -1 & -6 & | & -20 \end{bmatrix} \xrightarrow[R_3 + R_2]{R_1 + R_2}$

$\begin{bmatrix} 1 & 0 & 1 & | & 4 \\ 0 & 1 & -1 & | & -1 \\ 0 & 0 & -7 & | & -21 \end{bmatrix} \xrightarrow{-\frac{1}{7}R_3} \begin{bmatrix} 1 & 0 & 1 & | & 4 \\ 0 & 1 & -1 & | & -1 \\ 0 & 0 & 1 & | & 3 \end{bmatrix} \xrightarrow[R_2 + R_3]{R_1 - R_3} \begin{bmatrix} 1 & 0 & 0 & | & 1 \\ 0 & 1 & 0 & | & 2 \\ 0 & 0 & 1 & | & 3 \end{bmatrix}$. Therefore, $x = 1$, $y = 2$, and $z = 3$.

21. $\begin{bmatrix} 3 & -2 & 4 & | & 11 \\ 2 & -4 & 5 & | & 4 \\ 1 & 2 & -1 & | & 10 \end{bmatrix} \xrightarrow{R_1 - R_2} \begin{bmatrix} 1 & 2 & -1 & | & 7 \\ 2 & -4 & 5 & | & 4 \\ 1 & 2 & -1 & | & 10 \end{bmatrix} \xrightarrow[R_3 - R_1]{R_2 - 2R_1} \begin{bmatrix} 1 & 2 & -1 & | & 7 \\ 0 & -8 & 7 & | & -10 \\ 0 & 0 & 0 & | & 3 \end{bmatrix}.$

Since this last row implies that $0 = 3$, we conclude that the system has no solution.

23. $\begin{bmatrix} 3 & -2 & 1 & | & 4 \\ 1 & 3 & -4 & | & -3 \\ 2 & -3 & 5 & | & 7 \\ 1 & -8 & 9 & | & 10 \end{bmatrix} \xrightarrow{R_1 - R_3} \begin{bmatrix} 1 & 1 & -4 & | & -3 \\ 1 & 3 & -4 & | & -3 \\ 2 & -3 & 5 & | & 7 \\ 1 & -8 & 9 & | & 10 \end{bmatrix} \xrightarrow[\substack{R_2 - R_1 \\ R_3 - 2R_1 \\ R_4 - R_1}]{} \begin{bmatrix} 1 & 1 & -4 & | & -3 \\ 0 & 2 & 0 & | & 0 \\ 0 & -5 & 13 & | & 13 \\ 0 & -9 & 13 & | & 13 \end{bmatrix} \xrightarrow{\frac{1}{2}R_2}$

$\begin{bmatrix} 1 & 1 & -4 & | & -3 \\ 0 & 1 & 0 & | & 0 \\ 0 & -5 & 13 & | & 13 \\ 0 & -9 & 13 & | & 13 \end{bmatrix} \xrightarrow[\substack{R_1 - R_2 \\ R_3 + 5R_2 \\ R_4 + 9R_2}]{} \begin{bmatrix} 1 & 0 & -4 & | & -3 \\ 0 & 1 & 0 & | & 0 \\ 0 & 0 & 13 & | & 13 \\ 0 & 0 & 13 & | & 13 \end{bmatrix} \xrightarrow{\frac{1}{13}R_3} \begin{bmatrix} 1 & 0 & -4 & | & -3 \\ 0 & 1 & 0 & | & 0 \\ 0 & 0 & 1 & | & 1 \\ 0 & 0 & 13 & | & 13 \end{bmatrix} \xrightarrow[R_4 - 13R_3]{R_1 + 4R_3} \begin{bmatrix} 1 & 0 & 0 & | & 1 \\ 0 & 1 & 0 & | & 0 \\ 0 & 0 & 1 & | & 1 \\ 0 & 0 & 0 & | & 0 \end{bmatrix}.$

Thus, $x = 1$, $y = 0$, and $z = 1$.

25. $A^{-1} = \dfrac{1}{(3)(2) - (1)(1)} \begin{bmatrix} 2 & -1 \\ -1 & 3 \end{bmatrix} = \begin{bmatrix} \frac{2}{5} & -\frac{1}{5} \\ -\frac{1}{5} & \frac{3}{5} \end{bmatrix}.$

27. $A^{-1} = \dfrac{1}{(3)(2) - (2)(4)} \begin{bmatrix} 2 & -4 \\ -2 & 3 \end{bmatrix} = \begin{bmatrix} -1 & 2 \\ 1 & -\frac{3}{2} \end{bmatrix}.$

29. $\begin{bmatrix} 2 & 3 & 1 & | & 1 & 0 & 0 \\ 1 & -1 & 2 & | & 0 & 1 & 0 \\ 1 & 2 & 1 & | & 0 & 0 & 1 \end{bmatrix} \xrightarrow{R_1 - R_2} \begin{bmatrix} 1 & 4 & -1 & | & 1 & -1 & 0 \\ 1 & -1 & 2 & | & 0 & 1 & 0 \\ 1 & 2 & 1 & | & 0 & 0 & 1 \end{bmatrix} \xrightarrow[R_3 - R_1]{R_2 - R_1} \begin{bmatrix} 1 & 4 & -1 & | & 1 & -1 & 0 \\ 0 & -5 & 3 & | & -1 & 2 & 0 \\ 0 & -2 & 2 & | & -1 & 1 & 1 \end{bmatrix}$

$\xrightarrow{R_2 - 3R_3} \begin{bmatrix} 1 & 4 & -1 & | & 1 & -1 & 0 \\ 0 & 1 & -3 & | & 2 & -1 & -3 \\ 0 & -2 & 2 & | & -1 & 1 & 1 \end{bmatrix} \xrightarrow[R_3 + 2R_2]{R_1 - 4R_2} \begin{bmatrix} 1 & 0 & 11 & | & -7 & 3 & 12 \\ 0 & 1 & -3 & | & 2 & -1 & -3 \\ 0 & 0 & -4 & | & 3 & -1 & -5 \end{bmatrix} \xrightarrow{-\frac{1}{4}R_3}$

$\begin{bmatrix} 1 & 0 & 11 & | & -7 & 3 & 12 \\ 0 & 1 & -3 & | & 2 & -1 & -3 \\ 0 & 0 & 1 & | & -\frac{3}{4} & \frac{1}{4} & \frac{5}{4} \end{bmatrix} \xrightarrow[R_2 + 3R_3]{R_1 - 11R_3} \begin{bmatrix} 1 & 0 & 0 & | & \frac{5}{4} & \frac{1}{4} & -\frac{7}{4} \\ 0 & 1 & 0 & | & -\frac{1}{4} & -\frac{1}{4} & \frac{3}{4} \\ 0 & 0 & 1 & | & -\frac{3}{4} & \frac{1}{4} & \frac{5}{4} \end{bmatrix}$, so $A^{-1} = \begin{bmatrix} \frac{5}{4} & \frac{1}{4} & -\frac{7}{4} \\ -\frac{1}{4} & -\frac{1}{4} & \frac{3}{4} \\ -\frac{3}{4} & \frac{1}{4} & \frac{5}{4} \end{bmatrix}$.

31. $\begin{bmatrix} 1 & 2 & 4 & | & 1 & 0 & 0 \\ 3 & 1 & 2 & | & 0 & 1 & 0 \\ 1 & 0 & -6 & | & 0 & 0 & 1 \end{bmatrix} \xrightarrow[R_3 - R_1]{R_2 - 3R_1} \begin{bmatrix} 1 & 2 & 4 & | & 1 & 0 & 0 \\ 0 & -5 & -10 & | & -3 & 1 & 0 \\ 0 & -2 & -10 & | & -1 & 0 & 1 \end{bmatrix} \xrightarrow{R_2 - 3R_3} \begin{bmatrix} 1 & 2 & 4 & | & 1 & 0 & 0 \\ 0 & 1 & 20 & | & 0 & 1 & -3 \\ 0 & -2 & -10 & | & -1 & 0 & 1 \end{bmatrix} \xrightarrow[R_3 + 2R_2]{R_1 - 2R_2}$

$\begin{bmatrix} 1 & 0 & -36 & | & 1 & -2 & 6 \\ 0 & 1 & 20 & | & 0 & 1 & -3 \\ 0 & 0 & 30 & | & -1 & 2 & -5 \end{bmatrix} \xrightarrow{\frac{1}{30}R_3} \begin{bmatrix} 1 & 0 & -36 & | & 1 & -2 & 6 \\ 0 & 1 & 20 & | & 0 & 1 & -3 \\ 0 & 0 & 1 & | & -\frac{1}{30} & \frac{1}{15} & -\frac{1}{6} \end{bmatrix} \xrightarrow[R_2 - 20R_3]{R_1 + 36R_3} \begin{bmatrix} 1 & 0 & 0 & | & -\frac{1}{5} & \frac{2}{5} & 0 \\ 0 & 1 & 0 & | & \frac{2}{3} & -\frac{1}{3} & \frac{1}{3} \\ 0 & 0 & 1 & | & -\frac{1}{30} & \frac{1}{15} & -\frac{1}{6} \end{bmatrix}$, so

$A^{-1} = \begin{bmatrix} -\frac{1}{5} & \frac{2}{5} & 0 \\ \frac{2}{3} & -\frac{1}{3} & \frac{1}{3} \\ -\frac{1}{30} & \frac{1}{15} & -\frac{1}{6} \end{bmatrix}$.

33. $\left(A^{-1}B\right)^{-1} = B^{-1}\left(A^{-1}\right)^{-1} = B^{-1}A$. Now

$B^{-1} = \dfrac{1}{(3)(2) - 4(1)} \begin{bmatrix} 2 & -1 \\ -4 & 3 \end{bmatrix} = \begin{bmatrix} 1 & -\frac{1}{2} \\ -2 & \frac{3}{2} \end{bmatrix}$, so $B^{-1}A = \begin{bmatrix} 1 & -\frac{1}{2} \\ -2 & \frac{3}{2} \end{bmatrix}\begin{bmatrix} 1 & 2 \\ -1 & 2 \end{bmatrix} = \begin{bmatrix} \frac{3}{2} & 1 \\ -\frac{7}{2} & -1 \end{bmatrix}$.

35. $2A - C = \begin{bmatrix} 2 & 4 \\ -2 & 4 \end{bmatrix} - \begin{bmatrix} 1 & 1 \\ -1 & 2 \end{bmatrix} = \begin{bmatrix} 1 & 3 \\ -1 & 2 \end{bmatrix}$, so $(2A - C)^{-1} = \dfrac{1}{(1)(2) - (-1)(3)} \begin{bmatrix} 2 & -3 \\ 1 & 1 \end{bmatrix} = \begin{bmatrix} \frac{2}{5} & -\frac{3}{5} \\ \frac{1}{5} & \frac{1}{5} \end{bmatrix}$.

37. $A = \begin{bmatrix} 2 & 3 \\ 1 & -2 \end{bmatrix}$, $X = \begin{bmatrix} x \\ y \end{bmatrix}$, and $C = \begin{bmatrix} -8 \\ 3 \end{bmatrix}$, so $A^{-1} = \dfrac{1}{(-2)(2) - (1)(3)} \begin{bmatrix} -2 & -3 \\ -1 & 2 \end{bmatrix} = \begin{bmatrix} \frac{2}{7} & \frac{3}{7} \\ \frac{1}{7} & -\frac{2}{7} \end{bmatrix}$.

Thus, $\begin{bmatrix} x \\ y \end{bmatrix} = A^{-1}B = \begin{bmatrix} \frac{2}{7} & \frac{3}{7} \\ \frac{1}{7} & -\frac{2}{7} \end{bmatrix}\begin{bmatrix} -8 \\ 3 \end{bmatrix} = \begin{bmatrix} -1 \\ -2 \end{bmatrix}$.

39. Put $X = \begin{bmatrix} x \\ y \\ z \end{bmatrix}$, $A = \begin{bmatrix} 1 & -2 & 4 \\ 2 & 3 & -2 \\ 1 & 4 & -6 \end{bmatrix}$, and $C = \begin{bmatrix} 13 \\ 0 \\ -15 \end{bmatrix}$. Then $AX = C$ and $X = A^{-1}C$.

To find A^{-1}, we calculate $\begin{bmatrix} 1 & -2 & 4 & | & 1 & 0 & 0 \\ 2 & 3 & -2 & | & 0 & 1 & 0 \\ 1 & 4 & -6 & | & 0 & 0 & 1 \end{bmatrix} \xrightarrow[R_3 - R_1]{R_2 - 2R_1} \begin{bmatrix} 1 & -2 & 4 & | & 1 & 0 & 0 \\ 0 & 7 & -10 & | & -2 & 1 & 0 \\ 0 & 6 & -10 & | & -1 & 0 & 1 \end{bmatrix} \xrightarrow{R_2 - R_3}$

$\begin{bmatrix} 1 & -2 & 4 & | & 1 & 0 & 0 \\ 0 & 1 & 0 & | & -1 & 1 & -1 \\ 0 & 6 & -10 & | & -1 & 0 & 1 \end{bmatrix} \xrightarrow[R_3 - 6R_2]{R_1 + 2R_2} \begin{bmatrix} 1 & 0 & 4 & | & -1 & 2 & -2 \\ 0 & 1 & 0 & | & -1 & 1 & -1 \\ 0 & 0 & -10 & | & 5 & -6 & 7 \end{bmatrix} \xrightarrow{-\frac{1}{10}R_3} \begin{bmatrix} 1 & 0 & 4 & | & -1 & 2 & -2 \\ 0 & 1 & 0 & | & -1 & 1 & -1 \\ 0 & 0 & 1 & | & -\frac{1}{2} & \frac{3}{5} & -\frac{7}{10} \end{bmatrix} \xrightarrow{R_1 - 4R_3}$

$\begin{bmatrix} 1 & 0 & 0 & | & 1 & -\frac{2}{5} & \frac{4}{5} \\ 0 & 1 & 0 & | & -1 & 1 & -1 \\ 0 & 0 & 1 & | & -\frac{1}{2} & \frac{3}{5} & -\frac{7}{10} \end{bmatrix}$, so $A^{-1} = \begin{bmatrix} 1 & -\frac{2}{5} & \frac{4}{5} \\ -1 & 1 & -1 \\ -\frac{1}{2} & \frac{3}{5} & -\frac{7}{10} \end{bmatrix}$.

Therefore, $X = A^{-1}C = \begin{bmatrix} 1 & -\frac{2}{5} & \frac{4}{5} \\ -1 & 1 & -1 \\ -\frac{1}{2} & \frac{3}{5} & -\frac{7}{10} \end{bmatrix} \begin{bmatrix} 13 \\ 0 \\ -15 \end{bmatrix} = \begin{bmatrix} 1 \\ 2 \\ 4 \end{bmatrix}$; that is, $x = 1$, $y = 2$, and $z = 4$.

41. Let $S = \begin{matrix} \text{Premium} & \text{Super} & \text{Regular} & \text{Diesel} \\ \begin{bmatrix} 600 & 800 & 1000 & 700 \\ 700 & 600 & 1200 & 400 \\ 900 & 700 & 1400 & 800 \end{bmatrix} \end{matrix}$ and $T = \begin{matrix} \text{Premium} \\ \text{Super} \\ \text{Regular} \\ \text{Diesel} \end{matrix} \begin{bmatrix} 3.80 \\ 3.60 \\ 3.40 \\ 3.70 \end{bmatrix}$ be the matrices representing

the sales in the three gasoline stations and the unit prices for the various fuels. Then the total revenue at each station

is found by computing $ST = \begin{bmatrix} 600 & 800 & 1000 & 700 \\ 700 & 600 & 1200 & 400 \\ 900 & 700 & 1400 & 800 \end{bmatrix} \begin{bmatrix} 3.80 \\ 3.60 \\ 3.40 \\ 3.70 \end{bmatrix} = \begin{bmatrix} 11{,}150 \\ 10{,}380 \\ 13{,}660 \end{bmatrix}$.

We conclude that the total revenue of station A is \$11,150, that of station B is \$10,380, and that of station C is \$13,660.

43. a. $A = \begin{bmatrix} 800 & 1200 & 250 & 1500 \\ 600 & 1400 & 300 & 1200 \end{bmatrix}$ **b.** $B = \begin{bmatrix} 12.57 \\ 28.21 \\ 214.92 \\ 36.34 \end{bmatrix}$

c. $AB = \begin{bmatrix} 800 & 1200 & 250 & 1500 \\ 600 & 1400 & 300 & 1200 \end{bmatrix} \begin{bmatrix} 12.57 \\ 28.21 \\ 214.92 \\ 36.34 \end{bmatrix} = \begin{bmatrix} 152{,}148 \\ 155{,}120 \end{bmatrix}$

45. We wish to solve the system of equations

$$2x + 2y + 3z = 210$$
$$2x + 3y + 4z = 270$$
$$3x + 4y + 3z = 300$$

Using the Gauss–Jordan method, we find
$\begin{bmatrix} 2 & 2 & 3 & | & 210 \\ 2 & 3 & 4 & | & 270 \\ 3 & 4 & 3 & | & 300 \end{bmatrix} \xrightarrow{\frac{1}{2}R_1} \begin{bmatrix} 1 & 1 & \frac{3}{2} & | & 105 \\ 2 & 3 & 4 & | & 270 \\ 3 & 4 & 3 & | & 300 \end{bmatrix} \xrightarrow[R_3 - 3R_1]{R_2 - 2R_1}$

$\begin{bmatrix} 1 & 1 & \frac{3}{2} & | & 105 \\ 0 & 1 & 1 & | & 60 \\ 0 & 1 & -\frac{3}{2} & | & -15 \end{bmatrix} \xrightarrow[R_3 - R_2]{R_1 - R_2} \begin{bmatrix} 1 & 0 & \frac{1}{2} & | & 45 \\ 0 & 1 & 1 & | & 60 \\ 0 & 0 & -\frac{5}{2} & | & -75 \end{bmatrix} \xrightarrow{-\frac{2}{5}R_3} \begin{bmatrix} 1 & 0 & \frac{1}{2} & | & 45 \\ 0 & 1 & 1 & | & 60 \\ 0 & 0 & 1 & | & 30 \end{bmatrix} \xrightarrow[R_2 - R_3]{R_1 - \frac{1}{2}R_3} \begin{bmatrix} 1 & 0 & 0 & | & 30 \\ 0 & 1 & 0 & | & 30 \\ 0 & 0 & 1 & | & 30 \end{bmatrix}$. Thus,

$x = y = z = 30$, and so Desmond should produce 30 of each type of pendant.

47. a. The amount consumed is 200 (0.15), or \$30 million.

b. The amount consumed is 300 (0.1 + 0.15), or \$75 million.

c. The agricultural sector consumes the greatest amount of agricultural products, namely 0.2 units, in the production of each unit of goods in that sector. The manufacturing sector consumes the lesser, namely 0.15 units.

49. a. $I - A = \begin{bmatrix} 1 & 0 \\ 0 & 1 \end{bmatrix} - \begin{bmatrix} 0.2 & 0.15 \\ 0.1 & 0.15 \end{bmatrix} = \begin{bmatrix} 0.8 & -0.15 \\ -0.1 & 0.85 \end{bmatrix}$, so

$(I - A)^{-1} = \dfrac{1}{(0.8)(0.85) - (-0.1)(-0.15)} \begin{bmatrix} 0.85 & 0.15 \\ 0.1 & 0.8 \end{bmatrix} = \dfrac{1}{0.665} \begin{bmatrix} 0.85 & 0.15 \\ 0.1 & 0.8 \end{bmatrix}$ and

$X = (I - A)^{-1} D = \dfrac{1}{0.665} \begin{bmatrix} 0.85 & 0.15 \\ 0.1 & 0.8 \end{bmatrix} \begin{bmatrix} 100 \\ 80 \end{bmatrix} = \begin{bmatrix} 145.86 \\ 111.28 \end{bmatrix}$. Therefore, to fulfill demand, \$145.86 million

worth of agricultural products and \$111.28 million worth of manufactured goods should be produced.

b. To meet the gross output, the value of goods consumed in internal production is

$AX = X - D = \begin{bmatrix} 145.86 \\ 111.28 \end{bmatrix} - \begin{bmatrix} 100 \\ 80 \end{bmatrix} = \begin{bmatrix} 45.86 \\ 31.28 \end{bmatrix}$, or \$45.86 million worth of agricultural products and

\$31.28 million worth of manufactured goods.

CHAPTER 2   Before Moving On...   page 168

1. $\begin{bmatrix} 2 & 1 & -1 & | & -1 \\ 1 & 3 & 2 & | & 2 \\ 3 & 3 & -3 & | & -5 \end{bmatrix} \xrightarrow{R_1 \leftrightarrow R_2} \begin{bmatrix} 1 & 3 & 2 & | & 2 \\ 2 & 1 & -1 & | & -1 \\ 3 & 3 & -3 & | & -5 \end{bmatrix} \xrightarrow[R_2 - 2R_1]{R_3 - 3R_1} \begin{bmatrix} 1 & 3 & 2 & | & 2 \\ 0 & -5 & -5 & | & -5 \\ 0 & -6 & -9 & | & -11 \end{bmatrix} \xrightarrow{-\frac{1}{5}R_2}$

$\begin{bmatrix} 1 & 3 & 2 & | & 2 \\ 0 & 1 & 1 & | & 1 \\ 0 & -6 & -9 & | & -11 \end{bmatrix} \xrightarrow[R_3 + 6R_2]{R_1 - 3R_2} \begin{bmatrix} 1 & 0 & -1 & | & -1 \\ 0 & 1 & 1 & | & 1 \\ 0 & 0 & -3 & | & -5 \end{bmatrix} \xrightarrow{-\frac{1}{3}R_3} \begin{bmatrix} 1 & 0 & -1 & | & -1 \\ 0 & 1 & 1 & | & 1 \\ 0 & 0 & 1 & | & \frac{5}{3} \end{bmatrix} \xrightarrow[R_2 - R_3]{R_1 + R_3} \begin{bmatrix} 1 & 0 & 0 & | & \frac{2}{3} \\ 0 & 1 & 0 & | & -\frac{2}{3} \\ 0 & 0 & 1 & | & \frac{5}{3} \end{bmatrix}$.

The solution is $x = \frac{2}{3}$, $y = -\frac{2}{3}$, $z = \frac{5}{3}$.

2. a. $x = 2, y = -3, z = 1$

b. No solution.

c. $x = 2, y = 1 - 3t, z = t$, where t is a parameter.

d. $x = y = z = w = 0.$

e. $x = 2 + t, y = 3 - 2t, z = t$, where t is a parameter

3. a. $\begin{bmatrix} 1 & 2 & | & 3 \\ 3 & -1 & | & -5 \\ 4 & 1 & | & -2 \end{bmatrix} \xrightarrow[R_3 - 4R_1]{R_2 - 3R_1} \begin{bmatrix} 1 & 2 & | & 3 \\ 0 & -7 & | & -14 \\ 0 & -7 & | & -14 \end{bmatrix} \xrightarrow{-\frac{1}{7}R_2} \begin{bmatrix} 1 & 2 & | & 3 \\ 0 & 1 & | & 2 \\ 0 & -7 & | & -14 \end{bmatrix} \xrightarrow[R_3 + 7R_2]{R_1 - 2R_2} \begin{bmatrix} 1 & 0 & | & -1 \\ 0 & 1 & | & 2 \\ 0 & 0 & | & 0 \end{bmatrix}.$

The solution is $x = -1, y = 2.$

b. $\begin{bmatrix} 1 & -2 & 4 & | & 2 \\ 3 & 1 & -2 & | & 1 \end{bmatrix} \xrightarrow{R_2 - 3R_1} \begin{bmatrix} 1 & -2 & 4 & | & 2 \\ 0 & 7 & -14 & | & -5 \end{bmatrix} \xrightarrow{\frac{1}{7}R_2} \begin{bmatrix} 1 & -2 & 4 & | & 2 \\ 0 & 1 & -2 & | & -\frac{5}{7} \end{bmatrix} \xrightarrow{R_1 + 2R_2} \begin{bmatrix} 1 & 0 & 0 & | & \frac{4}{7} \\ 0 & 1 & -2 & | & -\frac{5}{2} \end{bmatrix}.$

The solution is $x = \frac{4}{7}, y = -\frac{5}{7} + 2t, z = t$, where t is a parameter.

4. a. $AB = \begin{bmatrix} 1 & -2 & 4 \\ 3 & 0 & 1 \end{bmatrix} \begin{bmatrix} 1 & -1 & 2 \\ 3 & 1 & -1 \\ 2 & 1 & 0 \end{bmatrix} = \begin{bmatrix} 3 & 1 & 4 \\ 5 & -2 & 6 \end{bmatrix}.$

b. $A + C^T = \begin{bmatrix} 1 & -2 & 4 \\ 3 & 0 & 1 \end{bmatrix} + \begin{bmatrix} 2 & 1 & 3 \\ -2 & 1 & 4 \end{bmatrix} = \begin{bmatrix} 3 & -1 & 7 \\ 1 & 1 & 5 \end{bmatrix}$ and

$(A + C^T) B = \begin{bmatrix} 3 & -1 & 7 \\ 1 & 1 & 5 \end{bmatrix} \begin{bmatrix} 1 & -1 & 2 \\ 3 & 1 & -1 \\ 2 & 1 & 0 \end{bmatrix} = \begin{bmatrix} 14 & 3 & 7 \\ 14 & 5 & 1 \end{bmatrix}.$

c. $C^T B - A B^T = \begin{bmatrix} 2 & 1 & 3 \\ -2 & 1 & 4 \end{bmatrix} \begin{bmatrix} 1 & -1 & 2 \\ 3 & 1 & -1 \\ 2 & 1 & 0 \end{bmatrix} \begin{bmatrix} 1 & -2 & 4 \\ 3 & 0 & 1 \end{bmatrix} \begin{bmatrix} 1 & 3 & 2 \\ -1 & 1 & 1 \\ 2 & -1 & 0 \end{bmatrix}$

$= \begin{bmatrix} 11 & 2 & 3 \\ 9 & 7 & -5 \end{bmatrix} - \begin{bmatrix} 11 & -3 & 0 \\ 5 & 8 & 6 \end{bmatrix} = \begin{bmatrix} 0 & 5 & 3 \\ 4 & -1 & -11 \end{bmatrix}.$

5. $\begin{bmatrix} 2 & 1 & 2 & | & 1 & 0 & 0 \\ 0 & -1 & 3 & | & 0 & 1 & 0 \\ 1 & 1 & 0 & | & 0 & 0 & 1 \end{bmatrix} \xrightarrow{R_1 \leftrightarrow R_3} \begin{bmatrix} 1 & 1 & 0 & | & 0 & 0 & 1 \\ 0 & -1 & 3 & | & 0 & 1 & 0 \\ 2 & 1 & 2 & | & 1 & 0 & 0 \end{bmatrix} \xrightarrow[R_3 - 2R_1]{\substack{R_1 + R_2 \\ -R_2}} \begin{bmatrix} 1 & 0 & 3 & | & 0 & 1 & 1 \\ 0 & 1 & -3 & | & 0 & -1 & 0 \\ 0 & -1 & 2 & | & 1 & 0 & -2 \end{bmatrix} \xrightarrow[R_1 - 3R_3]{R_2 + 3R_3}$

$\begin{bmatrix} 1 & 0 & 0 & | & 3 & -2 & -5 \\ 0 & 1 & 0 & | & -3 & 2 & 6 \\ 0 & 0 & 1 & | & -1 & 1 & 2 \end{bmatrix}$, so $A^{-1} = \begin{bmatrix} 3 & -2 & -5 \\ -3 & 2 & 6 \\ -1 & 1 & 2 \end{bmatrix}.$

6. $A = \begin{bmatrix} 2 & 0 & 1 \\ 2 & 1 & -1 \\ 3 & 1 & -1 \end{bmatrix}$, $B = \begin{bmatrix} 4 \\ -1 \\ 0 \end{bmatrix}$, and $X = \begin{bmatrix} x \\ y \\ z \end{bmatrix}$. To find A^{-1}, we calculate

$$\begin{bmatrix} 2 & 0 & 1 & | & 1 & 0 & 0 \\ 2 & 1 & -1 & | & 0 & 1 & 0 \\ 3 & 1 & -1 & | & 0 & 0 & 1 \end{bmatrix} \xrightarrow{R_1 \leftrightarrow R_3} \begin{bmatrix} 3 & 1 & -1 & | & 0 & 0 & 1 \\ 2 & 1 & -1 & | & 0 & 1 & 0 \\ 2 & 0 & 1 & | & 1 & 0 & 0 \end{bmatrix} \xrightarrow{R_1 - R_2} \begin{bmatrix} 1 & 0 & 0 & | & 0 & -1 & 1 \\ 2 & 1 & -1 & | & 0 & 1 & 0 \\ 2 & 0 & 1 & | & 1 & 0 & 0 \end{bmatrix} \xrightarrow[R_3 - 2R_1]{R_2 - 2R_1}$$

$$\begin{bmatrix} 1 & 0 & 0 & | & 0 & -1 & 1 \\ 0 & 1 & -1 & | & 0 & 3 & -2 \\ 0 & 0 & 1 & | & 1 & 2 & -2 \end{bmatrix} \xrightarrow{R_2 + R_3} \begin{bmatrix} 1 & 0 & 0 & | & 0 & -1 & 1 \\ 0 & 1 & 0 & | & 1 & 5 & -4 \\ 0 & 0 & 1 & | & 1 & 2 & -2 \end{bmatrix}.$$ Therefore, $A^{-1} = \begin{bmatrix} 0 & -1 & 1 \\ 1 & 5 & -4 \\ 1 & 2 & -2 \end{bmatrix}$, so

$$X = \begin{bmatrix} x \\ y \\ z \end{bmatrix} = A^{-1}B = \begin{bmatrix} 0 & -1 & 1 \\ 1 & 5 & -4 \\ 1 & 2 & -2 \end{bmatrix} \begin{bmatrix} 4 \\ -1 \\ 0 \end{bmatrix} = \begin{bmatrix} 1 \\ -1 \\ 2 \end{bmatrix},$$ and we conclude that $x = 1$, $y = -1$, and $z = 2$.

LINEAR PROGRAMMING: A GEOMETRIC APPROACH

3.1 Graphing Systems of Linear Inequalities in Two Variables

Problem-Solving Tips

1. When you graph an inequality, use a solid line to show inclusion in the solution set ($\leq$ or $\geq$) and a dashed line to show exclusion from the solution set ($<$ or $>$).

2. Pick the point $(0, 0)$ as your test point if it does not lie on the line you are considering. Otherwise, pick a point at which it is easy to evaluate the quantities you are working with.

3. If the solution set of a system of linear inequalities can be enclosed by a circle, the solutions set is **bounded**. Otherwise, it is **unbounded**.

Concept Questions   page 177

1. **a.** The solution set of $ax + by < c$ is a half-plane that does not include the line with equation $ax + by = c$. The solution set of $ax + by \leq c$, on the other hand, includes the line.

   b. It is the line with equation $ax + by = c$.

Exercises   page 178

1. $4x - 8 < 0$ implies $x < 2$.

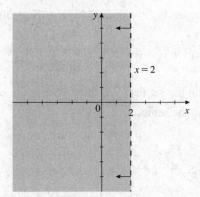

3. $x - y \leq 0$ implies $x \leq y$.

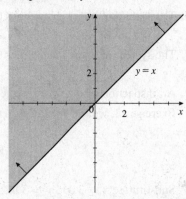

5. The graph of $x \leq -3$ is a half-plane.

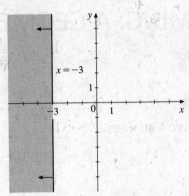

7. We first sketch the straight line with equation $2x + y = 4$. Next, picking the test point $(0, 0)$, we have $2(0) + (0) = 0 \leq 4$. We conclude that the half-plane containing the origin is the required half-plane.

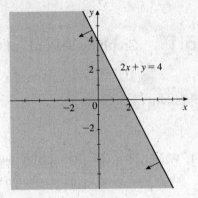

9. We first sketch the graph of the straight line $4x - 3y = -24$. Next, picking the test point $(0, 0)$, we see that $4(0) - 3(0) = 0 \not\leq -24$. We conclude that the half-plane not containing the origin is the required half-plane.

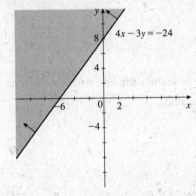

11. The system of linear inequalities that describes the shaded region is $x \geq 1, x \leq 5, y \geq 2, y \leq 4$. We may also combine the first and second inequalities and the third and fourth inequalities and write $1 \leq x \leq 5$ and $2 \leq y \leq 4$.

13. The system of linear inequalities that describes the shaded region is $2x - y \geq 2, 5x + 7y \geq 35, x \leq 4$.

15. The system of linear inequalities that describes the shaded region is $7x + 4y \leq 140, x + 3y \geq 30, x - y \geq -10$.

17. The system of linear inequalities that describes the shaded region is $x + y \geq 7, x \geq 2, y \geq 3, y \leq 7$.

19. An inspection of the figure suggests that the point $(3, 3)$ is in the set depicted there. Using the solution of Exercise 13, we see that the system of inequalities describing the set is

$$
\begin{aligned}
2x - y &\geq 2 \\
5x + 7y &\geq 35 \\
x &\leq 4
\end{aligned}
$$

Substituting $x = 3$ and $y = 3$ into this system gives

$$2\,(3) - \quad 3 = \; 3 \geq 2$$
$$5\,(3) + 7\,(3) = 36 \geq 35$$
$$3 \qquad \quad \leq \; 4$$

and so all the inequalities in the system are satisfied. Thus, $(3, 3)$ lies in S.

21. By inspection, the point $(10, 10)$ seems to lie in the solution set. Using the solution of Exercise 15, we see that the system of inequalities describing the set is

$$7x + 4y \leq \; 140$$
$$x + 3y \geq \quad 30$$
$$x - \quad y \geq -10$$

Substituting $x = y = 10$ into the system gives

$$7\,(10) + 4\,(10) = 110 \leq \; 140$$
$$10 + 3\,(10) = \quad 40 \geq \quad 30$$
$$10 - \quad 10 = \quad 0 \geq -10$$

Each inequality is satisfied, showing that $(10, 10)$ is in the solution set S.

23. The vertex $A\,(2, 3)$ is found by solving the system

$$\begin{cases} 2x + 4y = 16 \\ -x + 3y = \; 7 \end{cases}$$ Observe that a dashed line is

used to show that no point on the line constitutes a solution to the given problem. Observe also that this is an unbounded solution set.

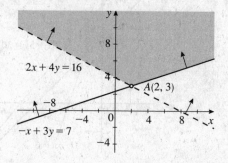

25. The vertex $A\,(2, 2)$ is found by solving the system

$$\begin{cases} x - \quad y = \; 0 \\ 2x + 3y = 10 \end{cases}$$ The solution set is unbounded.

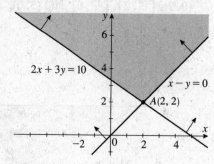

27. Because the two half-planes defined by the system of inequalities $\begin{cases} x + 2y \geq \;\;\; 3 \\ 2x + 4y \leq -2 \end{cases}$ have no point in common, we conclude that the system has no solution. The (empty) set is bounded.

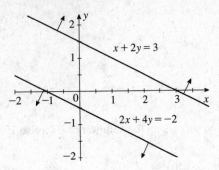

29. The vertex $A\,(3, 3)$ is found by solving the system $\begin{cases} x + y = 6 \\ x \;\;\;\;\;\;= 3 \end{cases}$ Observe that the solution set is bounded.

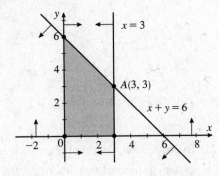

31. Observe that the two lines described by the equations $3x - 6y = 12$ and $-x + 2y = 4$ do not intersect because they are parallel. The solution set is unbounded.

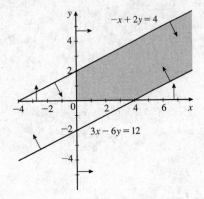

33. The vertices are $A\left(0, \frac{24}{7}\right)$, $B\left(0, \frac{8}{3}\right)$, and $C\,(8, 0)$. The solution set is unbounded.

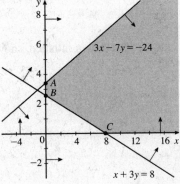

35. The corners of the bounded solution set are $\left(0, \frac{3}{2}\right)$, $(0, 2)$, $\left(\frac{24}{5}, 2\right)$, $\left(\frac{16}{5}, 0\right)$, and $(3, 0)$.

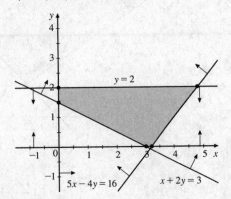

37. The bounded solution set has vertices $(0, 6)$, $(5, 0)$, $(4, 0)$, and $(1, 3)$.

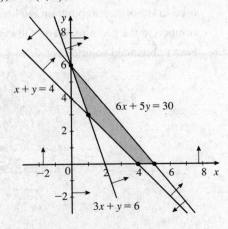

39. The unbounded solution set has vertices $(2, 8)$, $(0, 6)$, $(0, 3)$, and $(2, 2)$.

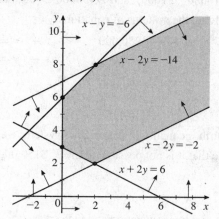

41. a. Let x and y denote the numbers of season ticket holders and other attendees. Then the total number of attendees is $x + y$. Since the capacity of the auditorium is 500, we must have $x + y \leq 500$. Also, the number of season ticket holders is at least 200, so $x \geq 200$. Similarly, we see that $y \geq 100$, so we have the system

$$
\begin{aligned}
x + y &\leq 500 \\
x \quad &\geq 200 \\
y &\geq 100
\end{aligned}
$$

b.

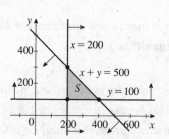

c. We substitute $x = 300$ and $y = 150$ into the system of inequalities:

$$
\begin{aligned}
300 + 150 = 400 &\leq 500 \\
300 \quad &\geq 200 \\
150 &\geq 100
\end{aligned}
$$

All the inequalities are satisfied, and we conclude that it is possible for 300 season ticket holders and 150 others to attend the concert.

43. a. Let x and y denote the amounts Louisa invests in Solaron and Windmill. Since she has at most \$250,000 to invest, we have $x + y \leq 250{,}000$. The condition that at least \$50,000 be invested in each company gives the inequalities $x \geq 50{,}000$ and $y \geq 50{,}000$. Finally, the condition that the amount invested in Solaron not exceed 120% of that invested in Windmill implies that $x \leq 1.2y$, or $x \leq \frac{6}{5}y$, or $5x - 6y \leq 0$. Thus, we have the system

$$
\begin{aligned}
x + \quad y &\leq 250{,}000 \\
x \quad\quad &\geq 50{,}000 \\
y &\geq 50{,}000 \\
5x - 6y &\leq \quad\quad 0
\end{aligned}
$$

b.

The vertices A (50000, 200000), $B\left(136363\frac{7}{11}, 113636\frac{4}{11}\right)$, C (60000, 50000), and D (50000, 50000) are found by solving the systems of equations

$$\begin{array}{ll} x + y = 250{,}000 & \quad x + y = 250{,}000 \\ x \phantom{{}+y} = 50{,}000 & \quad 5x - 6y = 0 \end{array}$$

$$\begin{array}{ll} y = 50{,}000 & \quad\quad\quad x \phantom{{}+y} = 50{,}000 \\ 5x - 6y = 0 & \text{and} \quad\;\; y = 50{,}000 \end{array}$$

c. Substituting $x = 150{,}000$ and $y = 100{,}000$ into the fourth inequality in the original system, we have $5(150{,}000) - 6(100{,}000) - 150{,}000 \nleq 0$. We conclude that it is not possible to invest \$150,000 in Solaron Corporation and \$100,000 in Windmill Corporation.

45. False. It is always a half-plane. A straight line is the graph of a linear equation and vice versa.

47. True. Since a circle can always be enclosed by a rectangle, the solution set of such a system is bounded if it can be enclosed by a rectangle.

3.2   Linear Programming Problems

Problem-Solving Tips

1. In a linear programming problem, a linear objective function is maximized or minimized subject to certain constraints. These constraints can be **linear equations** or **inequalities**.

2. When you solve an applied linear programming problem, it is important to understand the question in mathematical terms. If you are asked to maximize the profit, then the problem involves a linear objective function that should be maximized. If you are asked to minimize the cost, then the problem involves a linear objective function that should be minimized.

3. It is helpful to organize information in a table, as done in Examples 1–4, before you formulate a problem.

Concept Questions   page 185

1. See the definition on page 181 of the text.

3. In a maximization linear programming problem, we find the greatest value of the objective function. In a minimization problem, we find its least value.

Exercises   page 185

1. We tabulate the given information:

Machine	Product A	B	Time Available
I	6 minutes	9 minutes	300 minutes
II	5 minutes	4 minutes	180 minutes
Profit per unit	$3	$4	

Let x and y denote the numbers of units of Products A and B to be produced. Then the linear programming problem is:

$$\begin{aligned} \text{Maximize} \quad & P = 3x + 4y \\ \text{subject to} \quad & 6x + 9y \le 300 \\ & 5x + 4y \le 180 \\ & x \ge 0, y \ge 0. \end{aligned}$$

3. Let x and y denote the numbers of model A and model B grates to be produced. Since only 1000 pounds of cast iron are available, we must have $3x + 4y \le 1000$. The restriction that only 20 hours (or 1200 minutes) of labor are available per day implies that $6x + 3y \le 1200$. Then the profit on the production of these grates is given by $P = 2x + 1.5y$.

Summarizing, we have the following linear programming problem:

$$\begin{aligned} \text{Maximize} \quad & P = 2x + 1.5y \\ \text{subject to} \quad & 3x + 4y \le 1000 \\ & 6x + 3y \le 1200 \\ & x \ge 0, y \ge 0. \end{aligned}$$

5. Let x denote the number of tables and y the number of chairs to be manufactured. Since 3200 board feet are available, we have $40x + 16y \le 3200$. Next, since 520 hours of labor are available, we have $3x + 4y \le 520$. Then the profit for the production of tables and chairs is given by $P = 45x + 20y$.

Summarizing, we have the following linear programming problem:

$$\begin{aligned} \text{Maximize} \quad & P = 45x + 20y \\ \text{subject to} \quad & 40x + 16y \le 3200 \\ & 3x + 4y \le 520 \\ & x \ge 0, y \ge 0. \end{aligned}$$

7. Suppose the company extends x million dollars in homeowner loans and y million dollars in automobile loans. Then, the returns on these loans are given by $P = 0.1x + 0.12y$ million dollars. Since the company has a total of $20 million for these loans, we have $x + y \leq 20$. Furthermore, since the total amount of homeowner loans should be greater than or equal to four times the total amount of automobile loans, we have $x \geq 4y$. Therefore, the linear programming problem is:

$$\begin{aligned} \text{Maximize} \quad & P = 0.1x + 0.12y \\ \text{subject to} \quad & x + y \leq 20 \\ & x - 4y \geq 0 \\ & x \geq 0,\, y \geq 0. \end{aligned}$$

9. Let x and y denote the amounts that Justin should invest in medium- and high-risk stocks. Since he has at most $60,000 to invest, we have $x + y \leq 60{,}000$. The condition that medium-risk stocks should make up at least 40% of the total investment gives the inequality $x \geq \frac{2}{5}(x + y)$, so $\frac{3}{5}x \geq \frac{2}{5}y$ and $3x - 2y \geq 0$. Also, the stipulation that high-risk stocks make up at least 20% of the total investment leads to the condition $y \geq \frac{1}{5}(x + y)$, whence $x - 4y \leq 0$. Finally, Justin wants to make 12% from his medium-risk investments and 20% from his high-risk investments, so the linear programming problem is to maximize $P = 0.12x + 0.20y$ subject to

$$\begin{aligned} x + y &\leq 60{,}000 \\ 3x - 2y &\geq 0 \\ x - 4y &\leq 0 \\ x \geq 0,\, y &\geq 0 \end{aligned}$$

11. Let x and y denote the number of days the Saddle and Horseshoe mines are operated, respectively. Then the operating cost is $C = 14{,}000x + 16{,}000y$. The amount of gold produced in the two mines is $(50x + 75y)$ oz., and this amount must be at least 650 oz. Thus, we have $50x + 75y \geq 650$. Similarly, the requirement for silver production leads to the inequality $3000x + 1000y \geq 18{,}000$. So the linear programming problem is:

$$\begin{aligned} \text{Minimize} \quad & C = 14{,}000x + 16{,}000y \\ \text{subject to} \quad & 50x + 75y \geq 650 \\ & 3000x + 1000y \geq 18{,}000 \\ & x \geq 0,\, y \geq 0. \end{aligned}$$

13. Let x denote the number of fully assembled units to be produced daily and y the number of kits to be produced. Then the fraction of the day the fabrication department works on the fully assembled cabinets is $\frac{1}{200}x$. Similarly the fraction of the day the fabrication department works on kits is $\frac{1}{200}y$. Since the fraction of the day during which the fabrication department is busy cannot exceed one, we must have $\frac{1}{200}x + \frac{1}{200}y \leq 1$. Similarly, the restrictions on the assembly department lead to the inequality $\frac{1}{100}x + \frac{1}{300}y \leq 1$. The profit (objective) function is $P = 50x + 40y$. Summarizing, the linear programming problem is:

$$\begin{aligned} \text{Maximize} \quad & P = 50x + 40y \\ \text{subject to} \quad & \tfrac{1}{200}x + \tfrac{1}{200}y \leq 1 \\ & \tfrac{1}{100}x + \tfrac{1}{300}y \leq 1 \\ & x \geq 0,\, y \geq 0. \end{aligned}$$

15. Let x and y denote the numbers of gallons of water (in millions) obtained each day from the local reservoir and the pipeline, respectively. The requirement that at least 10 million gallons of water be supplied per day implies that $x + y \geq 10$. Next, because the maximum yield of the local reservoir is 5 million gallons per day, we have $x \leq 5$. The maximum yield of the pipeline is 10 million gallons per day and the pipeline has been contracted to supply at least 6 million gallons per day, so we have $6 \leq y \leq 10$. Then the cost function is given by $C = 300x + 500y$. Summarizing, we have the following linear programming problem:

$$\text{Minimize} \quad C = 300x + 500y$$
$$\text{subject to} \quad x + y \geq 10$$
$$0 \leq x \geq 5, 6 \leq y \leq 10.$$

17. Let x and y denote the amounts of food A and food B, respectively, used to prepare a meal. Then the requirement that the meal contain a minimum of 400 mg of calcium implies $30x + 25y \geq 400$. Similarly, the requirements that the meal contain at least 10 mg of iron and 40 mg of vitamin C imply that $x + 0.5y \geq 10$ and $2x + 5y \geq 40$. The cholesterol content is given by $C = 2x + 5y$. Therefore, the linear programming problem is:

$$\text{Minimize} \quad C = 2x + 5y$$
$$\text{subject to} \quad 30x + 25y \geq 400$$
$$x + 0.5y \geq 10$$
$$2x + 5y \geq 40$$
$$x \geq 0, y \geq 0.$$

19. Let x denote the number of LCD TVs shipped from location I to city A and y the number of TVs shipped from location I to city B. Since the number of TVs required by the two factories in cities A and B are 3000 and 4000, respectively, the number of TVs shipped from location II to cities A and B are $3000 - x$ and $4000 - y$, respectively. These numbers are shown in the diagram.

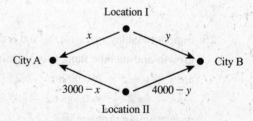

Referring to the schematic and the shipping schedule, we find that the total shipping costs incurred by the company are given by $C = 6x + 4y + 8(3000 - x) + 10(4000 - y) = 64{,}000 - 2x - 6y$. The production constraints on locations I and II lead to the inequalities $x + y \leq 6000$ and $(3000 - x) + (4000 - y) \leq 5000$. This last inequality simplifies to $x + y \geq 2000$. The requirements of the two factories lead to the inequalities $x \geq 0, y \geq 0$, $3000 - x \geq 0$, and $4000 - y \geq 0$. These last two inequalities may be written as $x \leq 3000$ and $y \leq 4000$. Summarizing, we have the following linear programming problem:

$$\text{Minimize} \quad C = 64{,}000 - 2x - 6y$$
$$\text{subject to} \quad x + y \leq 6000$$
$$x + y \geq 2000$$
$$0 \leq x \leq 3000, 0 \leq y \leq 4000.$$

21. Let x, y, and z denote the numbers of units produced of products A, B, and C, respectively. From the given information, we see that the time required by department I is given by $2x + y + 2z$, and this must not exceed 900 minutes. The time required by department II is given by $3x + y + 2z$, and this must not exceed 1080 minutes. The time required by department III is given by $2x + 2y + z$, and this must not exceed 840 minutes. The profit is given by $P = 18x + 12y + 15z$, and this is the quantity to be maximized. Thus, the linear programming problem is:

$$\begin{aligned} \text{Maximize} \quad & P = 18x + 12y + 15z \\ \text{subject to} \quad & 2x + y + 2z \leq 900 \\ & 3x + y + 2z \leq 1080 \\ & 2x + 2y + z \leq 840 \\ & x \geq 0, y \geq 0, z \geq 0. \end{aligned}$$

23. We first tabulate the given information:

Department	Model A	B	C	Time Available
Fabrication	$\frac{5}{4}$	$\frac{3}{2}$	$\frac{3}{2}$	310
Assembly	1	1	$\frac{3}{4}$	205
Finishing	1	1	$\frac{1}{2}$	190

Let x, y, and z denote the numbers of units of models A, B, and C to be produced, respectively. Then the required linear programming problem is:

$$\begin{aligned} \text{Maximize} \quad & P = 26x + 28y + 24z \\ \text{subject to} \quad & \tfrac{5}{4}x + \tfrac{3}{2}y + \tfrac{3}{2}z \leq 310 \\ & x + y + \tfrac{3}{4}z \leq 205 \\ & x + y + \tfrac{1}{2}z \leq 190 \\ & x \geq 0, y \geq 0, z \geq 0. \end{aligned}$$

25. Suppose Ashley invests x, y, and z dollars in the money market fund, the international equity fund, and the growth-and-income fund, respectively. Then the objective function is $P = 0.06x + 0.1y + 0.15z$. The constraints are $x + y + z \leq 250{,}000$, $z \leq 0.25(x + y + z)$, and $y \leq 0.5(x + y + z)$. The last two inequalities simplify to $-\tfrac{1}{4}x - \tfrac{1}{4}y + \tfrac{3}{4}z \leq 0$, or $-x - y + 3z \leq 0$, and $-\tfrac{1}{2}x + \tfrac{1}{2}y - \tfrac{1}{2}z \leq 0$, or $-x + y - z \leq 0$. So the required linear programming problem is:

$$\begin{aligned} \text{Maximize} \quad & P = 0.06x + 0.1y + 0.15z = \tfrac{3}{50}x + \tfrac{1}{10}y + \tfrac{3}{20}z \\ \text{subject to} \quad & x + y + z \leq 250{,}000 \\ & -x - y + 3z \leq 0 \\ & -x + y - z \leq 0 \\ & x \geq 0, y \geq 0, z \geq 0. \end{aligned}$$

27. The shipping costs per loudspeaker system in dollars are given in the first table. Letting x_1 denote the number of loudspeaker systems shipped from plant I to warehouse A, x_2 the number of loudspeaker systems shipped from plant I to warehouse B, and so on, we have the second table.

	Warehouse		
Plant	A	B	C
I	16	20	22
II	18	16	14

	Warehouse			
Plant	A	B	C	Max. Prod.
I	x_1	x_2	x_3	800
II	x_4	x_5	x_6	600
Min. Req.	500	400	400	

From the two tables, we see that the total monthly shipping cost incurred by Acrosonic is given by $C = 16x_1 + 20x_2 + 22x_3 + 18x_4 + 16x_5 + 14x_6$. Next, the production constraints on plants I and II lead to the inequalities $x_1 + x_2 + x_3 \le 800$ and $x_4 + x_5 + x_6 \le 600$. Also, the minimum requirements of each warehouse leads to the three inequalities $x_1 + x_4 \ge 500$, $x_2 + x_5 \ge 400$, $x_3 + x_6 \ge 400$.

Summarizing, we have the following linear programming problem:

$$\text{Minimize} \quad C = 16x_1 + 20x_2 + 22x_3 + 18x_4 + 16x_5 + 14x_6$$
$$\text{subject to} \quad x_1 + x_2 + x_3 \qquad\qquad\qquad \le 800$$
$$x_4 + x_5 + x_6 \le 600$$
$$x_1 \qquad\quad + x_4 \qquad\qquad \ge 500$$
$$x_2 \qquad\quad + x_5 \qquad \ge 400$$
$$x_3 \qquad\quad + x_6 \ge 400$$
$$x_1 \ge 0,\, x_2 \ge 0,\, x_3 \ge 0,\, x_4 \ge 0,\, x_5 \ge 0,\, x_6 \ge 0.$$

29. The given data can be summarized as follows:

	Concentrate			
Juice	Pineapple	Orange	Banana	Profit ($)
Pineapple-orange	8	8	0	1
Orange-banana	0	12	4	0.80
Pineapple-orange-banana	4	8	4	0.90
Maximum available (oz.)	16,000	24,000	5000	

Suppose x, y, and z cartons of pineapple-orange, orange-banana, and pineapple-orange-banana juice are to be produced, respectively. The linear programming problem is:

$$\text{Maximize} \quad P = x + 0.8y + 0.9z$$
$$\text{subject to} \quad 8x \qquad\quad + 4z \le 16{,}000$$
$$8x + 12y + 8z \le 24{,}000$$
$$4y + 4z \le 5000$$
$$x \ge 0,\, y \ge 0,\, 0 \le z \le 800.$$

31. False. The objective function $P = xy$ is not a linear function in x and y.

3.3 Graphical Solution of Linear Programming Problems

Problem-Solving Tips

1. To solve a linear programming problem using the method of corners, perform the following steps:

a. Graph the feasible set.

b. Find the coordinates of all corner points (vertices) of the feasible set.

c. Evaluate the objective function at each corner point.

Then check to see which vertex yields the maximum (or minimum). If only one vertex yields the maximum (or minimum) then the problem has a unique solution. If there are two adjacent vertices that yield the same maximum (or minimum), then any point lying on the line segment joining these two vertices is a solution.

2. It's helpful to set up a table, as done in Examples 1–3, to evaluate the objective function at each vertex.

Concept Questions   page 196

1. a. The feasible set is the set of points satisfying the constraints associated with the linear programming problem.

b. A feasible solution of a linear programming problem is a point in the feasible set.

c. An optimal solution of a linear programming problem is a feasible solution that also optimizes (maximizes or minimizes) the objective function.

Exercises   page 196

1. Evaluating the objective function at each of the corner points, we obtain the following table.

Vertex	$Z = 2x + 3y$
(1, 1)	5
(8, 5)	31
(4, 9)	35
(2, 8)	28

From the table, we conclude that the maximum value of Z is 35 and it occurs at the vertex (4, 9). The minimum value of Z is 5 and it occurs at the vertex (1, 1).

3. Evaluating the objective function at each of the corner points, we obtain the following table.

Vertex	$Z = 2x + 3y$
(0, 20)	60
(3, 10)	36
(4, 6)	26
(9, 0)	18

From the graph, we conclude that there is no maximum value since Z is unbounded. The minimum value of Z is 18 and it occurs at the vertex (9, 0).

5. Evaluating the objective function at each of the corner points, we obtain the following table.

Vertex	$Z = x + 4y$
$(0, 6)$	24
$(4, 10)$	44
$(12, 8)$	44
$(15, 0)$	15

From the table, we conclude that the maximum value of Z is 44 and it occurs at every point on the line segment joining the points $(4, 10)$ and $(12, 8)$. The minimum value of Z is 15 and it occurs at the vertex $(15, 0)$.

7. The problem is to maximize $P = 3x + 2y$ subject to $x + y \le 6$, $x \le 3$, $x \ge 0$, $y \ge 0$. The feasible set S for the problem is shown in the figure, and the values of the function P at the vertices of S are summarized in the table.

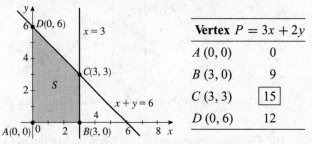

Vertex	$P = 3x + 2y$
$A (0, 0)$	0
$B (3, 0)$	9
$C (3, 3)$	15
$D (0, 6)$	12

We conclude that P attains a maximum value of 15 when $x = 3$ and $y = 3$.

9. The problem is to maximize $P = 2x + y$ subject to $x + y \le 4$, $2x + y \le 5$, $x \ge 0$, $y \ge 0$.

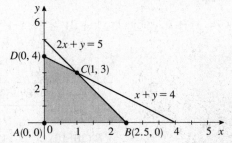

Vertex	$P = 2x + y$
$A (0, 0)$	0
$B (2.5, 0)$	5
$C (1, 3)$	5
$D (0, 4)$	4

From the figure and the table, we conclude that P attains a maximum value of 5 at any point (x, y) lying on the line segment joining $(1, 3)$ and $(2.5, 0)$.

11. The problem is: Maximize $P = x + 8y$ subject to $x + y \le 8$, $2x + y \le 10$, $x \ge 0$, $y \ge 0$.

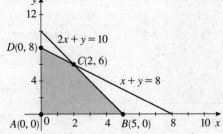

Vertex	$P = x + 8y$
$A (0, 0)$	0
$B (5, 0)$	5
$C (2, 6)$	50
$D (0, 8)$	64

From the figure and the table, we conclude that P attains a maximum value of 64 when $x = 0$ and $y = 8$.

13. The linear programming problem is: Maximize $P = x + 3y$ subject to $2x + y \le 6$, $x + y \le 4$, $x \le 1$, $x \ge 0$, $y \ge 0$.

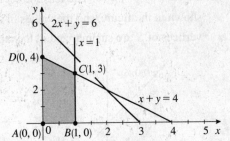

Vertex	$P = x + 3y$
$A\,(0, 0)$	0
$B\,(1, 0)$	1
$C\,(1, 3)$	10
$D\,(0, 4)$	$\boxed{12}$

From the figure and the table, we conclude that P attains a maximum value of 12 when $x = 0$ and $y = 4$.

15. The linear programming problem is: Minimize $C = 2x + 5y$ subject to $x + y \ge 3$, $x + 2y \ge 4$, $x \ge 0$, $y \ge 0$.

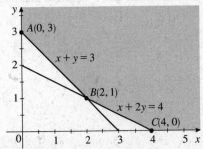

Vertex	$C = 2x + 5y$
$A\,(0, 3)$	15
$B\,(2, 1)$	9
$C\,(4, 0)$	$\boxed{8}$

From the figure and the table, we conclude that C attains a minimum value of 8 when $x = 4$ and $y = 0$.

17. The linear programming problem is: Minimize $C = 3x + 6y$ subject to $x + 2y \ge 40$, $x + y \ge 30$, $x \ge 0$, $y \ge 0$.

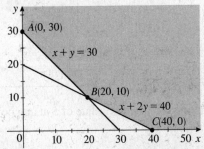

Vertex	$C = 3x + 6y$
$A\,(0, 30)$	180
$B\,(20, 10)$	$\boxed{120}$
$C\,(40, 0)$	$\boxed{120}$

From the figure and the table, we conclude that C attains a minimum value of 120 at any point on the line segment joining (20, 10) to (40, 0).

19. The problem is: Minimize $C = 2x + 10y$ subject to $5x + 2y \ge 40$, $x + 2y \ge 20$, $y \ge 3$, $x \ge 0$. The feasible set S for the problem is shown in the figure, and the values of the function C at the vertices of S are shown in the table.

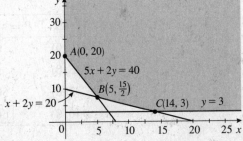

Vertex	$C = 2x + 10y$
$A\,(0, 20)$	200
$B\left(5, \frac{15}{2}\right)$	85
$C\,(14, 3)$	$\boxed{58}$

We conclude that C attains a minimum value of 58 when $x = 14$ and $y = 3$.

21. The problem is to minimize $C = 10x + 15y$ subject to $x + y \leq 10, 3x + y \geq 12, -2x + 3y \geq 3, x \geq 0, y \geq 0$. The feasible set is shown in the figure, and the values of C at each of the vertices of S are shown in the table.

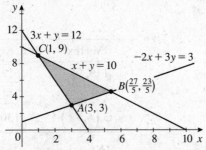

Vertex	$C = 10x + 15y$
$A\ (3, 3)$	$\boxed{75}$
$B\ \left(\frac{27}{5}, \frac{23}{5}\right)$	123
$C\ (1, 9)$	145

We conclude that C attains a minimum value of 75 when $x = 3$ and $y = 3$.

23. The problem is to maximize $P = 3x + 4y$ subject to $x + 2y \leq 50, 5x + 4y \leq 145, 2x + y \geq 25, y \geq 5, x \geq 0$. The feasible set S is shown in the figure, and the values of P at each of the vertices of S are shown in the table.

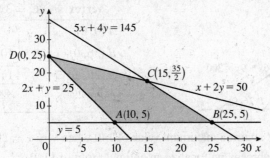

Vertex	$P = 3x + 4y$
$A\ (10, 5)$	50
$B\ (25, 5)$	95
$C\ \left(15, \frac{35}{2}\right)$	$\boxed{115}$
$D\ (0, 25)$	100

We conclude that P attains a maximum value of 115 when $x = 15$ and $y = \frac{35}{2}$.

25. The problem is to maximize $P = 2x + 3y$ subject to $x + y \leq 48, x + 3y \geq 60, 9x + 5y \leq 320, x \geq 10, y \geq 0$.

Vertex	$P = 2x + 3y$
$A\ \left(10, \frac{50}{3}\right)$	70
$B\ (30, 10)$	90
$C\ (20, 28)$	124
$D\ (10, 38)$	$\boxed{134}$

From the graph and the table, we conclude that P attains a maximum value of 134 when $x = 10$ and $y = 38$.

27. The problem is to find the maximum and minimum value of $P = 8x + 5y$ subject to $5x + 2y \geq 63$, $x + y \geq 18$, $3x + 2y \leq 51$, $x \geq 0$, $y \geq 0$.

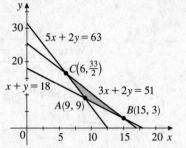

Vertex	$P = 8x + 5y$
$A\,(9, 9)$	$\boxed{117}$
$B\,(15, 3)$	$\boxed{135}$
$C\left(6, \frac{33}{2}\right)$	130.5

From the graph and the table, we see that P attains a maximum value of 135 when $x = 15$ and $y = 3$. The minimum value of P is 117, attained when $x = 9$ and $y = 9$.

29. Refer to the solution to Exercise 3.2.1 on page 79 of this manual. The problem is: Maximize $P = 3x + 4y$ subject to $6x + 9y \leq 300$, $5x + 4y \leq 180$, $x \geq 0$, $y \geq 0$.

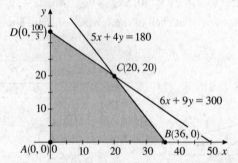

Vertex	$C = 3x + 4y$
$A\,(0, 0)$	0
$B\,(36, 0)$	108
$C\,(20, 20)$	$\boxed{140}$
$D\left(0, \frac{100}{3}\right)$	$\frac{400}{3}$

From the graph and the table, we see that P attains a maximum value of 140 when $x = y = 20$. Thus, by producing 20 units of each product in each shift, the company will realize an optimal profit of $140.

31. Refer to the solution to Exercise 3.2.3 on page 79 of this manual. The problem is: Maximize $P = 2x + 1.5y$ subject to $3x + 4y \leq 1000$, $6x + 3y \leq 1200$, $x \geq 0$, $y \geq 0$.

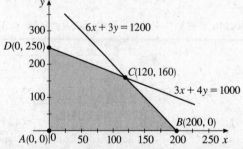

Vertex	$P = 2x + 1.5y$
$A\,(0, 0)$	0
$B\,(200, 0)$	400
$C\,(120, 160)$	$\boxed{480}$
$D\,(0, 250)$	375

From the graph and the table, we see that P attains a maximum value of 480 when $x = 120$ and $y = 160$. Thus, by producing 120 model A grates and 160 model B grates in each shift, the company will realize an optimal profit of $480.

33. Let x denote the number of tables and y denote the number of chairs to be manufactured. Then the linear programming problem is: Maximize $P = 45x + 20y$ subject to $40x + 16y \leq 3200$, $3x + 4y \leq 520$, $x \geq 0$, $y \geq 0$.

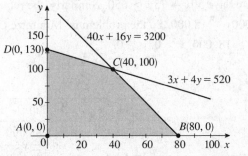

Vertex	$P = 45x + 20y$
$A\ (0, 0)$	0
$B\ (80, 0)$	3600
$C\ (40, 100)$	3800
$D\ (0, 130)$	2600

From the graph and the table, we see that Winston should manufacture 40 tables and 100 chairs for a maximum profit of $3800.

35. Refer to the solution to Exercise 3.2.7 on page 80 of this manual. The linear programming problem is: Maximize $P = 0.1x + 0.12y$ subject to $x + y \leq 20$, $x - 4y \geq 0$, $x \geq 0$, $y \geq 0$.

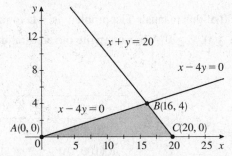

Vertex	$P = 0.1x + 0.12y$
$A\ (0, 0)$	0
$B\ (16, 4)$	2.08
$C\ (20, 0)$	2.00

The maximum value of P is attained when $x = 16$ and $y = 4$. Thus, by extending $16 million in housing loans and $4 million in automobile loans, the company will realize a return of $2.08 million on its loans.

37. Referring to the solution of Exercise 3.2.9 on page 80 of this manual, we see that the problem is: Maximize $P = 0.12x + 0.20y$ subject to

$$
\begin{aligned}
x + y &\leq 60{,}000 \\
3x - 2y &\geq 0 \\
x - 4y &\leq 0 \\
x \geq 0, \ y &\geq 0
\end{aligned}
$$

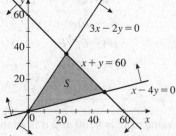

V	$P = 0.12x + 0.20y$
$(0, 0)$	0
$(48, 12)$	8160
$(24, 36)$	10,080

From the table, we see that the maximum of P occurs when $x = 24$ and $y = 36$, so Justin should invest $24,000 in medium-risk stock and $36,000 in high-risk stock. He can expect to make $10,080 in one year.

39. Let x and y denote the numbers of days the Saddle Mine and the Horseshoe Mine are operated, respectively. Then the operating cost is $C = 14{,}000x + 16{,}000y$. The amount of gold produced in the two mines is $(50x + 75y)$ oz., and this amount must be at least 650 oz., so we have $50x + 75y \geq 650$. Similarly, the requirement for silver production leads to the inequality $3000x + 1000y \geq 18{,}000$. So the problem is: Minimize $C = 14{,}000x + 16{,}000y$ subject to $50x + 75y \geq 650$, $3000x + 1000y \geq 18{,}000$, $x \geq 0$, $y \geq 0$.

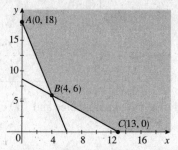

Vertex	$C = 14{,}000x + 16{,}000y$
$A\,(0, 18)$	288,000
$B\,(4, 6)$	152,000
$C\,(13, 0)$	182,000

From the figure and the table, we see that the minimum value of $C = 152{,}000$ is attained at $x = 4$ and $y = 6$. So, the Saddle Mine should be operated for 4 days and the Horseshoe Mine should be operated for 6 days at a minimum cost of $152,000/day.

41. Refer to the solution to Exercise 3.2.13 on page 80 of this manual. The problem is: Maximize $P = 50x + 40y$ subject to $\frac{1}{200}x + \frac{1}{200}y \leq 1$, $\frac{1}{100}x + \frac{1}{300}y \leq 1$, $x \geq 0$, $y \geq 0$. We can rewrite the constraints as $x + y \leq 200$, $3x + y \leq 300$, $x \leq 0$, $y \leq 0$.

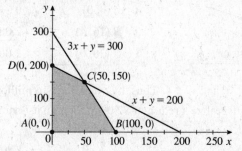

Vertex	$P = 50x + 40y$
$A\,(0, 0)$	0
$B\,(100, 0)$	5000
$C\,(50, 150)$	8500
$D\,(0, 200)$	8000

From the figure and the table, we conclude that the company should produce 50 fully assembled units and 150 kits daily in order to realize a profit of $8500.

43. Let x and y denote the numbers of gallons of water in millions obtained from the local reservoir and the pipeline per day, respectively. Then, we have the following linear programming problem: Minimize $C = 300x + 500y$ subject to $x + y \geq 10$, $x \leq 5$, $6 \leq y \leq 10$, $x \geq 0$.

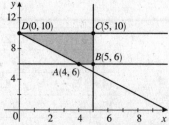

Vertex	$C = 300x + 500y$
$A\,(4, 6)$	4200
$B\,(5, 6)$	4500
$C\,(5, 10)$	6500
$D\,(0, 10)$	5000

From the figure and the table, we see that the minimum value of C is 4200 and it is attained at $x = 4$ and $y = 6$. Thus, 4 million gallons should be obtained from the reservoir and 6 million gallons from the pipeline at a minimum cost of $4200.

45. Refer to the solution to Exercise 3.2.17 on page 81 of this manual. The problem is: Minimize $C = 2x + 5y$ subject to $30x + 25y \geq 400$, $x + 0.5y \geq 10$, $2x + 5y \geq 40$, $x \geq 0$, $y \geq 0$.

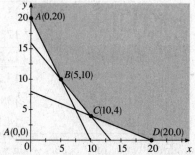

Vertex	$C = 2x + 5y$
$A\,(0, 20)$	100
$B\,(5, 10)$	60
$C\,(10, 4)$	40
$D\,(20, 0)$	40

From the figure and the table, we see that C attains a minimum value of 40 when $x = 10$ and $y = 4$ and when $x = 20$ and $y = 0$. This means that any point lying on the line joining the points $(10, 4)$ and $(20, 0)$ will satisfy these constraints. For example, we could use 10 ounces of food A and 4 ounces of food B, or we could use 20 ounces of food A and zero ounces of food B.

47. Refer to the solution to Exercise 3.2.19 on page 81 of this manual. Minimize $C = 64,000 - 2x - 6y$ subject to $x + y \leq 6000$, $x + y \geq 2000$, $x \leq 3000$, $y \leq 4000$, $x \geq 0$, $y \geq 0$.

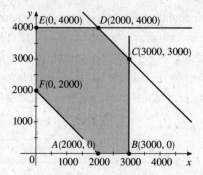

Vertex	$C = 64,000 - 2x - 6y$
$A\,(2000, 0)$	60,000
$B\,(3000, 0)$	58,000
$C\,(3000, 3000)$	40,000
$D\,(2000, 4000)$	36,000
$E\,(0, 4000)$	40,000
$F\,(0, 2000)$	52,000

Since x denotes the number of televisions shipped from location I to city A and y denotes the number of televisions shipped from location I to city B, we see that the company should ship 2000 televisions from location I to city A and 4000 televisions from location I to city B. Since the number of televisions required by the two factories in city A and city B are 3000 and 4000, respectively, the number of televisions shipped from location II to city A and city B are $3000 - x = 3000 - 2000 = 1000$ and $4000 - y = 4000 - 4000 = 0$, respectively. The minimum shipping cost is $36,000.

49. The problem is: Minimize $C = 14,500 - 20x - 10y$ subject to $x + y \geq 40$, $x + y \leq 100$, $0 \leq x \leq 80$, $0 \leq y \leq 70$.

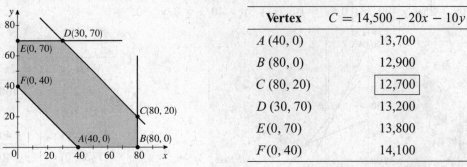

Vertex	$C = 14,500 - 20x - 10y$
$A\,(40, 0)$	13,700
$B\,(80, 0)$	12,900
$C\,(80, 20)$	12,700
$D\,(30, 70)$	13,200
$E\,(0, 70)$	13,800
$F\,(0, 40)$	14,100

From the figure and the table, we conclude that the minimum value of C occurs when $x = 80$ and $y = 20$. Thus, 80 engines should be shipped from plant I to assembly plant A, and 20 engines should be shipped from plant I to assembly plant B; whereas $80 - x = 80 - 80 = 0$ and $70 - y = 70 - 20 = 50$ engines should be shipped from plant II to assembly plants A and B, respectively, at a total cost of $12,700.

51. Let x and y denote Patricia's investments in growth and speculative stocks, respectively, where both x and y are measured in thousands of dollars. Then the return on her investments is given by $P = 0.15x + 0.25y$. Since her investment may not exceed $30,000, we have the constraint $x + y \leq 30$. The condition that her investment in growth stocks be at least 3 times as much as her investment in speculative stocks translates into the inequality $x \geq 3y$. Thus, we have the following linear programming problem: Maximize $P = 0.15x + 0.25y$ subject to $x + y \leq 30$, $x - 3y \geq 0$, $x \geq 0, y \geq 0$.

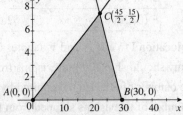

Vertex	$P = 0.15x + 0.25y$
$A\,(0, 0)$	0
$B\,(30, 0)$	4.5
$C\left(\frac{45}{2}, \frac{15}{2}\right)$	5.25

From the figure and the table, we conclude that the maximum value of P occurs when $x = \frac{45}{2}$ and $y = \frac{15}{2}$. Thus, by investing $22,500 in growth stocks and $7500 in speculative stocks. Patricia's maximum return is $5250.

53. Let x denote the number of urban families and let y denote the number of suburban families interviewed by the company. Then, the amount of money paid to Trendex will be $P = 6000 + 8(x + y) - 4.4x - 5y = 6000 + 3.6x + 3y$. Since a maximum of 1500 families are to be interviewed, we have $x + y \le 1500$. Next, the condition that at least 500 urban families are to be interviewed translates into the condition $x \ge 500$. Finally the condition that at least half of the families interviewed must be from the suburban area gives $y \ge \frac{1}{2}(x + y)$, or $y - x \ge 0$. Thus, we are led to the following programming problem: Maximize $P = 6000 + 3.6x + 3y$ subject to $x + y \le 1500$, $y - x \ge 0$, $x \ge 500$, $y \ge 0$.

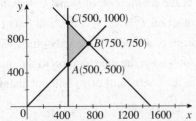

Vertex	$P = 6000 + 3.6x + 3y$
A (500, 500)	9300
B (750, 750)	10,950
C (500, 1000)	10,800

From the figure and the table, we conclude that the profit will be maximized when $x = 750$ and $y = 750$. Thus, a maximum profit of \$10,950 will be realized when 750 urban and 750 suburban families are interviewed.

55. False. It can have either one or infinitely many solutions.

57. a. True. Because $a > 0$, the term ax can be made as large as we please by taking x sufficiently large (because S is unbounded) and therefore P is unbounded as well.

b. True. Maximizing $P = ax + by$ on S is the same as minimizing
$Q = -P = -(ax + by) = -ax - by = Ax + By$, where $A \ge 0$ and $B \ge 0$. Since $x \ge 0$ and $y \ge 0$, the linear function Q (and therefore P) has at least one optimal solution.

59. Let $A(x_1, y_1)$ and $B(x_2, y_2)$. Then you can verify that $Q(\overline{x}, \overline{y})$, where $\overline{x} = x_1 + t(x_2 - x_1)$, $\overline{y} = y_1 + t(y_2 - y_1)$, and t is a number satisfying $0 < t < 1$. Therefore, the value of P at Q is $P = a\overline{x} + b\overline{y} = a[x_1 + t(x_2 - x_1)] + b[y_1 + t(y_2 - y_1)] = ax_1 + by_1 + [a(x_2 - x_1) + b(y_2 - y_1)]t$. Now if $c = a(x_2 - x_1) + b(y_2 - y_1) = 0$, then P has the (maximum) value $ax_1 + by_1$ on the line segment joining A and B; that is, the infinitely many solutions lie on this line segment. If $c > 0$, then a point a little to the right of Q will give a larger value of P. In this case, P is not maximal at Q. (Such a point can be found because Q lies in the interior of the line segment.) A similar statement holds for the case $c < 0$. Thus, the maximum of P cannot occur at Q unless it occurs in every point on the line segment joining A and B.

61. a.

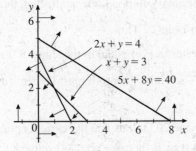

b. There is no point that satisfies all the given inequalities. Therefore, there is no solution.

3.4   Sensitivity Analysis

Problem-Solving Tips

In applied problems, it is important to interpret the mathematical solution of a problem in terms of the real-life problem that you are solving.

1. One consideration in sensitivity analysis is *how changes in the coefficients of the objective function* affect the optimal solution. In Example 1, we optimized the profit function $P = 2x + 1.5y$, and found that the optimal solution would not be changed if the contribution to the profit of a model A grate assumed values between $1.125 and $3.00 (changes in the coefficient of x); similarly, we found that the contribution to the profit of a model B grate could assume values between $1.00 and $2.67 (changes in the coefficient of y) without changing the optimal solution.

2. Another consideration is *how changes to the constants on the right-hand side of the constraint inequalities* affect the optimal solution. In the example discussed (production problem for Ace Novelty), we found that the profit could be improved from $148.80 to $152.40 by increasing the time available on Machine I by 10 minutes (a change in constraint 1).

Concept Questions   page 212

1. The coefficients of x and y represent the costs for producing one unit of Product A and one unit of product B, respectively.

3. **a.** The shadow price of the ith resource (associated with the ith constraint of a linear programming problem) is the amount by which the value of the objective function is improved if the right-hand side of the ith constraint is increased by 1 unit.

   b. Binding constraints are constraints which hold with equality at the optimal solution. They cannot be increased without increasing the resources.

Exercises   page 212

1. Refer to the discussion on pages 203–204 of the text.

   a. Suppose the contribution to the profit of each Type B souvenir is c, so that $P = x + cy$, or $y = -\dfrac{x}{c} + \dfrac{P}{c}$, and the slope of the isoprofit line is $-1/c$. In order for the current optimal solution to be unaffected, this slope must be less than or equal to the slope of the line associated with constraint 2. Thus, $-\dfrac{1}{c} \leq -\dfrac{1}{3}, \dfrac{1}{c} \geq \dfrac{1}{3}$ and $c \leq 3$. Similarly, we see that the slope of the isoprofit line must be greater than or equal to the slope of the line associated with constraint 1, so $-\dfrac{1}{c} \geq -2, \dfrac{1}{c} \leq 2$, and $c \geq \dfrac{1}{2}$. Therefore, the profit must lie between $0.50 and $3, as was to be shown.

   b. If the profit of a Type A souvenir is $1.50, then the results on page 205 tell us that the current optimal solution holds. That is, we produce 48 Type A (because $x = 48$) and 84 type B (because $y = 84$) souvenirs, giving a profit of $P = 1.5x + 1.2y = 1.5(48) + 1.2(84) = 172.8$, or $172.80.

   c. Here the results of part a show that the current optimal solution holds. Thus, with $x = 48$ and $y = 84$, we find $P = x + 2y = 48 + 2(84) = 216$, or $216.

3. Refer to the discussion on pages 203–204 in the text.

a. Suppose Resource 2 is changed from 1200 to $1200 + k$. Then the constraint is changed to $6x + 3y = 1200 + k$. The current optimal solution is shifted to the new optimal solution at the point D'. To find the coordinates of D', we solve the system $3x + 4y = 1000$, $6x + 3y = 1200 + k$, obtaining $x = \frac{4}{15}(450 + k)$ and $y = \frac{1}{5}(800 - k)$. The requirements that x and y be nonnegative require that $450 + k \geq 0$, so $k \geq -450$, and $800 - k \geq 0$, $k \leq 800$. Next, constraint 3 ($x \leq 180$) requires that $\frac{4}{15}(450 + k) \leq 180$, so

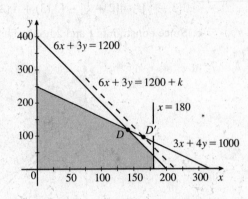

$450 + k \leq 675$, or $k \leq 225$. These three inequalities imply that $-450 \leq k \leq 225$. Therefore, Resource 2 must lie between $1200 - 450$ and $1200 + 225$; that is, between 750 and 1425.

b. Suppose Resource 3 is changed to $180 + \ell$. To find the coordinates of D' in the figure, we solve $6x + 3y = 1200$, $x = 180 + \ell$, obtaining $x = 180 + \ell$ and $y = 40 - 2\ell$. Now for the current optimal solution to hold, $x = 180 + \ell \geq 120$, or $\ell \geq -60$. Therefore, Resource 3 must be greater than or equal to $180 - 60 = 120$; that is, it cannot be decreased by more than 60.

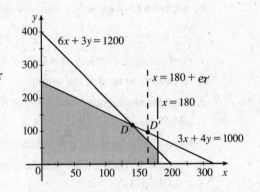

5. a.

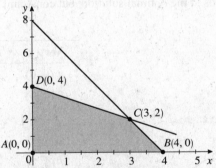

Vertex	$P = 3x + 4y$
$A\,(0, 0)$	0
$B\,(4, 0)$	12
$C\,(3, 2)$	$\boxed{17}$
$D\,(0, 4)$	16

From the figure and the table, we see that the optimal solution is $x = 3$, $y = 2$, giving $P = 17$.

b. Suppose $P = cx + 4y$. then $y = -\frac{c}{4}x + \frac{P}{4}$ and the slope of the isoprofit line is $-\frac{c}{4}$. In order for the current optimal solution to hold, we must have $-\frac{c}{4} \leq -\frac{2}{3}$, or $c \geq \frac{8}{3}$. (because the slope of constraint 1 is $-\frac{2}{3}$) and $-\frac{c}{4} \geq -2$, or $c \leq 8$ (because the slope of constraint 2 is -2). Therefore, $\frac{8}{3} \leq c \leq 8$.

c. Suppose resource 1 is changed from 12 to $12 + h$. Then the optimal solution is moved to a new optimal solution whose coordinates are found by solving the system $2x + 3y = 12 + h$, $2x + y = 8$. The solution is $x = 3 - \frac{h}{4}$, $y = 2 + \frac{h}{2}$. Next, the nonnegativity of x and y imply that $h \geq -4$ and $h \leq 12$. Thus, $-4 \leq h \leq 12$ and the resource can assume values between $12 - 4$ and $12 + 12$; that is, between 8 and 24.

d. If $h = 1$, then $x = 3 - \frac{1}{4}$ and $y = 2 + \frac{1}{2}$. The shadow price for resource I is

$$3[3 - \tfrac{1}{4}] + 4[2 + \tfrac{1}{2}] - [3(3) + 4(2)] = -\tfrac{3}{4} + 4\left(\tfrac{1}{2}\right) = \tfrac{5}{4}.$$

e. Since constraints 1 and 2 hold with equality at the optimal solution $(3, 2)$, they are both binding.

7. a.

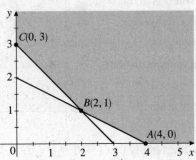

Vertex	$C = 2x + 5y$
$A\,(4, 0)$	$\boxed{8}$
$B\,(2, 1)$	9
$C\,(0, 3)$	15

From the figure and the table, we see that the optimal solution is $x = 4$, $y = 0$, giving $C = 8$.

b. Suppose $C = cx + 5y$. then $y = -\frac{c}{5} + \frac{C}{5}$. In order for the current optimal solution to hold, we must have

$-\frac{c}{5} \geq -\frac{1}{2}$, so $\frac{c}{5} \leq \frac{1}{2}$ and $c \leq \frac{5}{2}$; and $-\frac{c}{5} \geq -1$, so $\frac{c}{5} \leq 1$ and $c \leq 5$. Therefore, $0 \leq c \leq \frac{5}{2}$. Note that for a minimization problem, the optimal solution is the intersection of the corner point in the feasible set with the isoprofit line nearest the origin.

c. Suppose requirement 1 is changed from 4 to $4 + h$. Then solving the system $x + 2y = 4 + h$, $x + y = 3$, we obtain $x = 2 - h$ and $y = 1 + h$. We must have $y = 1 + h \geq 0$ or $h \geq -1$ (see the figure). Therefore, requirement 1 can assume values greater than or equal to $4 - 1$, or 3.

d. Put $h = 1$. Then we see that the shadow price for requirement 1 is $[2(5) + 5(0)] - [2(4) + 5(0)] = 2$ (because $C = 2x + 5y$).

e. Requirement 1 is binding since equality holds at the optimal solution, but constraint 2 is nonbinding.

9. a.

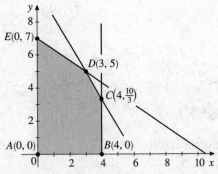

Vertex	$P = 4x + 3y$
$A\,(0, 0)$	0
$B\,(4, 0)$	16
$C\left(4, \frac{10}{3}\right)$	26
$D\,(3, 5)$	$\boxed{27}$
$E\,(0, 7)$	21

From the figure and the table, we see that the optimal solution is $x = 3$, $y = 5$, giving $P = 27$.

b. Suppose $P = cx + 3y$ so that $y = -\frac{c}{3}x + \frac{P}{3}$. For the current optimal solution to be unaffected, we must have

$-\frac{c}{3} \geq -\frac{5}{3}$, so $\frac{c}{3} \leq \frac{5}{3}$ and $c \leq 5$; and $-\frac{c}{3} \leq -\frac{2}{3}$, so $\frac{c}{3} \geq \frac{2}{3}$ and $c \geq 2$. Therefore, $2 \leq c \leq 5$.

c. Suppose the right-hand side of the constraint is replaced by $30 + h$. Then we have the system $5x + 3y = 30 + h$, $2x + 3y = 21$. The solutions are $x = 3 + \frac{h}{3}$ and $y = 5 - \frac{2h}{9}$. Since $x \geq 0$, we have $\frac{h}{3} + 3 \geq 0$, or $h \geq -9$.

Also, $y \geq 0$ implies $5 - \frac{2h}{9} \geq 0$, so $\frac{2h}{9} \leq 5$ and $h \leq \frac{45}{2}$. Finally, $x \leq 4$ implies $\frac{h}{3} + 3 \leq 4$, so $h \leq 3$. Therefore, $-9 \leq h \leq 3$, and we see that resource 1 can assume values between $30 - 9$ and $30 + 3$; that is, between 21 and 33.

d. Putting $h = 1$ in part (c) gives $x = 3 + \frac{1}{3} = \frac{10}{3}$ and $y - 5 - \frac{2}{9} = \frac{43}{9}$. Therefore, the shadow price of resource 1 is

$$\left[4\left(\tfrac{10}{3}\right) + 3\left(\tfrac{43}{9}\right)\right] - [4\,(3) + 3\,(5)] = \tfrac{2}{3} \ \text{(because } P = 4x + 3y).$$

e. The first two constraints are binding since equality holds at the optimal solution. Constraint 3 is nonbinding.

11. a. Let x and y denote the numbers of units of products A and B to be manufactured, respectively, Then the problem at hand is: Maximize $P = 3x + 4y$ subject to $6x + 9y \le 300$, $5x + 4y \le 180$, $x \ge 0$, $y \ge 0$.

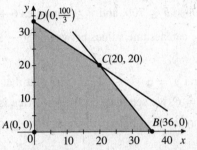

Vertex	$P = 3x + 4y$
$A\,(0, 0)$	0
$B\,(36, 0)$	108
$C\,(20, 20)$	140
$D\left(0, \frac{100}{3}\right)$	$\frac{400}{3}$

From the figure and the table, we see that the optimal solution is $x = y = 20$, giving $P = 140$. Thus, the company should produce 20 units of each product, for a maximum profit of \$140.

b. Suppose $P = cx + 4y$, so that $y = -\frac{c}{4}x + \frac{P}{4}$. In order for the current optimal solution to remain optimal, we

must have $-\frac{c}{4} \le -\frac{2}{3}$, so $\frac{c}{4} \ge \frac{2}{3}$ and $c \ge \frac{8}{3}$; and $-\frac{c}{4} \ge -\frac{5}{4}$, so $\frac{c}{4} \le \frac{5}{4}$ and $c \le 5$. Therefore, $\frac{8}{3} \le c \le 5$.

c. Suppose the right-hand side of the constraint is replaced by $300 + h$. Then we have the system

$6x + 9y = 300 + h$, $5x + 4y = 180$. The solution is $x = 20 - \frac{4}{21}h$, $y = 20 + \frac{5}{21}h$. Now $x \ge 0$ implies

$20 - \frac{4}{21}h \ge 0$, so $h \le 105$. Next, $y \ge 0$ implies $50 + \frac{5}{21}h \ge 0$, so $h \ge -84$. Therefore, $-84 \le h \le 105$. The resource can assume values between $300 - 84$ and $300 + 105$; that is, between 216 and 405.

d. Put $h = 1$ in part (c) to obtain $x = 20 - \frac{4}{21}$ and $y = 20 + \frac{5}{21}h$. Using this result and the result from part (a), we

see that the required shadow price is $3\left(20 - \frac{4}{21}\right) + 4\left(20 + \frac{5}{21}\right) - [3\,(20) + 4\,(20)] = \frac{8}{21}$.

13. a. Let x and y denote the numbers of days the Saddle and Horseshoe Mines are operated, respectively. Then the problem is: Minimize $C = 14{,}000x + 16{,}000y$ subject to $50x + 75y \ge 650$, $3000x + 1000y \ge 18{,}000$, $x \ge 0$, $y \ge 0$.

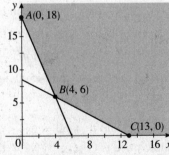

Vertex	$C = 14{,}000x + 16{,}000y$
$A\,(0, 18)$	288,000
$B\,(4, 6)$	152,000
$C\,(13, 0)$	182,000

From the figure and the table, we see that the optimal solution is $x = 4$, $y = 6$, giving $C = 152{,}000$. Therefore, the company should operate the Saddle Mine for 4 days and the Horseshoe Mine for 6 days at a minimal cost of \$152,000.

b. Suppose $C = cx + 16{,}000y$. Then $y = -\dfrac{c}{16{,}000}x + \dfrac{C}{16{,}000}$. In order for the current optimal solution to hold,

we must have $-\dfrac{c}{16{,}000} \le -\dfrac{50}{75}$, so $\dfrac{c}{16{,}000} \ge \dfrac{50}{75}$ and $c \ge \dfrac{32{,}000}{3}$; and $-\dfrac{c}{16{,}000} \ge -3$, so $\dfrac{c}{16{,}000} \le 3$ and

$c \le 48{,}000$. Therefore, $\dfrac{32{,}000}{3} \le c \le 48{,}000$.

c. Suppose the right-hand side of the first constraint (pertaining to gold requirements) is changed to $650 + h$. Then

we have the system $50x + 75y = 650 + h$, $2000x + 1000y = 18{,}000$. The solutions are $x = 4 - \dfrac{h}{175}$ and

$y = 6 + \dfrac{3h}{175}$. Now $x \ge 0$ implies $4 - \dfrac{h}{175} \ge 0$, so $h \le 700$; and $y \ge 0$ implies $6 + \dfrac{3h}{175} \ge 0$, so $h \ge -350$.

Therefore, $-350 \le h \le 700$ and so the resource can assume values between $650 - 350$ and $650 + 700$; that is,

between 300 and 1350.

d. Put $h = 1$ in part (a) to obtain $x = 4 - \tfrac{1}{175}$ and $y = 6 + \tfrac{3}{175}$. Using the fact that the

optimal solution found in part (a) is $x = 4$, $y = 6$, we see that the shadow price is

$14{,}000 \left(4 - \tfrac{1}{175}\right) + 16{,}000 \left(6 + \tfrac{3}{175}\right) - [14{,}000\,(4) + 16{,}000\,(6)] = 194.29$, or \$194.29.

15. a. Let x and y denote the numbers of Model A and Model B satellite radios to be produced. We are led to the

following problem: Maximize $P = 12x + 10y$ subject to $15x + 10y \le 1500$, $10x + 12y \le 1320$, $x \le 80$, $x \ge 0$,

$y \ge 0$.

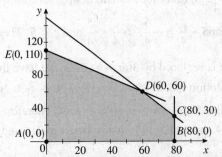

Vertex	$P = 12x + 10y$
$A\,(0, 0)$	0
$B\,(80, 0)$	960
$C\,(80, 30)$	1260
$D\,(60, 60)$	1320
$E\,(0, 110)$	1100

From the figure and the table, we see that the optimal solution is $x = y = 60$, giving $P = 1320$. So Soundex

should produce 60 of each model for a maximum profit of \$1320.

b. Suppose $P = cx + 10y$, so that $y = -\dfrac{c}{10}x + \dfrac{P}{10}$. In order for the current optimal solution to hold, we must have

$-\dfrac{c}{10} \ge -\dfrac{3}{2}$, so $\dfrac{c}{10} \le \dfrac{3}{2}$ and $c \le 15$; and $-\dfrac{c}{10} \le -\dfrac{5}{6}$, so $\dfrac{c}{10} \ge \dfrac{5}{6}$ and $c \ge \dfrac{25}{3}$. Therefore, $\tfrac{25}{3} \le c \le 15$.

c. Suppose the right-hand side of the first constraint (pertaining to the use of Machine I) is changed to $1500 + h$.

Then we have the system $15x + 10y = 1500 + h$, $10x + 12y = 1320$. The solution is $x = 60 + \tfrac{3}{20}h$,

$y = 60 - \tfrac{1}{8}h$. Now $x \ge 0$ implies $60 + \tfrac{3}{20}h \ge 0$, so $h \ge -400$; and $x \le 80$ implies that $60 + \tfrac{3}{20}h \le 80$,

$3h \le 400$, and $h \le 133\tfrac{1}{3}$. Also, $y \ge 0$ implies $60 - \tfrac{1}{8}h \ge 0$, so $h \le 480$. Therefore, $-400 \le h \le 133\tfrac{1}{3}$, and we

see that the values assumed by this resource must lie between $1500 - 400$ and $1500 + 133\tfrac{1}{3}$; that is, between

1100 and $1633\tfrac{1}{3}$.

d. Using the result of part (c) with $h = 1$, we find $x = 60 + \tfrac{3}{20}$ and $y = 60 - \tfrac{1}{8}$.

Next, using the result of part (a), we find that the shadow price is

$12 \left(60 + \tfrac{3}{20}\right) + 10 \left(60 - \tfrac{1}{8}\right) - [12\,(60) + 10\,(60)] = 12 \left(\tfrac{3}{20}\right) + 10 \left(-\tfrac{1}{8}\right) = 0.55$, or \$0.55.

e. Constraints 1 and 2 are binding because equality holds at the optimal solution. Constraint 3 is nonbinding.

17. a. Let x and y denote the numbers of model A and model B grates to be produced. We are led to the following problem: Maximize $P = 2x + 1.5y$ subject to $3x + 4y \leq 1000$ (constraint 1), $6x + 3y \leq 1200$ (Constraint 2), $y \leq 200$ (Constraint 3), $x \geq 0$, $y \geq 0$.

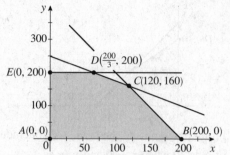

Vertex	$P = 2x + 1.5y$
$A\,(0, 0)$	0
$B\,(200, 0)$	400
$C\,(120, 160)$	480
$D\left(\frac{200}{3}, 200\right)$	433.33
$E\,(0, 200)$	300

From the figure and the table, we see that the optimal solution is $x = 120$, $y = 160$, giving $P = 480$. Therefore, Kane Manufacturing should produce 120 model A grates and 160 model B grates for a maximum daily profit of $480.

b. Suppose $P = cx + \frac{3}{2}y$, so that $y = -\frac{2}{3}cx + \frac{2}{3}P$. In order for the current optimal solution to hold, we must have $-\frac{2c}{3} \leq -\frac{3}{4}$, so $\frac{2c}{3} \geq \frac{3}{4}$ and $c \geq 1.125$; $-\frac{2c}{3} \geq -2$, so $\frac{2c}{3} \leq 2$ and $c \leq 3$; and $-\frac{3}{2}c \leq 0$, so $c \geq 0$. Therefore, $1.125 \leq c \leq 3$.

c. The constraint of interest here is constraint 1. Suppose we replace the right-hand side of this constraint by $1000 + h$. Then we have the system $3x + 4y = 1000 + h$, $6x + 3y = 1200$. The solution is $x = 120 - \frac{1}{5}h$, $y = 160 + \frac{2}{5}h$. Because $x \geq 0$, we have $120 - \frac{1}{5}h \geq 0$, so $h \leq 600$, and because $y \geq 0$, we have $160 + \frac{2}{5}h \geq 0$, so $h \geq -400$. Finally, $y \leq 200$ implies $160 + \frac{2}{5}h \leq 200$, so $h \leq 100$. Therefore, $-400 \leq h \leq 100$, and the values assumed by this resource must lie between $1000 - 400$ and $1000 + 100$; that is, between 600 and 1100.

d. Let $h = 1$. Then the results of part (c) give $x = 120 - \frac{1}{5}$ and $y = 160 + \frac{2}{5}$. Recall that the current optimal solution is $x = 120$ and $y = 160$. We see that the shadow price associated with the resource for cast iron (constraint 1) is $2\left(120 - \frac{1}{5}\right) + \frac{3}{2}\left(160 + \frac{2}{5}\right) - \left[2\,(120) + \frac{3}{2}\,(160)\right] = 2\left(-\frac{1}{5}\right) + \frac{3}{2}\left(\frac{2}{5}\right) = \frac{1}{5}$, or \$0.20.

e. Constraints 1 and 2 are binding because equality holds at the optimal solution. Constraint 3 is nonbinding because it reduces to $160 \leq 200$ when $x = 120$ and $y = 160$, and so equality does not hold.

| CHAPTER 3 | Concept Review Questions | page 215 |

1. a. half-plane, line

b. $ax + by \leq c$; $ax + by = c$

3. objective function, maximized, minimized, linear, inequalities

5. parameters, optimal

CHAPTER 3   Review Exercises   page 216

1. We evaluate Z at each of the corner points of the feasible set S.

Vertex	$Z = 2x + 3y$
$(0, 0)$	$\boxed{0}$
$(5, 0)$	10
$(3, 4)$	$\boxed{18}$
$(0, 6)$	$\boxed{18}$

From the table, we conclude that Z attains a minimum value of 0 when $x = 0$ and $y = 0$, and a maximum value of 18 when x and y lie on the line segment joining $(3, 4)$ and $(0, 6)$.

3.

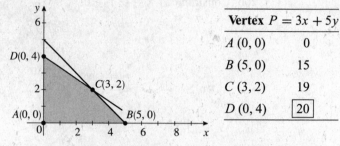

Vertex	$P = 3x + 5y$
$A\,(0, 0)$	0
$B\,(5, 0)$	15
$C\,(3, 2)$	19
$D\,(0, 4)$	$\boxed{20}$

From the graph and the table, we conclude that the maximum value of P is 20 when $x = 0$ and $y = 4$.

5.

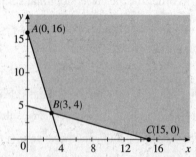

Vertex	$C = 2x + 5y$
$A\,(0, 16)$	80
$B\,(3, 4)$	$\boxed{26}$
$C\,(15, 0)$	30

From the graph and the table, we conclude that the minimum value of C is 26 when $x = 3$ and $y = 4$.

7.

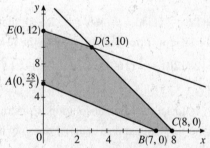

Vertex	$P = 3x + 2y$
$A\left(0, \frac{28}{5}\right)$	$\frac{56}{5}$
$B\,(7, 0)$	21
$C\,(8, 0)$	24
$D\,(3, 10)$	$\boxed{29}$
$E\,(0, 12)$	24

From the graph and the table, we conclude that P attains a maximum value of 29 when $x = 3$ and $y = 10$.

9.

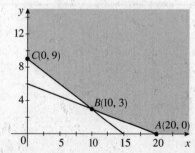

Vertex	$C = 2x + 7y$
$A\,(20, 0)$	$\boxed{40}$
$B\,(10, 3)$	41
$C\,(0, 9)$	63

From the graph and the table, we conclude that C attains a minimum value of 40 when $x = 20$ and $y = 0$.

11.

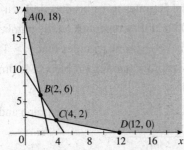

Vertex	$C = 4x + y$
$A\,(0, 18)$	18
$B\,(2, 6)$	$\boxed{14}$
$C\,(4, 2)$	18
$D\,(12, 0)$	48

From the graph and the table, we conclude that C attains a minimum value of 14 when $x = 2$ and $y = 6$.

13.

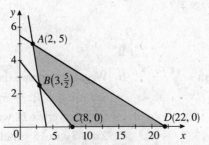

Vertex	$Q = x + y$
$A\,(2, 5)$	7
$B\left(3, \frac{5}{2}\right)$	$\boxed{\frac{11}{2}}$
$C\,(8, 0)$	8
$D\,(22, 0)$	$\boxed{22}$

From the graph and the table, we conclude that Q attains a maximum value of 22 when $x = 22$ and $y = 0$ and a minimum value of $\frac{11}{2}$ when $x = 3$ and $y = \frac{5}{2}$.

15. Suppose the investor puts x and y thousand dollars into the stocks of companies A and B, respectively. Then we have the following linear programming problem: Maximize $P = 0.14x + 0.20y$ subject to $x + y \leq 80$, $0.01x + 0.04y \leq 2$, $x \geq 0$, $y \geq 0$.

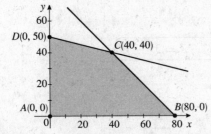

Vertex	$P = 0.14x + 0.20y$
$A\,(0, 0)$	0
$B\,(80, 0)$	11.2
$C\,(40, 40)$	$\boxed{13.6}$
$D\,(0, 50)$	10

From the graph and the table, we conclude that P attains a maximum value of 13.6 when $x = 40$ and $y = 40$. Thus, by investing \$40,000 in the stocks of each company, the investor will achieve a maximum return of \$13,600.

17. Let x and y denote the numbers of model A and model B grates to be produced. Then the constraint on the amount of cast iron available leads to $3x + 4y \leq 1000$ and the constraint on the number of minutes of labor used each day leads to $6x + 3y \leq 1200$. One additional constraint specifies that $y \geq 180$. The daily profit is $P = 2x + 1.5y$. Therefore, we have the following linear programming problem: Maximize $P = 2x + 1.5y$ subject to $3x + 4y \leq 1000$, $6x + 3y \leq 1200$, $x \geq 0$, $y \geq 180$.

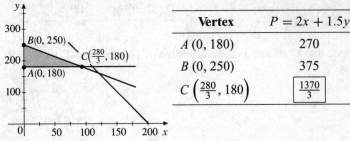

Vertex	$P = 2x + 1.5y$
$A\,(0, 180)$	270
$B\,(0, 250)$	375
$C\left(\frac{280}{3}, 180\right)$	$\boxed{\frac{1370}{3}}$

From the graph and the table, we conclude that the optimal profit of \$456 is realized when 93 model A grates and 180 model B grates are produced.

CHAPTER 3   Before Moving On...   page 217

1. a. We wish to graph the solution set for the system $2x + y \leq 10$, $x + 3y \leq 15$, $x \leq 4$, $x \geq 0$, $y \geq 0$. To find the vertex D, we solve $2x + y = 10$ and $x + 3y = 15$ simultaneously, obtaining $5y = 20$ (so $y = 4$) and $x = 15 - 3y = 15 - 3(4) = 3$.

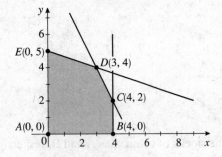

b. We wish to graph the solution set for the system $2x + y \geq 8$, $2x + 3y \geq 15$, $x \geq 0$, $y \geq 2$. To find the vertex B, we solve $2x + y = 8$ and $2x + 3y = 15$ simultaneously, obtaining $2y = 7$ (so $y = \frac{7}{2}$) and $2x + \frac{7}{2} = 8$ (so $2x = \frac{9}{2}$ and $x = \frac{9}{4}$).

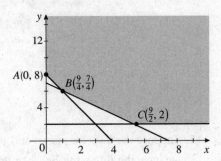

2.

Vertex	$Z = 3x - y$
$(8, 2)$	22
$(28, 8)$	$\boxed{76}$
$(16, 24)$	24
$(3, 16)$	$\boxed{-7}$

From the table, we see that the maximum value is $Z = 76$ and the minimum value is $Z = -7$.

3. Maximize $P = x + 3y$ subject to $2x + 3y \leq 11$, $3x + 7y \leq 24$, $x \geq 0$, $y \geq 0$. To find the coordinates of C, we solve $2x + 3y = 11$ and $3x + 7y = 24$ simultaneously, obtaining $6x + 9y = 33$, $6x + 14y = 48$, $5y = 15$, and $y = 3$, so $x = 1$.

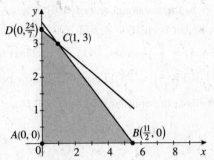

Vertex	$P = x + 3y$
$A\,(0, 0)$	0
$B\left(\frac{11}{2}, 0\right)$	$\frac{11}{2}$
$C\,(1, 3)$	10
$D\left(0, \frac{24}{7}\right)$	$\boxed{\dfrac{72}{7}}$

From the graph and the table, we see that the maximum value of P is $\frac{72}{7}$, attained at $x = 0$, $y = \frac{24}{7}$.

4. Minimize $C = 4x + y$ subject to $2x + y \geq 10$, $2x + 3y \geq 24$, $x + 3y \geq 15$, $x \geq 0$, $y \geq 0$. To find the coordinates of C, we solve $2x + y = 10$ and $2x + 3y = 24$ simultaneously, finding $2y = 14$, so $y = 7$ and $x = \frac{3}{2}$. To find the coordinates of B, we solve $2x + 3y = 24$ and $x + 3y = 15$ simultaneously, obtaining $x = 9$, so $y = 2$.

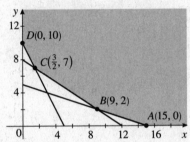

Vertex	$C = 4x + y$
$A\,(15, 0)$	70
$B\,(9, 2)$	38
$C\left(\frac{3}{2}, 7\right)$	13
$D\,(0, 10)$	$\boxed{10}$

From the graph and the table, we see that the minimum value of C is 10, attained at $x = 0$, $y = 10$.

5. Maximize $P = 2x + 3y$ subject to $x + 2y \leq 16$, $3x + 2y \leq 24$, $x \geq 0$, $y \geq 0$.

a. To find the coordinates of the vertex C, we solve $x + 2y = 16$ and $3x + 2y = 24$, obtaining $x = 4$, $y = 6$.

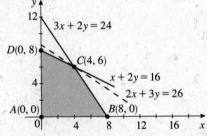

Vertex	$P = 2x + 3y$
$A\,(0, 0)$	0
$B\,(8, 0)$	16
$C\,(4, 6)$	$\boxed{26}$
$D\,(0, 8)$	24

From the graph and the table, we see that the solution is $x = 4$, $y = 6$, giving $P = 26$.

b. Let $P = cx + 3y$. Solving for y, we have $y = -\frac{1}{3}cx + \frac{1}{3}P$. If the slope of the isoprofit line is greater than the slope of the line associated with constraint 1, then the optimal solution will shift from point C to point D. But the slope of the line associated with constraint 1 is $-\frac{1}{2}$, so we have $-\frac{1}{3}c \leq -\frac{1}{2}$, $\frac{1}{3}c \geq \frac{1}{2}$, and $c \geq \frac{3}{2}$. Therefore, $\frac{3}{2} \leq c \leq \frac{9}{2}$.

c. Suppose the right-hand side of constraint 1 is replaced by $16 + h$. Then the new optimal solution occurs at a new point C'. To find the coordinate of C', we solve $x + 2y = 16 + h$ and $3x + 2y = 24$ simultaneously, obtaining $2x = 8 - h$, so $x = \frac{1}{2}(8 - h)$. Then $2y = 24 - 3x = 24 - \frac{3}{2}(8 - h) = 12 + \frac{3}{2}h$, so $y = 6 + \frac{3}{4}h$. The nonnegativity of x implies $\frac{1}{2}(8 - h) \geq 0$, or $h \leq 8$, and the nonnegativity of y implies $6 + \frac{3}{4}h \geq 0$, or $h \geq -8$. Therefore, $-8 \leq h \leq 8$, and so resource 1 must lie between $16 - 8$, and $16 + 8$; that is, between 8 and 24.

d. If we set $h = 1$ in part (c), we obtain $x = \frac{1}{2}(8 - 1) = \frac{7}{2}$, $y = 6 + \frac{3}{4}(1) = \frac{27}{4}$. Therefore, the profit realized at this level of production is $P = 2x + 3y = 2\left(\frac{7}{2}\right) + 3\left(\frac{27}{4}\right) = \frac{109}{4}$. Thus, the shadow price is $\frac{109}{4} - 26$, or $1.25.

e. Both constraints are binding because equality holds at the optimal solution $C(4, 6)$.

4 LINEAR PROGRAMMING: AN ALGEBRAIC APPROACH

4.1 The Simplex Method: Standard Maximization Problems

Problem-Solving Tips

1. Make sure that you set up the initial simplex tableau correctly:

   a. First, rewrite the linear inequalities as equalities by introducing slack variables.

   b. Next, rewrite the objective function so that all variables are on the left-hand side and the coefficient of P is 1. Then place this equation below the other equations.

2. In the simplex method, the optimal solution has been reached if *all the entries* in the last row to the left of the vertical line are *nonnegative*.

3. In the simplex method, you will know that the optimal solution has not been reached if there are *negative entries in the last row to the left of the vertical line*. If so, locate the most negative entry in that row. This gives you the pivot column. Proceed to find the pivot row by dividing each positive entry in the pivot column by its corresponding entry in the column of constants. Look for the smallest ratio. The corresponding entry will be in the pivot row. Proceed to make the pivot column a unit column by pivoting about the pivot element.

Concept Questions   page 235

1. a. The objective function is to be maximized.

   b. All the variables involved in the problem are nonnegative.

   c. Each linear constraint may be written so that the expression involving the variables is less than or equal to a nonnegative constant.

3. To find the *pivot column*, locate the most negative entry to the left of the vertical line in the last row. The column containing this entry is the pivot column. To find the *pivot row*, divide each positive entry in the pivot column into its corresponding entry in the column of constants. The pivot row is the row corresponding to the smallest ratio thus obtained. The *pivot element* is the element common to the pivot column and the pivot row.

Exercises   page 236

1. a. It is already in standard form.

   b.

x	y	u	v	P	Constant
1	4	1	0	0	12
1	3	0	1	0	10
−2	−4	0	0	1	0

3. a. We multiply the second inequality by -1, obtaining the following standard maximization problem:

Maximize $P = 2x + 3y$
subject to $x + y \leq 10$
$x + 2y \leq 12$
$2x + y \leq 12$
$x \geq 0, y \geq 0$

b.

x	y	u	v	w	P	Constant
1	1	1	0	0	0	10
1	2	0	1	0	0	12
2	1	0	0	1	0	12
-2	-3	0	0	0	1	0

5. a. We multiply the second inequality by -1, obtaining the following standard maximization problem:

Maximize $P = x + 3y + 4z$
subject to $x + 2y + z \leq 40$
$x + y + z \leq 30$
$x \geq 0, y \geq 0, z \geq 0$

b.

x	y	z	u	v	P	Constant
1	2	1	1	0	0	40
1	1	1	0	1	0	30
-1	-3	-4	0	0	1	0

7. All entries in the last row of the simplex tableau are nonnegative and an optimal solution has been reached. We find $x = \frac{30}{7}, y = \frac{20}{7}, u = 0, v = 0$, and $P = \frac{220}{7}$.

9. The simplex tableau is not in final form because there is a negative entry in the last row. The entry in the first row, second column, is the next pivot element and has a value of $\frac{1}{2}$.

11. The simplex tableau is in final form. We find $x = \frac{1}{3}, y = 0, z = \frac{13}{3}, u = 0, v = 6, w = 0$ and $P = 17$.

13. The simplex tableau is not in final form because there are two negative entries in the last row. The entry in the third row, second column, is the pivot element and has a value of 1.

15. The simplex tableau is in final form. The solutions are $x = 30, y = 10, z = 0, u = 0, v = 0, P = 60$ and $x = 30$, $y = 0, z = 0, u = 10, v = 0, P = 60$. (There are infinitely many solutions.)

17. We calculate the following sequence of tableaus:

	x	y	u	v	P	Constant	Ratio
Pivot row →	1	①	1	0	0	4	4
	2	1	0	1	0	5	5
	-3	-4	0	0	1	0	

↑ Pivot column

$\xrightarrow[R_3 + 4R_1]{R_2 - R_1}$

x	y	u	v	P	Constant
1	1	①	0	0	4
1	0	-1	1	0	1
1	0	4	0	1	16

The last tableau is in final form and we conclude that $x = 0, y = 4, u = 0, v = 1$, and $P = 16$.

19. We calculate the following sequence of tableaus:

	x	y	u	v	P	Constant	Ratio
Pivot row →	1	②	1	0	0	12	6
	3	2	0	1	0	24	12
	−10	−12	0	0	1	0	

↑ Pivot column

$\xrightarrow{\frac{1}{2}R_1}$

x	y	u	v	P	Constant
$\frac{1}{2}$	①	$\frac{1}{2}$	0	0	6
3	2	0	1	0	24
−10	−12	0	0	1	0

$\xrightarrow[R_3 + 12R_1]{R_2 - 2R_1}$

	x	y	u	v	P	Constant	Ratio
	$\frac{1}{2}$	1	$\frac{1}{2}$	0	0	6	12
Pivot row →	②	0	−1	1	0	12	6
	−4	0	6	0	1	72	

↑ Pivot column

$\xrightarrow{\frac{1}{2}R_2}$

x	y	u	v	P	Constant
$\frac{1}{2}$	1	$\frac{1}{2}$	0	0	6
①	0	$-\frac{1}{2}$	$\frac{1}{2}$	0	6
−4	0	6	0	1	72

$\xrightarrow[R_3 + 4R_2]{R_1 - \frac{1}{2}R_2}$

x	y	u	v	P	Constant
0	1	$\frac{3}{4}$	$-\frac{1}{4}$	0	3
1	0	$-\frac{1}{2}$	$\frac{1}{2}$	0	6
0	0	4	2	1	96

The last tableau is in final form. We conclude that $x = 6$, $y = 3$, $u = 0$, $v = 0$, and $P = 96$.

21. We calculate the following sequence of tableaus:

	x	y	u	v	w	P	Constant	Ratio
	3	1	1	0	0	0	24	24
	2	1	0	1	0	0	18	18
Pivot row →	1	③	0	0	1	0	24	8
	−4	−6	0	0	0	1	0	

↑ Pivot column

$\xrightarrow{\frac{1}{3}R_3}$

x	y	u	v	w	P	Constant
3	1	1	0	0	0	24
2	1	0	1	0	0	18
$\frac{1}{3}$	①	0	0	$\frac{1}{3}$	0	8
−4	−6	0	0	0	1	0

$\xrightarrow[R_4 + 6R_3]{\begin{array}{c}R_1 - R_3\\ R_2 - R_3\end{array}}$

	x	y	u	v	w	P	Constant	Ratio
Pivot row →	$\frac{8}{3}$	0	1	0	$-\frac{1}{3}$	0	16	6
	$\frac{5}{3}$	0	0	1	$-\frac{1}{3}$	0	10	6
	$\frac{1}{3}$	1	0	0	$\frac{1}{3}$	0	8	24
	−2	0	0	0	2	1	48	

↑ Pivot column

$\xrightarrow{\frac{3}{8}R_1}$

x	y	u	v	w	P	Constant
①	0	$\frac{3}{8}$	0	$-\frac{1}{8}$	0	6
$\frac{5}{3}$	0	0	1	$-\frac{1}{3}$	0	10
$\frac{1}{3}$	1	0	0	$\frac{1}{3}$	0	8
−2	0	0	0	2	1	48

$\xrightarrow[R_4 + 2R_1]{\begin{array}{c}R_2 - \frac{5}{3}R_1\\ R_3 - \frac{1}{3}R_1\end{array}}$

x	y	u	v	w	P	Constant
1	0	$\frac{3}{8}$	0	$-\frac{1}{8}$	0	6
0	0	$-\frac{5}{8}$	1	$-\frac{1}{8}$	0	0
0	1	$-\frac{1}{8}$	0	$\frac{3}{8}$	0	6
0	0	$\frac{3}{4}$	0	$\frac{7}{4}$	1	60

Observe that we have a choice after the third tableau, in which the ratios on the first and second lines are both 6. The last tableau is in final form, and we deduce that $x = 6$, $y = 6$, $u = 0$, $v = 0$, $w = 0$, and $P = 60$.

23. We calculate the following sequence of tableaus:

	x	y	z	u	v	w	P	Constant	Ratio
	3	10	5	1	0	0	0	120	12
Pivot row →	5	②	8	0	1	0	0	6	3
	8	10	3	0	0	1	0	105	$\frac{21}{2}$
	-3	-4	-1	0	0	0	1	0	

Pivot column (under y)

$\xrightarrow{\frac{1}{2}R_2}$

x	y	z	u	v	w	P	Constant
3	10	5	1	0	0	0	120
$\frac{5}{2}$	①	4	0	$\frac{1}{2}$	0	0	3
8	10	3	0	0	1	0	105
-3	-4	-1	0	0	0	1	0

$\xrightarrow[\ \ \ \ \ \ \ \ ]{\begin{array}{c}R_1 - 10R_2 \\ R_3 - 10R_2 \\ \hline R_4 + 4R_2\end{array}}$

x	y	z	u	v	w	P	Constant
-22	0	-35	1	-5	0	0	90
$\frac{5}{2}$	1	4	0	$\frac{1}{2}$	0	0	3
-17	0	-37	0	-5	1	0	75
7	0	15	0	2	0	1	12

The last tableau is in final form. We conclude that $x = 0$, $y = 3$, $z = 0$, $u = 90$, $v = 0$, $w = 75$, and $P = 12$.

25. We calculate the following sequence of tableaus:

	x	y	z	u	v	P	Constant	Ratio
	1	1	1	1	0	0	8	8
Pivot row →	3	2	④	0	1	0	24	6
	-3	-4	-5	0	0	1	0	

Pivot column (under z)

$\xrightarrow{\frac{1}{4}R_2}$

x	y	z	u	v	P	Constant
1	1	1	1	0	0	8
$\frac{3}{4}$	$\frac{1}{2}$	①	0	$\frac{1}{4}$	0	6
-3	-4	-5	0	0	1	0

$\xrightarrow[\ \ \ \ \ \ ]{\begin{array}{c}R_1 - R_2 \\ \hline R_3 + 5R_2\end{array}}$

	x	y	z	u	v	P	Constant	Ratio
Pivot row →	$\frac{1}{4}$	½	0	1	$-\frac{1}{4}$	0	2	4
	$\frac{3}{4}$	$\frac{1}{2}$	1	0	$\frac{1}{4}$	0	6	12
	$\frac{3}{4}$	$-\frac{3}{2}$	0	0	$\frac{5}{4}$	1	30	

Pivot column (under y)

$\xrightarrow{2R_1}$

x	y	z	u	v	P	Constant
$\frac{1}{2}$	①	0	2	$-\frac{1}{2}$	0	4
$\frac{3}{4}$	$\frac{1}{2}$	1	0	$\frac{1}{4}$	0	6
$\frac{3}{4}$	$-\frac{3}{2}$	0	0	$\frac{5}{4}$	1	30

$\xrightarrow[\ \ \ \ \ \ ]{\begin{array}{c}R_2 - \frac{1}{2}R_1 \\ \hline R_3 + \frac{3}{2}R_1\end{array}}$

x	y	z	u	v	P	Constant
$\frac{1}{2}$	1	0	2	$-\frac{1}{2}$	0	4
$\frac{1}{2}$	0	1	-1	$\frac{1}{2}$	0	4
$\frac{3}{4}$	0	0	3	$\frac{1}{2}$	1	36

The last tableau is in final form, and we deduce that $x = 0$, $y = 4$, $z = 4$, $u = 0$, $v = 0$, and $P = 36$.

27. We calculate the following sequence of tableaus:

	x	y	z	u	v	w	P	Constant	Ratio
	1	1	1	1	0	0	0	20	20
Pivot row →	2	④	3	0	1	0	0	42	$\frac{21}{2}$
	2	0	3	0	0	1	0	30	—
	−4	−6	−5	0	0	0	1	0	

↑ Pivot column

$\xrightarrow{\frac{1}{4}R_2}$

	x	y	z	u	v	w	P	Constant
	1	1	1	1	0	0	0	20
	$\frac{1}{2}$	①	$\frac{3}{4}$	0	$\frac{1}{4}$	0	0	$\frac{21}{2}$
	2	0	3	0	0	1	0	30
	−4	−6	−5	0	0	0	1	0

$\xrightarrow{\substack{R_1 - R_2 \\ R_4 + 6R_2}}$

	x	y	z	u	v	w	P	Constant	Ratio
	$\frac{1}{2}$	0	$\frac{1}{4}$	1	$-\frac{1}{4}$	0	0	$\frac{19}{2}$	19
	$\frac{1}{2}$	1	$\frac{3}{4}$	0	$\frac{1}{4}$	0	0	$\frac{21}{2}$	21
Pivot row →	②	0	3	0	0	1	0	30	15
	−1	0	$-\frac{1}{2}$	0	$\frac{3}{2}$	0	1	63	

↑ Pivot column

$\xrightarrow{\frac{1}{2}R_3}$

	x	y	z	u	v	w	P	Constant
	$\frac{1}{2}$	0	$\frac{1}{4}$	1	$-\frac{1}{4}$	0	0	$\frac{19}{2}$
	$\frac{1}{2}$	1	$\frac{3}{4}$	0	$\frac{1}{4}$	0	0	$\frac{21}{2}$
	①	0	$\frac{3}{2}$	0	0	$\frac{1}{2}$	0	15
	−1	0	$-\frac{1}{2}$	0	$\frac{3}{2}$	0	1	63

$\xrightarrow{\substack{R_1 - \frac{1}{2}R_3 \\ R_2 - \frac{1}{2}R_3 \\ R_4 + R_3}}$

x	y	z	u	v	w	P	Constant
0	0	$-\frac{1}{2}$	1	$-\frac{1}{4}$	$-\frac{1}{4}$	0	2
0	1	0	0	$\frac{1}{4}$	$-\frac{1}{4}$	0	3
1	0	$\frac{3}{2}$	0	0	$\frac{1}{2}$	0	15
0	0	1	0	$\frac{3}{2}$	$\frac{1}{2}$	1	78

The last tableau is in final form. We conclude that $x = 15$, $y = 3$, $z = 0$, $u = 2$, $v = 0$, $w = 0$, and $P = 78$.

29. We calculate the following sequence of tableaus:

	x	y	z	u	v	w	P	Constant	Ratio
Pivot row →	②	1	1	1	0	0	0	10	5
	3	5	1	0	1	0	0	45	15
	2	5	1	0	0	1	0	40	20
	−12	−10	−5	0	0	0	1	0	

↑ Pivot column

$\xrightarrow{\frac{1}{2}R_1}$

	x	y	z	u	v	w	P	Constant
	①	$\frac{1}{2}$	$\frac{1}{2}$	$\frac{1}{2}$	0	0	0	5
	3	5	1	0	1	0	0	45
	2	5	1	0	0	1	0	40
	−12	−10	−5	0	0	0	1	0

$\xrightarrow{\substack{R_2 - 3R_1 \\ R_3 - 2R_1 \\ R_4 + 12R_1}}$

	x	y	z	u	v	w	P	Constant	Ratio
	1	$\frac{1}{2}$	$\frac{1}{2}$	$\frac{1}{2}$	0	0	0	5	10
	0	$\frac{7}{2}$	$-\frac{1}{2}$	$-\frac{3}{2}$	1	0	0	30	$\frac{60}{7}$
Pivot row →	0	④	0	−1	0	1	0	30	$\frac{15}{2}$
	0	−4	1	6	0	0	1	60	

↑ Pivot column

$\xrightarrow{\frac{1}{4}R_3}$

	x	y	z	u	v	w	P	Constant
	1	$\frac{1}{2}$	$\frac{1}{2}$	$\frac{1}{2}$	0	0	0	5
	0	$\frac{7}{2}$	$-\frac{1}{2}$	$-\frac{3}{2}$	1	0	0	30
	0	①	0	$-\frac{1}{4}$	0	$\frac{1}{4}$	0	$\frac{15}{2}$
	0	−4	1	6	0	0	1	60

$\xrightarrow{\substack{R_1 - \frac{1}{2}R_3 \\ R_2 - \frac{7}{2}R_3 \\ R_4 + 4R_3}}$

x	y	z	u	v	w	P	Constant
1	0	$\frac{1}{2}$	$\frac{5}{8}$	0	$-\frac{1}{8}$	0	$\frac{5}{4}$
0	0	$-\frac{1}{2}$	$-\frac{5}{8}$	1	$-\frac{7}{8}$	0	$\frac{15}{4}$
0	1	0	$-\frac{1}{4}$	0	$\frac{1}{4}$	0	$\frac{15}{2}$
0	0	1	5	0	1	1	90

The last tableau is in final form, and we conclude that $x = \frac{5}{4}$, $y = \frac{15}{2}$, $z = 0$, $u = 0$, $v = \frac{15}{4}$, $w = 0$, and $P = 90$.

31. We calculate the following sequence of tableaus:

	x	y	z	u	v	w	P	Constant	Ratio
Pivot row →	②	1	2	1	0	0	0	7	$\frac{7}{2}$
	2	3	1	0	1	0	0	8	4
	1	2	3	0	0	1	0	7	7
	−24	−16	−23	0	0	0	1	0	

Pivot column

$\xrightarrow{\frac{1}{2}R_1}$

x	y	z	u	v	w	P	Constant
①	$\frac{1}{2}$	1	$\frac{1}{2}$	0	0	0	$\frac{7}{2}$
2	3	1	0	1	0	0	8
1	2	3	0	0	1	0	7
−24	−16	−23	0	0	0	1	0

$\xrightarrow[\substack{R_2 - 2R_1 \\ R_3 - R_1 \\ R_4 + 24R_1}]{}$

	x	y	z	u	v	w	P	Constant	Ratio
	1	$\frac{1}{2}$	1	$\frac{1}{2}$	0	0	0	$\frac{7}{2}$	7
Pivot row →	0	②	−1	−1	1	0	0	1	$\frac{1}{2}$
	0	$\frac{3}{2}$	2	$-\frac{1}{2}$	0	1	0	$\frac{7}{2}$	$\frac{7}{3}$
	0	−4	1	12	0	0	1	84	

Pivot column

$\xrightarrow{\frac{1}{2}R_2}$

x	y	z	u	v	w	P	Constant
1	$\frac{1}{2}$	1	$\frac{1}{2}$	0	0	0	$\frac{7}{2}$
0	①	$-\frac{1}{2}$	$-\frac{1}{2}$	$\frac{1}{2}$	0	0	$\frac{1}{2}$
0	$\frac{3}{2}$	2	$-\frac{1}{2}$	0	1	0	$\frac{7}{2}$
0	−4	1	12	0	0	1	84

$\xrightarrow[\substack{R_1 - \frac{1}{2}R_2 \\ R_3 - \frac{3}{2}R_2 \\ R_4 + 4R_2}]{}$

	x	y	z	u	v	w	P	Constant	Ratio
	1	0	$\frac{5}{4}$	$\frac{3}{4}$	$\frac{1}{4}$	0	0	$\frac{13}{4}$	$\frac{13}{5}$
	0	1	$-\frac{1}{2}$	$-\frac{1}{2}$	$\frac{1}{2}$	0	0	$\frac{1}{2}$	−
Pivot row →	0	0	⑪/₄	$\frac{1}{4}$	$-\frac{3}{4}$	1	0	$\frac{11}{4}$	1
	0	0	−1	10	2	0	1	86	

Pivot column

$\xrightarrow{\frac{4}{11}R_3}$

x	y	z	u	v	w	P	Constant
1	0	$\frac{5}{4}$	$\frac{3}{4}$	$\frac{1}{4}$	0	0	$\frac{13}{4}$
0	1	$-\frac{1}{2}$	$-\frac{1}{2}$	$\frac{1}{2}$	0	0	$\frac{1}{2}$
0	0	①	$\frac{1}{11}$	$-\frac{3}{11}$	$\frac{4}{11}$	0	1
0	0	−1	10	2	0	1	86

$\xrightarrow[\substack{R_1 - \frac{5}{4}R_3 \\ R_2 + \frac{1}{2}R_3 \\ R_4 + R_3}]{}$

x	y	z	u	v	w	P	Constant
1	0	0	$\frac{7}{11}$	$\frac{13}{22}$	$-\frac{5}{11}$	0	2
0	1	0	$-\frac{5}{11}$	$\frac{4}{11}$	$\frac{2}{11}$	0	1
0	0	1	$\frac{1}{11}$	$-\frac{3}{11}$	$\frac{4}{11}$	0	1
0	0	0	$\frac{111}{11}$	$\frac{19}{11}$	$\frac{4}{11}$	1	87

The last tableau is in final form, and we conclude that P attains a maximum value of 87 when $x = 2$, $y = 1$, $z = 1$, $u = 0$, $v = 0$, and $w = 0$.

33. Pivoting about the x-column in the initial simplex tableau, we have

	x	y	z	u	v	P	Constant	Ratio		x	y	z	u	v	P	Constant	
	3	3	-2	1	0	0	100	$\frac{100}{3}$	$\xrightarrow{\frac{1}{5}R_2}$	3	3	-2	1	0	0	100	$\xrightarrow{\substack{R_1-3R_2 \\ R_3+2R_2}}$
Pivot row $\rightarrow$	⑤	5	3	0	1	0	150	30		①	1	$\frac{3}{5}$	0	$\frac{1}{5}$	0	30	
	-2	-2	4	0	0	1	0			-2	-2	4	0	0	1	0	

Pivot column $\uparrow$

x	y	z	u	v	P	Constant
0	0	$-\frac{19}{5}$	1	$-\frac{3}{5}$	0	10
1	1	$\frac{3}{5}$	0	$\frac{1}{5}$	0	30
0	0	$\frac{26}{5}$	0	$\frac{2}{5}$	1	60

and we see that one optimal solution occurs when $x = 30$, $y = 0$, $z = 0$, and $P = 60$. Similarly, pivoting about the y-column, we obtain another optimal solution: $x = 0$, $y = 30$, $z = 0$, $P = 60$.

35. Let x and y denote the numbers of model A and model B portable printers produced each shift, respectively. Then we have the following linear programming problem:

$$\text{Maximize} \quad P = 30x + 40y$$
$$\text{subject to} \quad 100x + 150y \le 600{,}000$$
$$x + y \le 2500$$
$$x \ge 0, y \ge 0$$

Using the simplex method, we calculate the following sequence of tableaus:

	x	y	u	v	P	Constant	Ratio		x	y	u	v	P	Constant
	100	150	1	0	0	600,000	4000	$\xrightarrow{\substack{R_1-150R_2 \\ R_3+40R_2}}$	-50	0	1	-150	0	225,000
Pivot row $\rightarrow$	1	①	0	1	0	2500	2500		1	①	0	1	0	2500
	-30	-40	0	0	1	0			10	0	0	40	1	100,000

Pivot column $\uparrow$

We conclude that the maximum monthly profit is \$100,000, and this occurs when no model A and 2500 model B portable printers are produced.

37. Refer to the solution of Exercise 3.2.9 on page 80 of this manual. We rewrite the problem in standard form by multiplying the second inequality by -1, and obtain the following:

$$\text{Maximize} \quad P = 0.12x + 0.20y$$
$$\text{subject to} \quad x + y \le 60{,}000$$
$$-3x + 2y \le \quad 0$$
$$x - 4y \le \quad 0$$
$$x \ge 0, y \ge 0$$

We calculate the following sequence of tableaus:

	x	y	u	v	w	P	Constant	Ratio
	1	1	1	0	0	0	60,000	60,000
Pivot row →	-3	②	0	1	0	0	0	0
	1	-4	0	0	1	0	0	—
	$-\frac{3}{25}$	$-\frac{1}{5}$	0	0	0	1	0	

↑ Pivot column

$\xrightarrow{\frac{1}{2}R_2}$

x	y	u	v	w	P	Constant
1	1	1	0	0	0	60,000
$-\frac{3}{2}$	1	0	$\frac{1}{2}$	0	0	0
1	-4	0	0	1	0	0
$-\frac{3}{25}$	$-\frac{1}{5}$	0	0	0	1	0

$\xrightarrow[\substack{R_3 + 4R_2 \\ R_4 + \frac{1}{5}R_3}]{R_1 - R_2}$

	x	y	u	v	w	P	Constant	Ratio
Pivot row →	$\frac{5}{2}$	0	1	$-\frac{1}{2}$	0	0	60,000	24,000
	$-\frac{3}{2}$	1	0	$\frac{1}{2}$	0	0	0	—
	-5	0	0	2	1	0	0	—
	$-\frac{21}{50}$	0	0	$\frac{1}{10}$	0	1	0	

↑ Pivot column

$\xrightarrow{\frac{2}{5}R_1}$

x	y	u	v	w	P	Constant
1	0	$\frac{2}{5}$	$-\frac{1}{5}$	0	0	24,000
$-\frac{3}{2}$	1	0	$\frac{1}{2}$	0	0	0
-5	0	0	2	1	0	0
$-\frac{21}{50}$	0	0	$\frac{1}{10}$	0	1	0

$\xrightarrow[\substack{R_3 + 5R_1 \\ R_4 + \frac{21}{50}R_1}]{R_2 + \frac{3}{2}R_1}$

x	y	u	v	w	P	Constant
1	0	$\frac{2}{5}$	$-\frac{1}{5}$	0	0	24,000
0	1	$\frac{3}{5}$	$\frac{1}{5}$	0	0	36,000
0	0	2	1	1	0	120,000
0	0	$\frac{21}{125}$	$\frac{2}{125}$	0	1	10,080

The last tableau is in final form, and we conclude that the maximum value of 10,080 is obtained when $x = 24{,}000$ and $y = 36{,}000$. Thus, Justin should invest \$24,000 in medium-risk stocks and \$36,000 in high-risk stocks.

39. Refer to the solution to Exercise 3.3.33 on page 89 of this manual. The linear programming problem is:

$$\text{Maximize} \quad P = 45x + 20y$$
$$\text{subject to} \quad 40x + 16y \leq 3200$$
$$3x + 4y \leq 520$$
$$x \geq 0, y \geq 0$$

We calculate the following sequence of tableaus:

Pivot row →

x	y	u	v	P	Constant	Ratio
⑩ 40	16	1	0	0	3200	80
3	4	0	1	0	520	$\frac{520}{3}$
−45	−20	0	0	1	0	

↑ Pivot column

$\xrightarrow{\frac{1}{40}R_1}$

x	y	u	v	P	Constant
①	$\frac{2}{5}$	$\frac{1}{40}$	0	0	80
3	4	0	1	0	520
−45	−20	0	0	1	0

$\xrightarrow[R_3 + 45R_1]{R_2 - 3R_1}$

x	y	u	v	P	Constant	Ratio
1	$\frac{2}{5}$	$\frac{1}{40}$	0	0	80	200
0	⑭⁄₅ $\frac{14}{5}$	$-\frac{3}{40}$	1	0	280	100
0	−2	$\frac{9}{8}$	0	1	3600	

Pivot row →   ↑ Pivot column

$\xrightarrow{\frac{5}{14}R_2}$

x	y	u	v	P	Constant
1	$\frac{2}{5}$	$\frac{1}{40}$	0	0	80
0	①	$-\frac{3}{112}$	$\frac{5}{14}$	0	100
0	−2	$\frac{9}{8}$	0	1	3600

$\xrightarrow[R_3 + 2R_2]{R_1 - \frac{2}{5}R_2}$

x	y	u	v	P	Constant
1	0	$\frac{1}{28}$	$-\frac{1}{7}$	0	40
0	1	$-\frac{3}{112}$	$\frac{5}{14}$	0	100
0	0	$\frac{15}{14}$	$\frac{5}{7}$	1	3800

We conclude that Winston should manufacture 40 tables and 100 chairs for a maximum profit of $3800.

41. We have the following linear programming problem:

$$\text{Maximize} \quad P = 18x + 12y + 15z$$

$$\text{subject to} \quad \begin{array}{rcl} 2x + y + 2z &\leq& 900 \\ 3x + y + 2z &\leq& 1080 \\ 2x + 2y + z &\leq& 840 \end{array}$$

$$x \geq 0, y \geq 0, z \geq 0$$

Let u, v, and w be slack variables. We calculate the following tableaus:

	x	y	z	u	v	w	P	Constant	Ratio
	2	1	2	1	0	0	0	900	450
Pivot row →	③	1	2	0	1	0	0	1080	360
	2	2	1	0	0	1	0	840	420
	−18	−12	−15	0	0	0	1	0	

Pivot column ↑ (under x)

$\xrightarrow{\frac{1}{3}R_2}$

	x	y	z	u	v	w	P	Constant	
	2	1	2	1	0	0	0	900	$R_1 - 2R_2$
	①	$\frac{1}{3}$	$\frac{2}{3}$	0	$\frac{1}{3}$	0	0	360	$R_3 - 2R_2$
	2	2	1	0	0	1	0	840	$\overline{R_4 + 18R_2}$
	−18	−12	−15	0	0	0	1	0	

	x	y	z	u	v	w	P	Constant	Ratio
	0	$\frac{1}{3}$	$\frac{2}{3}$	1	$-\frac{2}{3}$	0	0	180	540
	1	$\frac{1}{3}$	$\frac{2}{3}$	0	$\frac{1}{3}$	0	0	360	1080
Pivot row →	0	④⁄₃	$-\frac{1}{3}$	0	$-\frac{2}{3}$	1	0	120	90
	0	−6	−3	0	6	0	1	6480	

Pivot column ↑ (under y)

$\xrightarrow{\frac{3}{4}R_3}$

	x	y	z	u	v	w	P	Constant	
	0	$\frac{1}{3}$	$\frac{2}{3}$	1	$-\frac{2}{3}$	0	0	180	$R_1 - \frac{1}{3}R_3$
	1	$\frac{1}{3}$	$\frac{2}{3}$	0	$\frac{1}{3}$	0	0	360	$R_2 - \frac{1}{3}R_3$
	0	①	$-\frac{1}{4}$	0	$-\frac{1}{2}$	$\frac{3}{4}$	0	90	$\overline{R_4 + 6R_3}$
	0	−6	−3	0	6	0	1	6480	

	x	y	z	u	v	w	P	Constant	Ratio
Pivot row →	0	0	③⁄₄	1	$-\frac{1}{2}$	$-\frac{1}{4}$	0	150	200
	1	0	$\frac{3}{4}$	0	$\frac{1}{2}$	$-\frac{1}{4}$	0	330	440
	0	1	$-\frac{1}{4}$	0	$-\frac{1}{2}$	$\frac{3}{4}$	0	90	–
	0	0	$-\frac{9}{2}$	0	6	$\frac{9}{2}$	1	7020	

Pivot column ↑ (under z)

$\xrightarrow{\frac{4}{3}R_1}$

	x	y	z	u	v	w	P	Constant	
	0	0	①	$\frac{4}{3}$	$-\frac{2}{3}$	$-\frac{1}{3}$	0	200	$R_2 - \frac{3}{4}R_1$
	1	0	$\frac{3}{4}$	0	$\frac{1}{2}$	$-\frac{1}{4}$	0	330	$R_3 + \frac{1}{4}R_1$
	0	1	$-\frac{1}{4}$	0	$-\frac{1}{2}$	$\frac{3}{4}$	0	90	$\overline{R_4 + \frac{9}{2}R_1}$
	0	0	$-\frac{9}{2}$	0	6	$\frac{9}{2}$	1	7020	

x	y	z	u	v	w	P	Constant
0	0	1	$\frac{4}{3}$	$-\frac{2}{3}$	$-\frac{1}{3}$	0	200
1	0	0	−1	1	0	0	180
0	1	0	$\frac{1}{3}$	$-\frac{2}{3}$	$\frac{2}{3}$	0	140
0	0	0	6	3	3	1	7920

The last tableau is in final form, and we conclude that the company will realize a maximum profit of $7920 by producing 180 units of product A, 140 units of product B, and 200 units of product C. Since $u = v = w = 0$, there are no resources left over.

43. Let x and y denote the numbers of Pandas and Saint Bernards produced. Then the linear programming problem is:

$$\text{Maximize} \quad P = 10x + 15y$$

$$\text{subject to} \quad 1.5x + 2y \leq 3600$$

$$30x + 35y \leq 66{,}000$$

$$5x + 8y \leq 13{,}600$$

$$x \geq 0, y \geq 0$$

We calculate the following tableaus:

	x	y	u	v	w	P	Constant	Ratio
	$\frac{3}{2}$	2	1	0	0	0	3600	1800
	30	35	0	1	0	0	66,000	$\frac{13{,}200}{7}$
Pivot row →	5	⑧	0	0	1	0	13,600	1700
	−10	−15	0	0	0	1	0	

Pivot column (under y)

$\xrightarrow{\frac{1}{8}R_3}$

x	y	u	v	w	P	Constant	
$\frac{3}{2}$	2	1	0	0	0	3600	
6	7	0	1	0	0	66,000	$\begin{array}{l} R_1 - 2R_3 \\ R_2 - 35R_3 \\ \hline R_4 + 15R_3 \end{array}$
$\frac{5}{8}$	①	0	0	$\frac{1}{8}$	0	1700	
−10	−15	0	0	0	1	0	

$\xrightarrow{}$

	x	y	u	v	w	P	Constant	Ratio
Pivot row →	①	0	1	0	$-\frac{1}{4}$	0	200	800
	$\frac{65}{8}$	0	0	1	$-\frac{35}{8}$	0	6500	800
	$\frac{5}{8}$	1	0	0	$\frac{1}{8}$	0	1700	2720
	$-\frac{5}{8}$	0	0	0	$\frac{15}{8}$	1	25,500	

(① is $\frac{1}{4}$)

Pivot column (under x)

$\xrightarrow{4R_1}$

x	y	u	v	w	P	Constant	
①	0	4	0	−1	0	800	
$\frac{65}{8}$	0	0	1	$-\frac{35}{8}$	0	6500	$\begin{array}{l} R_2 - \frac{65}{8}R_1 \\ R_3 - \frac{5}{8}R_1 \\ \hline R_4 + \frac{5}{8}R_1 \end{array}$
$\frac{5}{8}$	1	0	0	$\frac{1}{8}$	0	1700	
$-\frac{5}{8}$	0	0	0	$\frac{15}{8}$	1	25,500	

x	y	u	v	w	P	Constant
1	0	4	0	−1	0	800
0	0	$-\frac{65}{2}$	1	$\frac{15}{4}$	0	52,800
0	1	$-\frac{5}{2}$	0	$\frac{3}{4}$	0	1200
0	0	$\frac{5}{2}$	0	$\frac{5}{4}$	1	26,000

The last tableau is in final form, and the solution is $x = 800$, $y = 1200$, $P = 26{,}000$. We conclude that 800 pandas and 1200 Saint Bernards should be produced for a maximum profit of $26,000.

45. We first tabulate the given information:

| | **Model** | | | |
Department	A	B	C	Time Available
Fabrication	$\frac{5}{4}$	$\frac{3}{2}$	$\frac{3}{2}$	310
Assembly	1	1	$\frac{3}{4}$	205
Finishing	1	1	$\frac{1}{2}$	190
Profit	26	28	24	—

Let x, y, and z denote the numbers of units of models A, B, and C to be produced, respectively. Then the required linear programming problem is:

$$\text{Maximize} \quad P = 26x + 28y + 24z$$
$$\text{subject to} \quad \tfrac{5}{4}x + \tfrac{3}{2}y + \tfrac{3}{2}z \le 310$$
$$x + y + \tfrac{3}{4}z \le 205$$
$$x + y + \tfrac{1}{2}z \le 190$$
$$x \ge 0, y \ge 0, z \ge 0$$

Using the simplex method, we obtain the following tableaus:

	x	y	z	u	v	w	P	Constant	Ratio	
	$\frac{5}{4}$	$\frac{3}{2}$	$\frac{3}{2}$	1	0	0	0	310	$\frac{620}{3}$	$R_1 - \frac{3}{2}R_3$
	1	1	$\frac{3}{4}$	0	1	0	0	205	205	$R_2 - R_3$
Pivot row →	1	①	$\frac{1}{2}$	0	0	1	0	190	190	$\xrightarrow{R_4 + 28R_3}$
	−26	−28	−24	0	0	0	1	0		

Pivot column ↑ (under y)

	x	y	z	u	v	w	P	Constant	Ratio	
Pivot row →	$-\frac{1}{4}$	0	③$\frac{3}{4}$	1	0	$-\frac{3}{2}$	0	25	$\frac{100}{3}$	$\xrightarrow{\frac{4}{3}R_1}$
	0	0	$\frac{1}{4}$	0	1	−1	0	15	60	
	1	1	$\frac{1}{2}$	0	0	1	0	190	380	
	2	0	−10	0	0	28	1	5320		

Pivot column ↑ (under z)

x	y	z	u	v	w	P	Constant	
$-\frac{1}{3}$	0	①	$\frac{4}{3}$	0	−2	0	$\frac{100}{3}$	$R_2 - \frac{1}{4}R_1$
0	0	$\frac{1}{4}$	0	1	−1	0	15	$\xrightarrow{\substack{R_3 - \frac{1}{2}R_1 \\ R_4 + 10R_1}}$
1	1	$\frac{1}{2}$	0	0	1	0	190	
2	0	−10	0	0	28	1	5320	

	x	y	z	u	v	w	P	Constant	Ratio	
	$-\frac{1}{3}$	0	1	$\frac{4}{3}$	0	−2	0	$\frac{100}{3}$	—	$\xrightarrow{12R_2}$
Pivot row →	①$\frac{1}{12}$	0	0	$-\frac{1}{3}$	1	$-\frac{1}{2}$	0	$\frac{20}{3}$	80	
	$\frac{7}{6}$	1	0	$-\frac{2}{3}$	0	2	0	$\frac{520}{3}$	$\frac{1040}{7}$	
	$-\frac{4}{3}$	0	0	$\frac{40}{3}$	0	8	1	$\frac{16,960}{3}$		

Pivot column ↑ (under x)

x	y	z	u	v	w	P	Constant
$-\frac{1}{3}$	0	1	$\frac{4}{3}$	0	-2	0	$\frac{100}{3}$
①	0	0	-4	12	-6	0	80
$\frac{7}{6}$	1	0	$-\frac{2}{3}$	0	2	0	$\frac{520}{3}$
$-\frac{4}{3}$	0	0	$\frac{40}{3}$	0	8	1	$\frac{16{,}960}{3}$

$$\xrightarrow[\substack{R_1 + \frac{1}{3}R_2 \\ R_3 - \frac{7}{6}R_2 \\ R_4 + \frac{4}{3}R_2}]{}$$

x	y	z	u	v	w	P	Constant
0	0	1	0	4	-4	0	60
1	0	0	-4	12	-6	0	80
0	1	0	4	-14	9	0	80
0	0	0	8	16	0	1	5760

The last tableau is in final form. We see that $x = 80$, $y = 80$, $z = 60$, $u = 0$, $v = 0$, $w = 0$, and $P = 5760$. Thus, by producing 80 units each of Models A and B and 60 units of Model C, the company stands to make a maximum profit of \$5760. Because $u = v = w = 0$, there are no resources left over.

47. Rewriting and simplifying the constraint for the average risk factor for the investment, we have $10x + 6y + 2z \le 5x + 5y + 5z$, or $5x + y - 3z \le 0$. The linear programming problem is:

$$\text{Maximize} \quad P = 0.12x + 0.10y + 0.06z$$
$$\text{subject to} \quad x + y - z \le 0$$
$$x - 3y + z \le 0$$
$$5x + y - 3z \le 0$$
$$x + y + z \le 200{,}000$$
$$x \ge 0, y \ge 0, z \ge 0$$

We calculate the following tableaus:

	x	y	z	t	u	v	w	P	Constant	Ratio
Pivot row →	①	1	-1	1	0	0	0	0	0	0
	1	-3	1	0	1	0	0	0	0	0
	5	1	-3	0	0	1	0	0	0	0
	1	1	1	0	0	0	1	0	200,000	200,000
	-0.12	-0.10	-0.06	0	0	0	0	1	0	

$$\xrightarrow[\substack{R_2 - R_1 \\ R_3 - 5R_1 \\ R_4 - R_1 \\ R_5 + 0.12R_1}]{}$$

Pivot column (↑ under x)

	x	y	z	t	u	v	w	P	Constant	Ratio
	1	1	-1	1	0	0	0	0	0	–
Pivot row →	0	-4	②	-1	1	0	0	0	0	0
	0	-4	2	-5	0	1	0	0	0	0
	0	0	2	-1	0	0	1	0	200,000	100,000
	0	0.02	-0.18	0.12	0	0	0	1	0	

$$\xrightarrow{\frac{1}{2}R_2}$$

Pivot column (↑ under z)

	x	y	z	t	u	v	w	P	Constant
	1	1	-1	1	0	0	0	0	0
	0	-2	①	$-\frac{1}{2}$	$\frac{1}{2}$	0	0	0	0
	0	-4	2	-5	0	1	0	0	0
	0	0	2	-1	0	0	1	0	200,000
	0	0.02	-0.18	0.12	0	0	0	1	0

$$\xrightarrow[\substack{R_1 + R_2 \\ R_3 - 2R_2 \\ R_4 - 2R_2 \\ R_5 + 0.18R_2}]{}$$

	x	y	z	t	u	v	w	P	Constant	Ratio
	1	-1	0	$\frac{1}{2}$	$\frac{1}{2}$	0	0	0	0	—
	0	-2	1	$-\frac{1}{2}$	$\frac{1}{2}$	0	0	0	0	—
	0	0	0	-4	-1	1	0	0	0	—
Pivot row →	0	④	0	0	-1	0	1	0	200,000	50,000
	0	-0.34	0	0.03	0.09	0	0	1	0	

$\xrightarrow{\frac{1}{4}R_4}$

↑ Pivot column

x	y	z	t	u	v	w	P	Constant
1	-1	0	$\frac{1}{2}$	$\frac{1}{2}$	0	0	0	0
0	-2	1	$-\frac{1}{2}$	$\frac{1}{2}$	0	0	0	0
0	0	0	-4	-1	1	0	0	0
0	①	0	0	$-\frac{1}{4}$	0	$\frac{1}{4}$	0	50,000
0	-0.34	0	0.03	0.09	0	0	1	0

$\xrightarrow[R_5+0.34R_4]{\begin{array}{c}R_1+R_4\\ R_3+2R_4\end{array}}$

x	y	z	t	u	v	w	P	Constant
1	0	0	$\frac{1}{2}$	$\frac{1}{4}$	0	$\frac{1}{4}$	0	50,000
0	0	1	$-\frac{1}{2}$	0	0	$\frac{1}{2}$	0	100,000
0	0	0	-4	-1	1	0	0	0
0	1	0	0	$-\frac{1}{4}$	0	$\frac{1}{4}$	0	50,000
0	0	0	0.03	0.005	0	0.085	1	17,000

The last tableau is in final form, and so the maximum return on the investment each year is $17,000 if Sharon invests \$50,000 in growth funds, \$50,000 in balanced funds, and \$100,000 in income funds.

49. Let x, y, and z denote the numbers (in thousands) of bottles of formulas I, II, and III produced, respectively. The resulting linear programming problem is:

$$\text{Maximize} \quad P = 180x + 200y + 300z$$
$$\text{subject to} \quad \tfrac{5}{2}x + 3y + 4z \le 70$$
$$x \le 9, y \le 12, z \le 6$$
$$x \ge 0, y \ge 0, z \ge 0$$

Using the simplex method, we have

	x	y	z	s	t	u	v	P	Constant	Ratio
	$\frac{5}{2}$	3	4	1	0	0	0	0	70	$\frac{35}{2}$
	1	0	0	0	1	0	0	0	9	—
	0	1	0	0	0	1	0	0	12	—
Pivot row →	0	0	①	0	0	0	1	0	6	6
	-180	-200	-300	0	0	0	0	1	0	

$\xrightarrow[R_5+300R_4]{R_1-4R_4}$

↑ Pivot column

	x	y	z	s	t	u	v	P	Constant	Ratio
	$\frac{5}{2}$	3	0	1	0	0	-4	0	46	$\frac{46}{3}$
	1	0	0	0	1	0	0	0	9	—
Pivot row →	0	①	0	0	0	1	0	0	12	12
	0	0	1	0	0	0	1	0	6	—
	-180	-200	0	0	0	0	300	1	1800	

$\xrightarrow[R_5+200R_3]{R_1-3R_3}$

↑ Pivot column

	x	y	z	s	t	u	v	P	Constant	Ratio
Pivot row →	$\left(\tfrac{5}{2}\right)$	0	0	1	0	−3	−4	0	10	4
	1	0	0	0	1	0	0	0	9	9
	0	1	0	0	0	1	0	0	12	—
	0	0	1	0	0	0	1	0	6	—
	−180	0	0	0	0	200	300	1	4200	

$\xrightarrow{\;\tfrac{2}{5}R_1\;}$

↑ Pivot column

x	y	z	s	t	u	v	P	Constant
(1)	0	0	$\tfrac{2}{5}$	0	$-\tfrac{6}{5}$	$-\tfrac{8}{5}$	0	4
1	0	0	0	1	0	0	0	9
0	1	0	0	0	1	0	0	12
0	0	1	0	0	0	1	0	6
−180	0	0	0	0	200	300	1	4200

$\xrightarrow[R_5 + 180R_1]{R_2 - R_1}$

	x	y	z	s	t	u	v	P	Constant	Ratio
	1	0	0	$\tfrac{2}{5}$	0	$-\tfrac{6}{5}$	$-\tfrac{8}{5}$	0	4	—
Pivot row →	0	0	0	$-\tfrac{2}{5}$	1	$\left(\tfrac{6}{5}\right)$	$\tfrac{8}{5}$	0	5	$\tfrac{25}{6}$
	0	1	0	0	0	1	0	0	12	12
	0	0	1	0	0	0	1	0	6	—
	0	0	0	72	0	−16	12	1	4920	

↑ Pivot column

$\xrightarrow{\;\tfrac{5}{6}R_2\;}$

x	y	z	s	t	u	v	P	Constant
1	0	0	$\tfrac{2}{5}$	0	$-\tfrac{6}{5}$	$-\tfrac{8}{5}$	0	4
0	0	0	$-\tfrac{1}{3}$	$\tfrac{5}{6}$	(1)	$\tfrac{4}{3}$	0	$\tfrac{25}{6}$
0	1	0	0	0	1	0	0	12
0	0	1	0	0	0	1	0	6
0	0	0	72	0	−16	12	1	4920

$\xrightarrow[\substack{R_3 - R_2 \\ R_5 + 16R_2}]{R_1 + \tfrac{6}{5}R_2}$

x	y	z	s	t	u	v	P	Constant
1	0	0	0	1	0	0	0	9
0	0	0	$-\tfrac{1}{3}$	$\tfrac{5}{6}$	1	$\tfrac{4}{3}$	0	$\tfrac{25}{6}$
0	1	0	$\tfrac{1}{3}$	$-\tfrac{5}{6}$	0	$-\tfrac{4}{3}$	0	$\tfrac{47}{6}$
0	0	1	0	0	0	1	0	6
0	0	0	$\tfrac{200}{3}$	$\tfrac{40}{3}$	0	$\tfrac{100}{3}$	1	$\tfrac{14{,}960}{3}$

The last tableau is in final form. We conclude that $x = 9$, $y = \tfrac{47}{6}$, $z = 6$, $s = 0$, $t = 0$, $u = \tfrac{25}{6}$, and $P = \tfrac{14{,}960}{3} \approx 4986.67$; that is, the company should manufacture 9000 bottles of formula I, 7833 bottles of formula II, and 6000 bottles of formula III for a maximum profit of \$4986.60, leaving ingredients for 4167 bottles of formula II unused.

51. The linear programming problem is:

Maximize $P = 3x + 2y$

subject to $x - y \le 3$

$\qquad\qquad\; x \quad\; \le 2$

$\qquad\; x \ge 0,\, y \ge 0$

a.

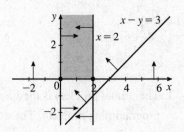

b. It is evident from part (a) that the feasible set is unbounded.

c. Letting u and v be slack variables, we calculate the following sequence of tableaus:

	x	y	u	v	P	Constant	Ratio
	1	-1	1	0	0	3	3
Pivot row →	①	0	0	1	0	2	2
	-3	-2	0	0	1	0	

$$\xrightarrow[R_3 + 3R_2]{R_1 - R_2}$$

x	y	u	v	P	Constant
0	-1	1	-1	0	1
1	0	0	1	0	2
0	-2	0	3	1	6

↑ Pivot column   ↑ Pivot column

In the second tableau, the entries in the pivot column are 0, and -1, and so the ratios cannot be computed.

d. The first line in the second tableau in part (c) tells us that $-y + u - v = 1$ or $y = u - v - 1$. The last equation shows that y can be made as large as we please by taking u sufficiently large. Thus, P can be made as large as we please, and so the linear programming problem has no solution.

53. True. See the explanation of the simplex method on page 225 of the text.

55. True.

Technology Exercises   page 246

1. $x = 1.2, y = 0, z = 1.6, w = 0; P = 8.8$.

3. $x = 1.6, y = 0, z = 0, w = 3.6; P = 12.4$.

4.2   The Simplex Method: Standard Minimization Problems

Problem-Solving Tips

1. The given problem is called the **primal problem** and the problem related to it is called the **dual problem**. Dual problems are standard minimization problems.

2. To solve a dual problem, first *write down the tableau* for the primal problem. (Note that this is not a simplex tableau as there are no slack variables.) Then *interchange the columns and rows* of this tableau. Use this tableau to *write the dual problem* and then use the simplex method to complete the solution to the problem. The minimum value of C will appear in the lower right corner of the final simplex tableau.

Concept Questions   page 256

1. Maximize $P = -C = 3x + 5y$ subject to

$$5x + 2y \le 30$$
$$x + 3y \le 21$$
$$x \ge 0, y \ge 0$$

3. The primal problem is the linear programming (maximization) problem associated with a minimization linear programming problem. The dual problem is the linear programming (minimization) problem associated with the maximization linear programming problem.

Exercises page 257

1. We solve the associated regular problem:

$$\begin{aligned} \text{Maximize} \quad & P = -C = 2x - y \\ \text{subject to} \quad & x + 2y \le 6 \\ & 3x + 2y \le 12 \\ & x \ge 0, y \ge 0 \end{aligned}$$

Using the simplex method with slack variables u and v, we have

	x	y	u	v	P	Constant	Ratio
	1	2	1	0	0	6	6
	(3)	2	0	1	0	12	4
Pivot row →	−2	1	0	0	1	0	
	↑ Pivot column						

$\xrightarrow{\frac{1}{3}R_2}$

x	y	u	v	P	Constant
1	2	1	0	0	6
(1)	$\frac{2}{3}$	0	$\frac{1}{3}$	0	4
−2	1	0	0	1	0

$\xrightarrow[R_3 + 2R_2]{R_1 - R_2}$

x	y	u	v	P	Constant
0	$\frac{4}{3}$	1	$-\frac{1}{3}$	0	2
1	$\frac{2}{3}$	0	$\frac{1}{3}$	0	4
0	$\frac{7}{3}$	0	$\frac{2}{3}$	1	8

Therefore, $x = 4$, $y = 0$, and $C = -P = -8$.

3. We maximize $P = -C = 3x + 2y$. Using the simplex method, we obtain

	x	y	u	v	P	Constant	Ratio
	3	4	1	0	0	24	8
Pivot row →	(7)	−4	0	1	0	16	$\frac{16}{7}$
	−3	−2	0	0	1	0	
	↑ Pivot column						

$\xrightarrow{\frac{1}{7}R_2}$

x	y	u	v	P	Constant
3	4	1	0	0	24
(1)	$-\frac{4}{7}$	0	$\frac{1}{7}$	0	$\frac{16}{7}$
−3	−2	0	0	1	0

$\xrightarrow[R_3 + 3R_2]{R_1 - 3R_2}$

	x	y	u	v	P	Constant	Ratio
Pivot row →	0	$\left(\frac{40}{7}\right)$	1	$-\frac{3}{7}$	0	$\frac{120}{7}$	3
	1	$-\frac{4}{7}$	0	$\frac{1}{7}$	0	$\frac{16}{7}$	–
	0	$-\frac{26}{7}$	0	$\frac{3}{7}$	1	$\frac{48}{7}$	
		↑ Pivot column					

$\xrightarrow{\frac{7}{40}R_1}$

x	y	u	v	P	Constant
0	(1)	$\frac{7}{40}$	$-\frac{3}{40}$	0	3
1	$-\frac{4}{7}$	0	$\frac{1}{7}$	0	$\frac{16}{7}$
0	$-\frac{26}{7}$	0	$\frac{3}{7}$	1	$\frac{48}{7}$

$\xrightarrow[R_3 + \frac{26}{7}R_1]{R_2 + \frac{4}{7}R_1}$

x	y	u	v	P	Constant
0	1	$\frac{7}{40}$	$-\frac{3}{40}$	0	3
1	0	$\frac{1}{10}$	$\frac{1}{10}$	0	4
0	0	$\frac{13}{20}$	$\frac{3}{20}$	1	18

The last tableau is in final form. We find $x = 4$, $y = 3$, and $C = -P = -18$.

5. We maximize $P = -C = -2x + 3y + 4z$ subject to the given constraints. Using the simplex method, we obtain

	x	y	z	u	v	w	P	Constant	Ratio
	-1	2	-1	1	0	0	0	8	$-$
Pivot row $\rightarrow$	1	-2	②	0	1	0	0	10	5
	2	4	-3	0	0	1	0	12	$-$
	2	-3	-4	0	0	0	1	0	

$\xrightarrow{\frac{1}{2}R_2}$

	x	y	z	u	v	w	P	Constant	
	-1	2	-1	1	0	0	0	8	$R_1 + R_2$
	$\frac{1}{2}$	-1	①	0	$\frac{1}{2}$	0	0	5	$R_3 + 3R_2$
	2	4	-3	0	0	1	0	12	$\xrightarrow{\hspace{1cm}}$
	2	-3	-4	0	0	0	1	0	$R_4 + 4R_2$

Pivot column ↑ (under z)

	x	y	z	u	v	w	P	Constant	Ratio
Pivot row $\rightarrow$	$-\frac{1}{2}$	①	0	1	$\frac{1}{2}$	0	0	13	13
	$\frac{1}{2}$	-1	1	0	$\frac{1}{2}$	0	0	5	$-$
	$\frac{7}{2}$	1	0	0	$\frac{3}{2}$	1	0	27	27
	4	-7	0	0	2	0	1	20	

$\xrightarrow[\begin{array}{c}R_3 - R_1\\ \hline R_4 + 7R_1\end{array}]{R_2 + R_1}$

	x	y	z	u	v	w	P	Constant
	$-\frac{1}{2}$	①	0	1	$\frac{1}{2}$	0	0	13
	0	0	1	1	1	0	0	18
	4	0	0	-1	1	1	0	14
	$\frac{1}{2}$	0	0	7	$\frac{11}{2}$	0	1	111

Pivot column ↑ (under y)

The last tableau is in final form. We see that $x = 0$, $y = 13$, $z = 18$, $u = 0$, $v = 0$, $w = 14$, and $C = -P = -111$.

7. $x = \frac{5}{4}$, $y = \frac{1}{4}$, $u = 2$, $v = 3$, and $C = P = 13$.

9. $x = 5$, $y = 10$, $z = 0$, $u = 1$, $v = 2$, and $C = P = 80$.

11. We first write out the primal tableau:

x	y	Constant
2	3	90
3	2	120
3	2	

Then we obtain a dual tableau by interchanging rows and columns:

u	v	Constant
2	3	3
3	2	2
90	120	

From this table we construct the dual problem:

$$\begin{aligned}\text{Maximize} \quad & P = 90u + 120v \\ \text{subject to} \quad & 2u + 3v \leq 3 \\ & 3u + 2v \leq 2 \\ & u \geq 0, v \geq 0\end{aligned}$$

Solving the dual problem using the simplex method with x and y as the slack variables, we obtain

	u	v	x	y	P	Constant	Ratio
	2	3	1	0	0	3	1
Pivot row →	3	②	0	1	0	2	1
	−90	−120	0	0	1	0	

$\xrightarrow{\frac{1}{2}R_2}$

	u	v	x	y	P	Constant
	2	3	1	0	0	3
	$\frac{3}{2}$	①	0	$\frac{1}{2}$	0	1
	−90	−120	0	0	1	0

$\xrightarrow[R_3 + 120R_2]{R_1 - 3R_2}$

↑ Pivot column

u	v	x	y	P	Constant
$-\frac{5}{2}$	0	1	$-\frac{3}{2}$	0	0
$\frac{3}{2}$	1	0	$\frac{1}{2}$	0	1
90	0	0	60	1	120

We conclude that $x = 0$, $y = 60$, $u = 0$, $v = 1$, and $C = P = 120$.

13. We write the primal tableau, then obtain a dual tableau by interchanging rows and columns.

x	y	Constant
6	1	60
2	1	40
1	1	30
6	4	

u	v	w	Constant
6	2	1	6
1	1	1	4
60	40	30	

From this table we construct the dual problem:

$$\text{Maximize} \quad P = 60u + 40v + 30w$$
$$\text{subject to} \quad 6u + 2v + w \le 6$$
$$u + v + w \le 4$$
$$u \ge 0, v \ge 0, w \ge 0$$

We solve the problem as follows:

	u	v	w	x	y	P	Constant	Ratio
Pivot row →	⑥	2	1	1	0	0	6	1
	1	1	1	0	1	0	4	4
	−60	−40	−30	0	0	1	0	−

$\xrightarrow{\frac{1}{6}R_1}$

	u	v	w	x	y	P	Constant
	①	$\frac{1}{3}$	$\frac{1}{6}$	$\frac{1}{6}$	0	0	1
	1	1	1	0	1	0	4
	−60	−40	−30	0	0	1	0

$\xrightarrow[R_3 + 60R_1]{R_2 - R_1}$

↑ Pivot column

	u	v	w	x	y	P	Constant	Ratio
	1	$\frac{1}{3}$	$\frac{1}{6}$	$\frac{1}{6}$	0	0	1	6
Pivot row →	0	$\frac{2}{3}$	⑤⁄₆	$-\frac{1}{6}$	1	0	3	$\frac{18}{5}$
	0	−20	−20	10	0	1	60	−

$\xrightarrow{\frac{6}{5}R_2}$

	u	v	w	x	y	P	Constant
	1	$\frac{1}{3}$	$\frac{1}{6}$	$\frac{1}{6}$	0	0	1
	0	$\frac{4}{5}$	①	$-\frac{1}{5}$	$\frac{6}{5}$	0	$\frac{18}{5}$
	0	−20	−20	10	0	1	60

$\xrightarrow[R_3 + 20R_2]{R_1 - \frac{1}{6}R_2}$

↑ Pivot column

	u	v	w	x	y	P	Constant	Ratio
Pivot row →	1	$\left(\frac{1}{5}\right)$	0	$\frac{1}{5}$	$-\frac{1}{5}$	0	$\frac{2}{5}$	2
	0	$\frac{4}{5}$	1	$-\frac{1}{5}$	$\frac{6}{5}$	0	$\frac{18}{5}$	$\frac{9}{2}$
	0	-4	0	6	24	1	132	—

↑ Pivot column

$\xrightarrow{5R_1}$

u	v	w	x	y	P	Constant
5	①	0	1	-1	0	2
0	$\frac{4}{5}$	1	$-\frac{1}{5}$	$\frac{6}{5}$	0	$\frac{18}{5}$
0	-4	0	6	24	1	132

$\xrightarrow[R_3+4R_1]{R_2-\frac{4}{5}R_1}$

u	v	w	x	y	P	Constant
5	1	0	1	-1	0	2
-4	0	1	-1	2	0	2
20	0	0	10	20	1	140

The last tableau is in final form. We find that $x = 10$, $y = 20$, $u = 0$, $v = 2$, $w = 0$, and $C = 140$.

15. We write the primal tableau, then obtain a dual tableau by interchanging rows and columns.

x	y	z	Constant
20	10	1	10
1	1	2	20
200	150	120	

u	v	Constant
20	1	200
10	1	150
1	2	120
10	20	

From this table we construct the dual problem:

Maximize $P = 10u + 20v$

subject to $20u + v \le 200$

$10u + v \le 150$ Solving the dual problem using the simplex method with x, y, and z as slack

$u + 2v \le 120$

$u \ge 0, v \ge 0$

variables, we obtain the following tableaus:

	u	v	x	y	z	P	Constant	Ratio
	20	1	1	0	0	0	200	200
	10	1	0	1	0	0	150	150
Pivot row →	1	②	0	0	1	0	120	60
	-10	-20	0	0	0	1	0	

↑ Pivot column

$\xrightarrow{\frac{1}{2}R_3}$

u	v	x	y	z	P	Constant
20	1	1	0	0	0	200
10	1	0	1	0	0	150
$\frac{1}{2}$	①	0	0	$\frac{1}{2}$	0	60
-10	-20	0	0	0	1	0

$\xrightarrow[R_4+20R_3]{\begin{array}{c}R_1-R_3\\R_2-R_3\end{array}}$

u	v	x	y	z	P	Constant
$\frac{39}{2}$	0	1	0	$-\frac{1}{2}$	0	140
$\frac{19}{2}$	0	0	1	$-\frac{1}{2}$	0	90
$\frac{1}{2}$	1	0	0	$\frac{1}{2}$	0	60
0	0	0	0	10	1	1200

This last tableau is in final form. We find that $x = 0$, $y = 0$, $z = 10$, and $C = 1200$.

17. We write the primal tableau, then obtain a dual tableau by interchanging rows and columns.

x	y	z	Constant
1	2	2	10
2	1	1	24
1	1	1	16
6	8	4	

u	v	w	Constant
1	2	1	6
2	1	1	8
2	1	1	4
10	24	16	

From this table we construct the dual problem:

$$\text{Maximize} \quad P = 10u + 24v + 16w$$
$$\text{subject to} \quad u + 2v + w \le 6$$
$$2u + v + w \le 8$$
$$2u + v + w \le 4$$
$$u \ge 0, v \ge 0, w \ge 0$$

Solving the dual problem using the simplex method with x, y, and z as slack variables, we obtain the following tableaus:

	u	v	w	x	y	z	P	Constant	Ratio
Pivot row →	1	②	1	1	0	0	0	6	3
	2	1	1	0	1	0	0	8	8
	2	1	1	0	0	1	0	4	4
	−10	−24	−16	0	0	0	1	0	

$\xrightarrow{\frac{1}{2}R_1}$

↑ Pivot column

u	v	w	x	y	z	P	Constant
$\frac{1}{2}$	①	$\frac{1}{2}$	$\frac{1}{2}$	0	0	0	3
2	1	1	0	1	0	0	8
2	1	1	0	0	1	0	4
−10	−24	−16	0	0	0	1	0

$\xrightarrow[\substack{R_3 - R_1 \\ R_4 + 24R_1}]{R_2 - R_1}$

u	v	w	x	y	z	P	Constant
$\frac{1}{2}$	1	$\frac{1}{2}$	$\frac{1}{2}$	0	0	0	3
$\frac{3}{2}$	0	$\frac{1}{2}$	$-\frac{1}{2}$	1	0	0	5
3	0	1	−1	0	2	0	2
2	0	−4	12	0	0	1	72

$\xrightarrow[\substack{R_2 - \frac{1}{2}R_3 \\ R_4 + 4R_3}]{R_1 - \frac{1}{2}R_3}$

u	v	w	x	y	z	P	Constant
−1	1	0	1	0	−1	0	2
0	0	0	0	1	−1	0	4
3	0	1	−1	0	2	0	2
14	0	0	8	0	8	1	80

The solution to the primal problem is thus $x = 8$, $y = 0$, $z = 8$, $u = 0$, $v = 2$, $w = 2$, and $C = 80$.

19. We write the primal tableau, then obtain a dual tableau by interchanging rows and columns.

x	y	z	Constant
2	4	3	6
6	0	1	2
0	6	2	4
30	12	20	

u	v	w	Constant
2	6	0	30
4	0	6	12
3	1	2	20
6	2	4	

From this table we construct the dual problem:

$$\text{Maximize} \quad P = 6u + 2v + 4w$$
$$\text{subject to} \quad 2u + 6v \quad\quad \leq 30$$
$$4u \quad\quad + 6w \leq 12$$
$$3u + v + 2w \leq 20$$
$$u \geq 0, v \geq 0, w \geq 0$$

Solving the dual problem using the simplex method with x, y, and z as slack variables, we obtain the following tableaus:

	u	v	w	x	y	z	P	Constant	Ratio
	2	6	0	1	0	0	0	30	15
Pivot row →	④	0	6	0	1	0	0	12	3
	3	1	2	0	0	1	0	20	$\frac{20}{3}$
	−6	−2	−4	0	0	0	1	0	

↑ Pivot column

$\xrightarrow{\frac{1}{4}R_2}$

u	v	w	x	y	z	P	Constant
2	6	0	1	0	0	0	30
①	0	$\frac{3}{2}$	0	$\frac{1}{4}$	0	0	3
3	1	2	0	0	1	0	20
−6	−2	−4	0	0	0	1	0

$\begin{array}{c} R_1 - 2R_2 \\ R_3 - 3R_2 \\ \xrightarrow{} \\ R_4 + 6R_2 \end{array}$

	u	v	w	x	y	z	P	Constant	Ratio
Pivot row →	0	⑥	−3	1	$-\frac{1}{2}$	0	0	24	4
	1	0	$\frac{3}{2}$	0	$\frac{1}{4}$	0	0	3	—
	0	1	$-\frac{5}{2}$	0	$-\frac{3}{4}$	1	0	11	11
	0	−2	5	0	$\frac{3}{2}$	0	1	18	

↑ Pivot column

$\xrightarrow{\frac{1}{6}R_1}$

u	v	w	x	y	z	P	Constant
0	①	$-\frac{1}{2}$	$\frac{1}{6}$	$-\frac{1}{12}$	0	0	4
1	0	$\frac{3}{2}$	0	$\frac{1}{4}$	0	0	3
0	1	$-\frac{5}{2}$	0	$-\frac{3}{4}$	1	0	11
0	−2	5	0	$\frac{3}{2}$	0	1	18

$\begin{array}{c} R_3 - R_1 \\ \xrightarrow{} \\ R_4 + 2R_1 \end{array}$

u	v	w	x	y	z	P	Constant
0	1	$-\frac{1}{2}$	$\frac{1}{6}$	$-\frac{1}{12}$	0	0	4
1	0	$\frac{3}{2}$	0	$\frac{1}{4}$	0	0	3
0	0	−2	$-\frac{1}{6}$	$-\frac{2}{3}$	1	0	7
0	0	4	$\frac{1}{3}$	$\frac{4}{3}$	0	1	26

The last tableau is in final form. We find $x = \frac{1}{3}$, $y = \frac{4}{3}$, $z = 0$, $u = 3$, $v = 4$, $w = 0$, and $C = 26$.

21. Let x denote the number of type A vessels and y the number of type B vessels to be operated. Then the problem is:

$$\text{Maximize} \quad C = 44{,}000x + 54{,}000y$$
$$\text{subject to} \quad 60x + 80y \geq 360$$
$$160x + 120y \geq 680$$
$$x \geq 0, y \geq 0$$

We first write a tableau for the primal problem, then obtain a dual tableau by interchanging rows and columns.

x	y	Constant
60	80	360
160	120	680
44,000	54,000	

u	v	Constant
60	160	44,000
80	120	54,000
360	680	

Proceeding, we are led to the dual problem:

$$\text{Maximize} \quad P = 360u + 680v$$
$$\text{subject to} \quad 60u + 160y \leq 44{,}000$$
$$80x + 120v \leq 54{,}000$$
$$u \geq 0, v \geq 0$$

Let x and y be slack variables. We obtain the following tableaus:

	u	v	x	y	P	Constant	Ratio
Pivot row →	60	(160)	1	0	0	44,000	275
	80	120	0	1	0	54,000	450
	−360	−680	0	0	1	0	

$\xrightarrow{\frac{1}{160}R_1}$

Pivot column (under v)

	u	v	x	y	P	Constant	
	$\frac{3}{8}$	(1)	$\frac{1}{160}$	0	0	275	$\xrightarrow[R_3 + 680R_1]{R_2 - 120R_1}$
	80	120	0	1	0	54,000	
	−360	−680	0	0	1	0	

	u	v	x	y	P	Constant	Ratio
	$\frac{3}{8}$	1	$\frac{1}{160}$	0	0	275	$\frac{2200}{3}$
Pivot row →	(35)	0	$-\frac{3}{4}$	1	0	21,000	600
	−105	0	$\frac{17}{4}$	0	1	187,000	

$\xrightarrow{\frac{1}{35}R_2}$

Pivot column (under u)

	u	v	x	y	P	Constant	
	$\frac{3}{8}$	1	$\frac{1}{160}$	0	0	275	$\xrightarrow[R_3 + 105R_2]{R_1 - \frac{3}{8}R_2}$
	(1)	0	$-\frac{3}{140}$	$\frac{1}{35}$	0	600	
	−105	0	$\frac{17}{4}$	0	1	187,000	

u	v	x	y	P	Constant
0	1	$\frac{1}{70}$	$-\frac{3}{280}$	0	50
1	0	$-\frac{3}{140}$	$\frac{1}{35}$	0	600
0	0	2	3	1	250,000

The last tableau is in final form. The fundamental theorem of duality tells us that the solution to the primal problem is $x = 2$, $y = 3$ with a minimum value for C of 250,000. Thus, Deluxe River Cruises should use two type A vessels and three type B vessels. The minimum operating cost is $250,000.

23. The given data may be summarized as follows:

	Orange Juice	Grapefruit Juice
Vitamin A	60 I.U.	120 I.U.
Vitamin C	16 I.U.	12 I.U.
Calories	14	11

Suppose x ounces of orange juice and y ounces of pink grapefruit juice are required for each glass of the blend. Then the problem is:

$$\text{Minimize} \quad C = 14x + 11y$$
$$\text{subject to} \quad 60x + 120y \geq 1200$$
$$16x + 12y \geq 200$$
$$x \geq 0, y \geq 0$$

We write down a tableau for the primal problem, then interchange columns and rows to obtain a duplex tableau.

x	y	Constant
60	120	1200
16	12	200
14	11	

u	v	Constant
60	16	14
120	12	11
1200	200	

Thus, the dual problem is:

$$\text{Maximize} \quad P = 1200u + 200v$$
$$\text{subject to} \quad 60u + 16v \leq 14$$
$$120u + 12v \leq 11$$
$$u \geq 0, v \geq 0$$

Using the slack variables x and y, we obtain the following sequence of tableaus:

	u	v	x	y	P	Constant	Ratio
	60	16	1	0	0	14	$\frac{7}{30}$
Pivot row →	(120)	12	0	1	0	11	$\frac{11}{120}$
	-1200	-200	0	0	1	0	

$\xrightarrow{\frac{1}{120}R_2}$

↑ Pivot column

	u	v	x	y	P	Constant
	60	16	1	0	0	14
	(1)	$\frac{1}{10}$	0	$\frac{1}{120}$	0	$\frac{11}{120}$
	-1200	-200	0	0	1	0

$\xrightarrow[R_3 + 1200R_2]{R_1 - 60R_2}$

	u	v	x	y	P	Constant	Ratio
Pivot row →	0	(10)	1	$-\frac{1}{2}$	0	$\frac{17}{2}$	$\frac{17}{20}$
	1	$\frac{1}{10}$	0	$\frac{1}{120}$	0	$\frac{11}{120}$	$\frac{11}{12}$
	0	-80	0	10	1	110	

$\xrightarrow{\frac{1}{10}R_1}$

↑ Pivot column

	u	v	x	y	P	Constant
	0	(1)	$\frac{1}{10}$	$-\frac{1}{20}$	0	$\frac{17}{20}$
	1	$\frac{1}{10}$	0	$\frac{1}{120}$	0	$\frac{11}{120}$
	0	-80	0	10	1	110

$\xrightarrow[R_3 + 80R_1]{R_2 - \frac{1}{10}R_1}$

u	v	x	y	P	Constant
0	1	$\frac{1}{10}$	$-\frac{1}{20}$	0	$\frac{17}{20}$
1	0	$-\frac{1}{100}$	$\frac{1}{75}$	0	$\frac{1}{150}$
0	0	8	6	1	178

We conclude that the owner should use 8 ounces of orange juice and 6 ounces of pink grapefruit juice per glass of the blend for a minimal calorie count of 178.

25. This problem was formulated in Exercise 3.2.27 on page 82 of this manual. We rewrite the constraints in the form

$$-x_1 - x_2 - x_3 \geq -800$$
$$-x_4 - x_5 - x_6 \geq -600$$
$$x_1 \qquad + x_4 \qquad \geq 500$$
$$x_2 \qquad + x_5 \qquad \geq 400$$
$$x_3 \qquad + x_6 \geq 400$$

We solve this problem using duality, first writing a tableau for the primal problem, then interchanging rows and columns to obtain a dual tableau.

x_1	x_2	x_3	x_4	x_5	x_6	Constant
−1	−1	−1	0	0	0	−800
0	0	0	−1	−1	−1	−600
1	0	0	1	0	0	500
0	1	0	0	1	0	400
0	0	1	0	0	1	400
16	20	22	18	16	14	

c_1	c_2	c_3	c_4	c_5	Constant
−1	0	1	0	0	16
−1	0	0	1	0	20
−1	0	0	0	1	22
0	−1	1	0	0	18
0	−1	0	1	0	16
0	−1	0	0	1	14
−800	−600	−500	400	400	

Thus, the dual problem is:

$$\text{Maximize } P = -800u_1 - 600u_2 + 500u_3 + 400u_4 + 400u_5$$
$$\text{subject to } -u_1 \qquad + u_3 \qquad \leq 16$$
$$-u_1 \qquad + u_4 \qquad \leq 20$$
$$-u_1 \qquad + u_5 \leq 22$$
$$-u_2 + u_3 \qquad \leq 18$$
$$-u_2 \qquad + u_4 \qquad \leq 16$$
$$-u_2 \qquad + u_5 \leq 14$$
$$u_1, u_2, u_3, u_4, u_5 \geq 0$$

We obtain the following sequence of tableaus:

	u_1	u_2	u_3	u_4	u_5	x_1	x_2	x_3	x_4	x_5	x_6	P	Constant	Ratio
Pivot row →	−1	0	①	0	0	1	0	0	0	0	0	0	16	16
	−1	0	0	1	0	0	1	0	0	0	0	0	20	−
	−1	0	0	0	1	0	0	1	0	0	0	0	22	−
	0	−1	1	0	0	0	0	0	1	0	0	0	18	18
	0	−1	0	1	0	0	0	0	0	1	0	0	16	−
	0	−1	0	0	1	0	0	0	0	0	1	0	14	−
	800	600	−500	−400	−400	0	0	0	0	0	0	1	0	

$$\xrightarrow{\begin{array}{c} R_4 - R_1 \\ R_7 + 500R_1 \end{array}}$$

↑ Pivot column

u_1	u_2	u_3	u_4	u_5	x_1	x_2	x_3	x_4	x_5	x_6	P	Constant
−1	0	①	0	0	1	0	0	0	0	0	0	16
−1	0	0	1	0	0	1	0	0	0	0	0	20
−1	0	0	0	1	0	0	1	0	0	0	0	22
1	−1	0	0	0	−1	0	0	1	0	0	0	2
0	−1	0	1	0	0	0	0	0	1	0	0	16
0	−1	0	0	1	0	0	0	0	0	1	0	14
300	600	0	−400	−400	500	0	0	0	0	0	1	8000

$$\xrightarrow[R_7 + 400R_5]{R_2 - R_5}$$

u_1	u_2	u_3	u_4	u_5	x_1	x_2	x_3	x_4	x_5	x_6	P	Constant	Ratio
−1	0	1	0	0	1	0	0	0	0	0	0	16	−
−1	1	0	0	0	0	1	0	0	−1	0	0	4	−
−1	0	0	0	1	0	0	1	0	0	0	0	22	22
1	−1	0	0	0	−1	0	0	1	0	0	0	2	−
0	−1	0	1	0	0	0	0	0	1	0	0	16	−
0	−1	0	0	①	0	0	0	0	0	1	0	14	14
300	200	0	0	−400	500	0	0	0	400	0	1	14,400	

Pivot row → (row 6)

↑ Pivot column

$$\xrightarrow[R_7 + 400R_6]{R_3 - R_6}$$

u_1	u_2	u_3	u_4	u_5	x_1	x_2	x_3	x_4	x_5	x_6	P	Constant	Ratio
−1	0	1	0	0	1	0	0	0	0	0	0	16	−
−1	①	0	0	0	0	1	0	0	−1	0	0	4	4
−1	1	0	0	0	0	0	1	0	0	−1	0	8	8
1	−1	0	0	0	−1	0	0	1	0	0	0	2	−
0	−1	0	1	0	0	0	0	0	1	0	0	16	−
0	−1	0	0	1	0	0	0	0	0	1	0	14	−
300	−200	0	0	0	500	0	0	0	400	400	1	20,000	

Pivot row → (row 2)

↑ Pivot column

$$\xrightarrow[\substack{R_3 - R_2 \\ R_4 + R_2 \\ R_5 + R_2 \\ R_6 + R_2 \\ R_7 + 200R_2}]{}$$

u_1	u_2	u_3	u_4	u_5	x_1	x_2	x_3	x_4	x_5	x_6	P	Constant
−1	0	1	0	0	1	0	0	0	0	0	0	16
−1	①	0	0	0	0	1	0	0	−1	0	0	4
0	0	0	0	0	0	−1	1	0	1	−1	0	4
0	0	0	0	0	−1	1	0	1	−1	0	0	6
−1	0	0	1	0	0	1	0	0	0	0	0	20
−1	0	0	0	1	0	1	0	0	−1	1	0	18
100	0	0	0	0	500	200	0	0	200	400	1	20,800

We find $x_1 = 500$, $x_2 = 200$, $x_3 = 0$, $x_4 = 0$, $x_5 = 200$, $x_6 = 400$, and $C = 20{,}800$. Thus, Plant I should ship 500 units to Warehouse A and 200 units to Warehouse B, and Plant II should ship 200 units to Warehouse B and 400 units to Warehouse C. The minimum shipping cost is \$20,800.

27. We first tabulate the data as follows:

	Output			
	Low	Medium	High	Daily operating cost ($)
Refinery I	200	100	100	200
Refinery II	100	200	600	300
Minimum requirement (barrels)	1000	1400	3000	

Let x and y denote the number of days Refineries I and II should be operated. Then we have the following linear programming problem:

$$\text{Minimize} \quad C = 200x + 300y$$
$$\text{subject to} \quad 200x + 100y \geq 1000$$
$$100x + 200y \geq 1400$$
$$100x + 600y \geq 3000$$
$$x \geq 0, y \geq 0$$

We write down a tableau for the primal problem, then interchange columns and rows to obtain a duplex tableau.

x	y	Constant
200	100	1000
100	200	1400
100	600	3000
200	300	

u	v	w	Constant
200	100	100	200
100	200	600	300
1000	1400	3000	

Thus, the dual problem is:

$$\text{Maximize} \quad P = 1000u + 1400v + 3000w$$
$$\text{subject to} \quad 200u + 100v + 100w \leq 200$$
$$100u + 200v + 600w \leq 300$$
$$u \geq 0, v \geq 0, w \geq 0$$

Using the slack variables x and y, we obtain the following sequence of tableaus:

	u	v	w	x	y	P	Constant	Ratio	
	200	100	100	1	0	0	200	2	$\xrightarrow{\frac{1}{600}R_2}$
Pivot row →	100	200	(600)	0	1	0	300	$\frac{1}{2}$	
	−1000	−1400	−3000	0	0	1	0		

Pivot column

	u	v	w	x	y	P	Constant	
	200	100	100	1	0	0	200	$\xrightarrow[R_3 + 3000R_2]{R_1 - 100R_2}$
	$\frac{1}{6}$	$\frac{1}{3}$	①	0	$\frac{1}{600}$	0	$\frac{1}{2}$	
	−1000	−1400	−3000	0	0	1	0	

	u	v	w	x	y	P	Constant	Ratio	
Pivot row →	$\left(\frac{550}{3}\right)$	$\frac{200}{3}$	0	1	$-\frac{1}{6}$	0	150	$\frac{9}{11}$	$\xrightarrow{\frac{3}{550}R_1}$
	$\frac{1}{6}$	$\frac{1}{3}$	1	0	$\frac{1}{600}$	0	$\frac{1}{2}$	3	
	−500	−400	0	0	5	1	1500		

Pivot column

u	v	w	x	y	P	Constant	Ratio
①	$\frac{4}{11}$	0	$\frac{3}{550}$	$-\frac{1}{1100}$	0	$\frac{9}{11}$	$\frac{9}{11}$
$\frac{1}{6}$	$\frac{1}{3}$	1	0	$\frac{1}{600}$	0	$\frac{1}{2}$	3
-500	-400	0	0	5	1	1500	

$$\xrightarrow[R_3 + 500R_1]{R_2 - \frac{1}{6}R_1}$$

u	v	w	x	y	P	Constant	Ratio
1	$\frac{4}{11}$	0	$\frac{3}{550}$	$-\frac{1}{1100}$	0	$\frac{9}{11}$	$\frac{9}{4}$
0	③$/11$	1	$-\frac{1}{1100}$	$\frac{1}{550}$	0	$\frac{4}{11}$	$\frac{4}{3}$
0	$-\frac{2400}{11}$	0	$\frac{30}{11}$	$\frac{50}{11}$	1	$\frac{21{,}000}{11}$	

Pivot row → (row 2)   Pivot column ↑ (v)

$$\xrightarrow{\frac{11}{3}R_2}$$

u	v	w	x	y	P	Constant
1	$\frac{4}{11}$	0	$\frac{3}{550}$	$-\frac{1}{1100}$	0	$\frac{9}{11}$
0	1	$\frac{11}{3}$	$-\frac{1}{300}$	$\frac{1}{150}$	0	$\frac{4}{3}$
0	$-\frac{2400}{11}$	0	$\frac{30}{11}$	$\frac{50}{11}$	1	$\frac{21{,}000}{11}$

$$\xrightarrow[R_3 + \frac{2400}{11}R_2]{R_1 - \frac{4}{11}R_2}$$

u	v	w	x	y	P	Constant
1	0	$-\frac{4}{3}$	$\frac{1}{150}$	$-\frac{1}{300}$	0	$\frac{1}{3}$
0	0	$\frac{11}{3}$	$-\frac{1}{300}$	$\frac{1}{150}$	0	$\frac{4}{3}$
0	0	800	2	6	1	2200

We conclude that $x = 2$, $y = 6$, and $C = P = 2200$. So the company should operate Refinery I for 2 days and Refinery II for 6 days for a minimum cost of $2200.

29. False. The optimal value is the same for both the primal and the dual problem. This follows from the Fundamental Theorem of Duality.

Technology Exercises   page 263

1. $x = \frac{4}{3}$, $y = \frac{10}{3}$, $z = 0$, and $C = \frac{14}{3}$ **3.** $x = 0.9524$, $y = 4.2857$, $z = 0$, and $C = 6.0952$

5. a. $x = 3$, $y = 2$, and $P = 17$ **b.** $\frac{8}{3} \le c_1 \le 8$, $\frac{3}{2} \le c_2 \le \frac{9}{2}$ **c.** $8 \le b_1 \le 24$, $4 \le b_2 \le 12$

  d. $\frac{5}{4}, \frac{1}{4}$ **e.** Both constraints are binding.

7. a. $x = 4$, $y = 0$, and $C = 8$ **b.** $0 \le c_1 \le \frac{5}{2}$, $4 \le c_2 < \infty$ **c.** $3 \le b_1 < \infty$, $-\infty < b_2 \le 4$

  d. $2, 0$ **e.** First constraint binding, second constraint non-binding

4.3   The Simplex Method: Nonstandard Problems (Optional)

Problem-Solving Tips

1. If you are solving a problem involving mixed constraints, make sure that the problem is written as a maximization problem. All constraints in a mixed constraint problem except $x \ge 0$, $y \ge 0$, and $z \ge 0$ should be written using $\le$. If a constraint is in the form of an equality, rewrite it in the form of two equivalent inequalities. For example, $x = 1$ can be written as the two inequalities $x \le 1$ and $-x \le -1$.

2. To find the *pivot element in a mixed constraint problem*, first check to see if there are any negative entries in the column of constants. If so, pick any negative entry in the row in which a negative entry in the column of constants occurs. (If there are no negative entries, use the simplex method to solve the problem.) The column containing this entry is the pivot column. Now locate the pivot row by computing the *positive* ratios of the numbers in the column of constants to the corresponding numbers in the pivot column (excluding the last row). The smallest ratio corresponds to the pivot row. The pivot element occurs at the intersection of the pivot row and the pivot column.

Concept Questions   page 272

1. It is not a standard maximization problem because the second inequality in the system of constraints cannot be written in a form in which the expression involving the variables is less than or equal to a nonnegative constant.

3. It is not a standard maximization problem because the second constraint in the system of constraints is an equation. It cannot be rewritten as a restricted minimization problem because if the problem is written as a minimization problem, the objective function $C = -P = -x - 3y$ has coefficients that are not all nonnegative.

Exercises   page 273

1. Maximize $P = -C = -2x + 3y$
    subject to $-3x - 5y \le -20$
    $$3x + y \le 16$$
    $$-2x + y \le 1$$
    $$x \ge 0, y \ge 0$$

3. Maximize $P = -C = -5x - 10y - z$
    subject to $-2x - y - z \le -4$
    $$-x - 2y - 2z \le -2$$
    $$2x + 4y + 3z \le 12$$
    $$x \ge 0, y \ge 0, z \ge 0$$

5. We set up the tableau and solve the problem using the simplex method:

	x	y	u	v	P	Constant	Ratio
Pivot row →	2	⑤	1	0	0	20	4
	1	−5	0	1	0	−5	1
	−1	−2	0	0	1	0	

Pivot column (under y)

$\xrightarrow{-\frac{1}{5}R_2}$

x	y	u	v	P	Constant
2	5	1	0	0	20
$-\frac{1}{5}$	1	0	$-\frac{1}{5}$	0	1
−1	−2	0	0	1	0

$\xrightarrow[R_3 + 2R_2]{R_1 - 5R_2}$

	x	y	u	v	P	Constant	Ratio
Pivot row →	③	0	1	1	0	15	5
	$-\frac{1}{5}$	1	0	$-\frac{1}{5}$	0	1	−
	$-\frac{7}{5}$	0	0	$-\frac{2}{5}$	1	2	

Pivot column (under x)

$\xrightarrow{\frac{1}{3}R_1}$

x	y	u	v	P	Constant
1	0	$\frac{1}{3}$	$\frac{1}{3}$	0	5
$-\frac{1}{5}$	1	0	$-\frac{1}{5}$	0	1
$-\frac{7}{5}$	0	0	$-\frac{2}{5}$	1	2

$\xrightarrow[R_3 + \frac{7}{5}R_1]{R_2 + \frac{1}{5}R_1}$

x	y	u	v	P	Constant
1	0	$\frac{1}{3}$	$\frac{1}{3}$	0	5
0	1	$\frac{1}{15}$	$-\frac{2}{15}$	0	2
0	0	$\frac{7}{15}$	$\frac{1}{15}$	1	9

The maximum value of P is 9 when $x = 5$ and $y = 2$.

7. We first rewrite the problem as a maximization problem with inequality constraints using $\leq$, obtaining the following equivalent problem:

$$\text{Maximize} \quad P = -C = 2x - y$$
$$\text{subject to} \quad x + 2y \leq 6$$
$$3x + 2y \leq 12$$
$$x \geq 0, y \geq 0$$

Following the procedure outlined for nonstandard problems, we obtain the following sequence of tableaus.

	x	y	u	v	P	Constant	Ratio
	1	2	1	0	0	6	6
Pivot row →	③	2	0	1	0	12	4
	-2	1	0	0	1	0	

$\xrightarrow{\frac{1}{3}R_2}$

x	y	u	v	P	Constant
1	2	1	0	0	6
1	$\frac{2}{3}$	0	$\frac{1}{3}$	0	4
-2	1	0	0	1	0

$\xrightarrow[R_3 + 2R_2]{R_1 - R_2}$

Pivot column (under x in first tableau)

x	y	u	v	P	Constant
0	$\frac{4}{3}$	1	$-\frac{1}{3}$	0	2
1	$\frac{2}{3}$	0	$\frac{1}{3}$	0	4
0	$\frac{7}{3}$	0	$\frac{2}{3}$	1	8

We conclude that C attains a minimum value of -8 when $x = 4$ and $y = 0$.

9. Using the simplex method, we have

	x	y	u	v	P	Constant	Ratio
	1	3	1	0	0	6	6
Pivot row →	⊝2	3	0	1	0	-6	3
	-1	-4	0	0	1	0	

$\xrightarrow{-\frac{1}{2}R_2}$

x	y	u	v	P	Constant
1	3	1	0	0	6
①	$-\frac{3}{2}$	0	$-\frac{1}{2}$	0	3
-1	-4	0	0	1	0

$\xrightarrow[R_3 + R_2]{R_1 - R_2}$

Pivot column (under x in first tableau)

x	y	u	v	P	Constant
0	$\frac{9}{2}$	1	$\frac{1}{2}$	0	3
1	$-\frac{3}{2}$	0	$-\frac{1}{2}$	0	3
0	$-\frac{11}{2}$	0	$-\frac{1}{2}$	1	3

$\xrightarrow{\frac{2}{9}R_1}$

x	y	u	v	P	Constant
0	1	$\frac{2}{9}$	$\frac{1}{9}$	0	$\frac{2}{3}$
1	$-\frac{3}{2}$	0	$-\frac{1}{2}$	0	3
0	$-\frac{11}{2}$	0	$-\frac{1}{2}$	1	3

$\xrightarrow[R_3 + \frac{11}{2}R_1]{R_2 + \frac{3}{2}R_1}$

x	y	u	v	P	Constant
0	1	$\frac{2}{9}$	$\frac{1}{9}$	0	$\frac{2}{3}$
1	0	$\frac{1}{3}$	$-\frac{1}{3}$	0	4
0	0	$\frac{11}{9}$	$\frac{1}{9}$	1	$\frac{20}{3}$

We conclude that P attains a maximum value of $\frac{20}{3}$ when $x = 4$ and $y = \frac{2}{3}$.

11. We rewrite the problem as follows:

$$\text{Maximize} \quad P = x + 2y$$
$$\text{subject to} \quad 2x + 3y \le 12$$
$$-x + 3y \le 3$$
$$-x + 3y \ge 3$$
$$x \ge 0, y \ge 0$$

We calculate the following sequence of tableaus.

	x	y	u	v	w	P	Constant	Ratio
	2	3	1	0	0	0	12	4
	-1	3	0	1	0	0	3	1
Pivot row →	1	$\boxed{-3}$	0	0	1	0	-3	1
	-1	-2	0	0	0	1	0	

↑ Pivot column

$\xrightarrow{-\frac{1}{3}R_3}$

x	y	u	v	w	P	Constant	
2	3	1	0	0	0	12	$R_1 - 3R_3$
-1	3	0	1	0	0	3	$R_2 - 3R_3$
$-\frac{1}{3}$	1	0	0	$-\frac{1}{3}$	0	1	$\xrightarrow{R_4 + 2R_3}$
-1	-2	0	0	0	1	0	

	x	y	u	v	w	P	Constant	Ratio
Pivot row →	$\boxed{3}$	0	1	0	1	0	9	3
	0	0	0	1	1	0	0	–
	$-\frac{1}{3}$	1	0	0	$-\frac{1}{3}$	0	1	–
	$-\frac{5}{3}$	0	0	0	$-\frac{2}{3}$	1	2	

↑ Pivot column

$\xrightarrow{\frac{1}{3}R_1}$

x	y	u	v	w	P	Constant	
1	0	$\frac{1}{3}$	0	$\frac{1}{3}$	0	3	$R_3 + \frac{1}{3}R_1$
0	0	0	1	1	0	0	$\xrightarrow{R_4 + \frac{5}{3}R_1}$
$-\frac{1}{3}$	1	0	0	$-\frac{1}{3}$	0	1	
$-\frac{5}{3}$	0	0	0	$-\frac{2}{3}$	1	2	

x	y	u	v	w	P	Constant
1	0	$\frac{1}{3}$	0	$\frac{1}{3}$	0	3
0	0	0	1	1	0	0
0	1	$\frac{1}{9}$	0	$-\frac{2}{9}$	0	2
0	0	5	0	1	1	7

We conclude that P attains a maximum value of 7 when $x = 3$ and $y = 2$.

13. We rewrite the problem as follows:

$$\text{Maximize} \quad P = x - 2y + z$$
$$\text{subject to} \quad 2x + 3y + 2z \le 12$$
$$-x - 2y + 3z \le -6$$
$$x \ge 0, y \ge 0, z \ge 0$$

We calculate the following sequence of tableaus.

	x	y	z	u	v	P	Constant	Ratio
Pivot row →	2	3	2	1	0	0	12	6
	$\boxed{-1}$	-2	3	0	1	0	-6	6
	-1	2	-1	0	0	1	0	

↑ Pivot column

$\xrightarrow{-R_2}$

x	y	z	u	v	P	Constant	
2	3	2	1	0	0	12	$R_1 - 2R_2$
$\boxed{1}$	2	-3	0	-1	0	6	$\xrightarrow{R_3 + R_2}$
-1	2	-1	0	0	1	0	

	x	y	z	u	v	P	Constant	Ratio
Pivot row →	0	-1	(8)	1	2	0	0	0
	1	2	-3	0	-1	0	6	-
	0	4	-4	0	-1	1	6	

Pivot column (↑ under z)

$\frac{1}{8}R_1 \longrightarrow$

x	y	z	u	v	P	Constant
0	$-\frac{1}{8}$	(1)	$\frac{1}{8}$	$\frac{1}{4}$	0	0
1	2	-3	0	-1	0	6
0	4	-4	0	-1	1	6

$\begin{array}{c} R_2 + 3R_1 \\ \longrightarrow \\ R_3 + 4R_1 \end{array}$

x	y	z	u	v	P	Constant
0	$-\frac{1}{8}$	1	$\frac{1}{8}$	$\frac{1}{4}$	0	0
1	$\frac{13}{8}$	0	$\frac{3}{8}$	$-\frac{1}{4}$	0	6
0	$\frac{7}{2}$	0	$\frac{1}{2}$	0	1	6

Thus, P attains a maximum value of 6 when $x = 6$, $y = 0$, and $z = 0$.

15. The problem is:

$$\begin{aligned} \text{Maximize} \quad & P = -C = -2x + 3y - 4z \\ \text{subject to} \quad & -x + 2y - z \le 8 \\ & x - 2y + 2z \le 10 \\ & 2x + 4y - 3z \le 12 \\ & x \ge 0, \, y \ge 0, \, z \ge 0 \end{aligned}$$

We calculate the following sequence of tableaus.

	x	y	z	u	v	w	P	Constant	Ratio
	-1	2	-1	1	0	0	0	8	4
	1	-2	2	0	1	0	0	10	-
Pivot row →	2	(4)	-3	0	0	1	0	12	3
	2	-3	4	0	0	0	1	0	

Pivot column (↑ under y)

$\frac{1}{4}R_3 \longrightarrow$

x	y	z	u	v	w	P	Constant
-1	2	-1	1	0	0	0	8
1	-2	2	0	1	0	0	10
$\frac{1}{2}$	(1)	$-\frac{3}{4}$	0	0	$\frac{1}{4}$	0	3
2	-3	4	0	0	0	1	0

$\begin{array}{c} R_1 - 2R_3 \\ R_2 + 2R_3 \\ \longrightarrow \\ R_4 + 3R_3 \end{array}$

x	y	z	u	v	w	P	Constant
-2	0	$\frac{1}{2}$	1	0	$-\frac{1}{2}$	0	2
2	0	$\frac{1}{2}$	0	1	$\frac{1}{2}$	0	16
$\frac{1}{2}$	1	$-\frac{3}{4}$	0	0	$\frac{1}{4}$	0	3
$\frac{7}{2}$	0	$\frac{7}{4}$	0	0	$\frac{3}{4}$	1	9

We deduce that C attains a minimum value of -9 when $x = 0$, $y = 3$, and $z = 0$.

17. Rewriting the third constraint as $-x + 2y - z \le -4$, we have the following problem:

$$\text{Maximize} \quad P = 2x + y + z$$
$$\text{subject to} \quad x + 2y + 3z \le 28$$
$$2x + 3y - z \le 6$$
$$-x + 2y - z \le -4$$
$$x \ge 0, y \ge 0, z \ge 0$$

We calculate the following sequence of tableaus.

	x	y	z	u	v	w	P	Constant	Ratio
	1	2	3	1	0	0	0	28	28
Pivot row →	(2)	3	-1	0	1	0	0	6	3
	-1	2	-1	0	0	1	0	-4	4
	-2	-1	-1	0	0	0	1	0	

↑ Pivot column

$\xrightarrow{\frac{1}{2} R_2}$

	x	y	z	u	v	w	P	Constant	
	1	2	3	1	0	0	0	28	$R_1 - R_2$
	(1)	$\frac{3}{2}$	$-\frac{1}{2}$	0	$\frac{1}{2}$	0	0	3	$\xrightarrow{\substack{R_3 + R_2 \\ R_4 + 2R_2}}$
	-1	2	-1	0	0	1	0	-4	
	-2	-1	-1	0	0	0	1	0	

	x	y	z	u	v	w	P	Constant	Ratio
	0	$\frac{1}{2}$	$\frac{7}{2}$	1	$-\frac{1}{2}$	0	0	25	$\frac{50}{7}$
	1	$\frac{3}{2}$	$-\frac{1}{2}$	0	$\frac{1}{2}$	0	0	3	—
Pivot row →	0	$\frac{7}{2}$	$\left(-\frac{3}{2}\right)$	0	$\frac{1}{2}$	1	0	-1	$\frac{2}{3}$
	0	2	-2	0	1	0	1	6	

↑ Pivot column

$\xrightarrow{-\frac{2}{3} R_3}$

	x	y	z	u	v	w	P	Constant	
	0	$\frac{1}{2}$	$\frac{7}{2}$	1	$-\frac{1}{2}$	0	0	25	$R_1 - \frac{7}{2} R_3$
	1	$\frac{3}{2}$	$-\frac{1}{2}$	0	$\frac{1}{2}$	0	0	3	$\xrightarrow{\substack{R_2 + \frac{1}{2} R_3 \\ R_4 + 2R_3}}$
	0	$-\frac{7}{3}$	(1)	0	$-\frac{1}{3}$	$-\frac{2}{3}$	0	$\frac{2}{3}$	
	0	2	-2	0	1	0	1	6	

	x	y	z	u	v	w	P	Constant	Ratio
Pivot row →	0	$\left(\frac{26}{3}\right)$	0	1	$\frac{2}{3}$	$\frac{7}{3}$	0	$\frac{68}{3}$	$\frac{34}{13}$
	1	$\frac{1}{3}$	0	0	$\frac{1}{3}$	$-\frac{1}{3}$	0	$\frac{10}{3}$	10
	0	$-\frac{7}{3}$	1	0	$-\frac{1}{3}$	$-\frac{2}{3}$	0	$\frac{2}{3}$	—
	0	$-\frac{8}{3}$	0	0	$\frac{1}{3}$	$-\frac{4}{3}$	1	$\frac{22}{3}$	

↑ Pivot column

$\xrightarrow{\frac{3}{26} R_1}$

	x	y	z	u	v	w	P	Constant	
	0	(1)	0	$\frac{3}{26}$	$\frac{1}{13}$	$\frac{7}{26}$	0	$\frac{34}{13}$	$R_2 - \frac{1}{3} R_1$
	1	$\frac{1}{3}$	0	0	$\frac{1}{3}$	$-\frac{1}{3}$	0	$\frac{10}{3}$	$\xrightarrow{\substack{R_3 + \frac{7}{3} R_1 \\ R_4 + \frac{8}{3} R_1}}$
	0	$-\frac{7}{3}$	1	0	$-\frac{1}{3}$	$-\frac{2}{3}$	0	$\frac{2}{3}$	
	0	$-\frac{8}{3}$	0	0	$\frac{1}{3}$	$-\frac{4}{3}$	1	$\frac{22}{3}$	

	x	y	z	u	v	w	P	Constant	Ratio
Pivot row →	0	1	0	$\frac{3}{26}$	$\frac{1}{13}$	$\left(\frac{7}{26}\right)$	0	$\frac{34}{13}$	$\frac{68}{7}$
	1	0	0	$-\frac{1}{26}$	$\frac{12}{39}$	$-\frac{11}{26}$	0	$\frac{32}{13}$	—
	0	0	1	$\frac{7}{26}$	$-\frac{2}{13}$	$-\frac{1}{26}$	0	$\frac{88}{13}$	—
	0	0	0	$\frac{4}{13}$	$\frac{7}{13}$	$-\frac{8}{13}$	1	$\frac{186}{13}$	

↑ Pivot column

$\xrightarrow{\frac{26}{7} R_1}$

	x	y	z	u	v	w	P	Constant	
	0	$\frac{26}{7}$	0	$\frac{3}{7}$	$\frac{2}{7}$	(1)	0	$\frac{68}{7}$	$R_2 + \frac{11}{26} R_1$
	1	0	0	$-\frac{1}{26}$	$\frac{12}{39}$	$-\frac{11}{26}$	0	$\frac{32}{13}$	$\xrightarrow{\substack{R_3 + \frac{1}{26} R_1 \\ R_4 + \frac{8}{13} R_1}}$
	0	0	1	$\frac{7}{26}$	$-\frac{2}{13}$	$-\frac{1}{26}$	0	$\frac{88}{13}$	
	0	0	0	$\frac{4}{13}$	$\frac{7}{13}$	$-\frac{8}{13}$	1	$\frac{186}{13}$	

x	y	z	u	v	w	P	Constant
0	$\frac{26}{7}$	0	$\frac{3}{7}$	$\frac{2}{7}$	1	0	$\frac{68}{7}$
1	$\frac{11}{7}$	0	$\frac{1}{7}$	$\frac{3}{7}$	0	0	$\frac{46}{7}$
0	$\frac{1}{7}$	1	$\frac{2}{7}$	$-\frac{1}{7}$	0	0	$\frac{50}{7}$
0	$\frac{16}{7}$	0	$\frac{4}{7}$	$\frac{5}{7}$	0	1	$\frac{142}{7}$

We deduce that P attains a maximum value of $\frac{142}{7}$ when $x = \frac{46}{7}$, $y = 0$, $z = \frac{50}{7}$, $u = 0$, $v = 0$, and $w = 0$.

19. Rewriting the third constraint $2x + y + z = 10$ as the two inequalities $2x + y + z \le 10$ and $-2x - y - z \le -10$ we have the following problem:

$$\begin{array}{ll} \text{Maximize} & P = x + 2y + 3z \\ \text{subject to} & x + 2y + z \le 20 \\ & 3x + y \le 30 \\ & 2x + y + z \le 10 \\ & -2x - y - z \le -10 \\ & x \ge 0,\ y \ge 0,\ z \ge 0 \end{array}$$

We calculate the following sequence of tableaus.

x	y	z	t	u	v	w	P	Constant
1	2	1	1	0	0	0	0	20
3	1	0	0	1	0	0	0	30
2	1	1	0	0	1	0	0	10
-2	-1	-1	0	0	0	1	0	-10
-1	-2	-3	0	0	0	0	1	0

$\xrightarrow{-R_4}$

x	y	z	t	u	v	w	P	Constant
1	2	1	1	0	0	0	0	20
3	1	0	0	1	0	0	0	30
2	1	1	0	0	1	0	0	10
2	1	1	0	0	0	-1	0	10
-1	-2	-3	0	0	0	0	1	0

$\begin{array}{c} R_1 - R_4 \\ R_3 - R_4 \\ \xrightarrow{\hspace{1cm}} \\ R_5 + 3R_4 \end{array}$

x	y	z	t	u	v	w	P	Constant	Ratio
-1	1	0	1	0	0	1	0	10	10
3	1	0	0	1	0	0	0	30	—
0	0	0	0	0	1	①	0	0	0
2	1	1	0	0	0	-1	0	10	—
5	1	0	0	0	0	-3	1	30	

Pivot row →

↑
Pivot column

$\begin{array}{c} R_1 - R_3 \\ R_4 + R_3 \\ \xrightarrow{\hspace{1cm}} \\ R_5 + 3R_3 \end{array}$

x	y	z	t	u	v	w	P	Constant
1	2	1	1	0	0	0	0	20
3	1	0	0	1	0	0	0	30
0	0	0	0	0	1	①	0	0
2	1	1	0	0	1	0	0	10
5	1	0	0	0	3	0	1	30

We conclude that P attains a maximum value of 30 when $x = 0$, $y = 0$, $z = 10$, $t = 20$, $u = 30$, $v = 0$, and $w = 0$.

21. Let x and y denote the dollar amounts invested in Companies A and B. Then the problem is:

$$\text{Maximize} \quad P = 0.10x + 0.20y$$
$$\text{subject to} \quad x + y \leq 50{,}000$$
$$-x + y \leq -20{,}000$$
$$x \geq 0, y \geq 0$$

We calculate the following sequence of tableaus.

	x	y	u	v	P	Constant
	1	1	1	0	0	50,000
Pivot row →	(−1)	1	0	1	0	−20,000
	−0.1	−0.2	0	0	1	0

↑ Pivot column

$\xrightarrow{-R_2}$

	x	y	u	v	P	Constant
	1	1	1	0	0	50,000
	1	−1	0	−1	0	20,000
	−0.1	−0.2	0	0	1	0

$\xrightarrow[R_3 + 0.1R_2]{R_1 - R_2}$

	x	y	u	v	P	Constant
Pivot row →	0	2	1	1	0	30,000
	1	−1	0	−1	0	20,000
	0	−0.3	0	−0.1	1	2000

↑ Pivot column

$\xrightarrow{\frac{1}{2}R_1}$

x	y	u	v	P	Constant
0	1	$\frac{1}{2}$	$\frac{1}{2}$	0	15,000
1	−1	0	−1	0	20,000
0	−0.3	0	−0.1	1	2000

$\xrightarrow[R_3 + 0.3R_2]{R_2 + R_1}$

x	y	u	v	P	Constant
0	1	$\frac{1}{2}$	$\frac{1}{2}$	0	15,000
1	0	$\frac{1}{2}$	$-\frac{1}{2}$	0	35,000
0	0	0.15	0.01	1	6500

Thus, Natsano should invest \$35,000 in Company A stock and \$15,000 in Company B stock for a maximum return of \$6,500 on his investment.

23. Let x and y denote the dollar amounts invested in home and commercial development loans, respectively. Then the problem is:

$$\text{Maximize} \quad P = 0.08x + 0.06y$$
$$\text{subject to} \quad -x + 3y \leq 0$$
$$y \geq 10{,}000{,}000$$
$$x + y = 60{,}000{,}000$$
$$x \geq 0, y \geq 0$$

Substituting $x = 60{,}000{,}000 - y$ into the first equation and the first and second inequalities, we simplify the problem:

$$\text{Maximize} \quad P = 0.08\,(60{,}000{,}000 - y) + 0.06y = 4{,}800{,}000 - 0.02y$$
$$\text{subject to} \quad y \leq 15{,}000{,}000$$
$$y \geq 10{,}000{,}000$$
$$x \geq 0, y \geq 0$$

We calculate the following sequence of tableaus.

	y	u	v	P	Constant
	1	1	0	0	15,000,000
Pivot row →	−1	0	1	0	−10,000,000
	0.02	0	0	1	4,800,000

$\xrightarrow{-R_2}$

y	u	v	P	Constant
1	1	0	0	15,000,000
1	0	−1	0	10,000,000
0.02	0	0	1	4,800,000

$\xrightarrow[R_3 - 0.02R_2]{R_1 - R_2}$

(Pivot column: ↑ under the y column)

y	u	v	P	Constant
0	1	1	0	5,000,000
1	0	−1	0	10,000,000
0	0	0.02	1	4,600,000

Then $y = 10{,}000{,}000$ and $x = 60{,}000{,}000 - 10{,}000{,}000 = 50{,}000{,}000$. We conclude that the bank should extend $50 million in home loans and $10 million in commercial development loans to attain a maximum return of $4.6 million.

25. Let x, y, and z denote the numbers of units of Products A, B, and C manufactured by the company. Then the linear programming problem is:

$$\text{Maximize} \quad P = 18x + 12y + 15z$$
$$\text{subject to} \quad 2x + y + 2z \leq 900$$
$$3x + y + 2z \leq 1080$$
$$2x + 2y + z \leq 840$$
$$x - y + z \leq 0$$
$$x \geq 0, y \geq 0, z \geq 0$$

We calculate the following sequence of tableaus.

	x	y	z	t	u	v	w	P	Constant	Ratio	
	2	1	2	1	0	0	0	0	900	450	
	3	1	2	0	1	0	0	0	1080	360	$R_1 - 2R_4$
	2	2	1	0	0	1	0	0	840	420	$R_2 - 3R_4$ $\longrightarrow$
Pivot row $\rightarrow$	①	−1	1	0	0	0	1	0	0	0	$R_3 - 2R_4$ $R_5 + 18R_4$
	−18	−12	−15	0	0	0	0	1	0		

↑ Pivot column

	x	y	z	t	u	v	w	P	Constant	Ratio	
	0	3	0	1	0	0	−2	0	900	300	
	0	4	−1	0	1	0	−3	0	1080	270	$\frac{1}{4}R_3$
Pivot row $\rightarrow$	0	④	−1	0	0	1	−2	0	840	210	$\longrightarrow$
	1	−1	1	0	0	0	1	0	0	−	
	0	−30	3	0	0	0	18	1	0		

↑ Pivot column

x	y	z	t	u	v	w	P	Constant	
0	3	0	1	0	0	−2	0	900	
0	4	−1	0	1	0	−3	0	1080	$R_1 - 3R_3$
0	1	$-\frac{1}{4}$	0	0	$\frac{1}{4}$	$-\frac{1}{2}$	0	210	$R_2 - 4R_3$ $\longrightarrow$
1	−1	1	0	0	0	1	0	0	$R_4 + R_3$ $R_5 + 30R_3$
0	−30	3	0	0	0	18	1	0	

	x	y	z	t	u	v	w	P	Constant	Ratio	
	0	0	$\frac{3}{4}$	1	0	$-\frac{3}{4}$	$-\frac{1}{2}$	0	270	$\frac{1080}{3}$	
	0	0	0	0	1	−1	−1	0	240	−	$\frac{4}{3}R_4$
	0	1	$-\frac{1}{4}$	0	0	$\frac{1}{4}$	$-\frac{1}{2}$	0	210	−	$\longrightarrow$
Pivot row $\rightarrow$	1	0	③⁄₄	0	0	$\frac{1}{4}$	$\frac{1}{2}$	0	210	−280	
	0	0	$-\frac{9}{2}$	0	0	$\frac{15}{2}$	3	1	6300		

↑ Pivot column

x	y	z	t	u	v	w	P	Constant	
0	0	$\frac{3}{4}$	1	0	$-\frac{3}{4}$	$-\frac{1}{2}$	0	270	
0	0	0	0	1	−1	−1	0	240	$R_1 - \frac{3}{4}R_4$
0	1	$-\frac{1}{4}$	0	0	$\frac{1}{4}$	$-\frac{1}{2}$	0	210	$R_3 + \frac{1}{4}R_4$ $\longrightarrow$
$\frac{4}{3}$	0	①	0	0	$\frac{1}{3}$	$\frac{2}{3}$	0	280	$R_5 + \frac{9}{2}R_4$
0	0	$-\frac{9}{2}$	0	0	$\frac{15}{2}$	3	1	6300	

x	y	z	t	u	v	w	P	Constant
−1	0	0	1	0	−1	−1	0	60
0	0	0	0	1	−1	−1	0	240
$\frac{1}{3}$	1	0	0	0	$\frac{1}{3}$	$-\frac{1}{3}$	0	280
$\frac{4}{3}$	0	1	0	0	$\frac{1}{3}$	$\frac{2}{3}$	0	240
6	0	0	0	0	9	6	1	7560

We conclude that the company should produce 0 units of Product A, 280 units of Product B, and 280 units of Product C to realize a maximum profit of \$7560.

27. Let x denote the number of ounces of Food A and y the number of ounces of Food B used in the meal. The problem is to minimize the amount of cholesterol in the meal, so the linear programming problem is:

$$\text{Maximize} \quad P = -C = -2x - 5y$$
$$\text{subject to} \quad 30x + 25y \geq 400$$
$$x + \tfrac{1}{2}y \geq 10$$
$$2x + 5y \geq 40$$
$$x \geq 0, y \geq 0$$

We calculate the following sequence of tableaus.

x	y	u	v	w	P	Constant
-30	-25	1	0	0	0	-400
-1	$-\tfrac{1}{2}$	0	1	0	0	-10
-2	-5	0	0	1	0	-40
2	5	0	0	0	1	0

$\xrightarrow{-R_2}$

x	y	u	v	w	P	Constant
-30	-25	1	0	0	0	-400
1	$\tfrac{1}{2}$	0	-1	0	0	10
-2	-5	0	0	1	0	-40
2	5	0	0	0	1	0

$\xrightarrow[\substack{R_3 + 2R_2 \\ R_4 - 2R_2}]{R_1 + 30R_2}$

x	y	u	v	w	P	Constant
0	-10	1	-30	0	0	-100
1	$\tfrac{1}{2}$	0	-1	0	0	10
0	-4	0	-2	1	0	-20
0	4	0	2	0	1	-20

$\xrightarrow{-\tfrac{1}{4}R_3}$

x	y	u	v	w	P	Constant
0	-10	1	-30	0	0	-100
1	$\tfrac{1}{2}$	0	-1	0	0	10
0	1	0	$\tfrac{1}{2}$	$-\tfrac{1}{4}$	0	5
0	4	0	2	0	1	-20

$\xrightarrow[\substack{R_2 - \tfrac{1}{2}R_3 \\ R_4 - 4R_3}]{R_1 + 10R_3}$

x	y	u	v	w	P	Constant
0	0	1	-25	$-\tfrac{5}{2}$	0	-50
1	0	0	$-\tfrac{5}{4}$	$\tfrac{1}{8}$	0	$\tfrac{15}{2}$
0	1	0	$\tfrac{1}{2}$	$-\tfrac{1}{4}$	0	5
0	0	0	0	1	1	-40

$\xrightarrow{-\tfrac{1}{25}R_1}$

x	y	u	v	w	P	Constant
0	0	$-\tfrac{1}{25}$	1	$\tfrac{1}{10}$	0	2
1	0	0	$-\tfrac{5}{4}$	$\tfrac{1}{8}$	0	$\tfrac{15}{2}$
0	1	0	$\tfrac{1}{2}$	$-\tfrac{1}{4}$	0	5
0	0	0	0	1	1	-40

$\xrightarrow[\substack{R_3 - \tfrac{1}{2}R_1}]{R_2 + \tfrac{5}{4}R_1}$

x	y	u	v	w	P	Constant
0	0	$-\tfrac{1}{25}$	1	$\tfrac{1}{10}$	0	2
1	0	$-\tfrac{1}{20}$	0	$\tfrac{1}{4}$	0	10
0	1	$\tfrac{1}{50}$	0	$\tfrac{3}{10}$	0	4
0	0	0	0	1	1	-40

Thus, the minimum content of cholesterol is 40 mg when 10 ounces of Food A and 4 ounces of Food B are used. Note that because the u-column is not in unit form, the problem has multiple solutions.

CHAPTER 4   Concept Review Questions   page 276

1. maximized, nonnegative, less than, equal to

3. minimized, nonnegative, greater than, equal to

1. This is a regular linear programming problem. Using the simplex method with u and v as slack variables, we obtain the following sequence of tableaus:

	x	y	u	v	P	Constant	Ratio
Pivot row →	1	③	1	0	0	15	5
	4	1	0	1	0	16	16
	−3	−4	0	0	1	0	

↑ Pivot column

$\xrightarrow{\frac{1}{3}R_1}$

x	y	u	v	P	Constant
$\frac{1}{3}$	①	$\frac{1}{3}$	0	0	5
4	1	0	1	0	16
−3	−4	0	0	1	0

$\xrightarrow[R_3+4R_1]{R_2-R_1}$

	x	y	u	v	P	Constant	Ratio
	$\frac{1}{3}$	1	$\frac{1}{3}$	0	0	5	15
Pivot row →	$\frac{11}{3}$	0	$-\frac{1}{3}$	1	0	11	3
	$-\frac{5}{3}$	0	$\frac{4}{3}$	0	1	20	

↑ Pivot column

$\xrightarrow{\frac{3}{11}R_2}$

x	y	u	v	P	Constant
$\frac{1}{3}$	1	$\frac{1}{3}$	0	0	5
①	0	$-\frac{1}{11}$	$\frac{3}{11}$	0	3
$-\frac{5}{3}$	0	$\frac{4}{3}$	0	1	20

$\xrightarrow[R_3+\frac{5}{3}R_2]{R_1-\frac{1}{3}R_2}$

x	y	u	v	P	Constant
0	1	$\frac{4}{11}$	$-\frac{1}{11}$	0	4
1	0	$-\frac{1}{11}$	$\frac{3}{11}$	0	3
0	0	$\frac{13}{11}$	$\frac{5}{11}$	1	25

We conclude that $x = 3$, $y = 4$, $u = 0$, $v = 0$, and $P = 25$.

3. This is a regular linear programming problem. Using the simplex method with u, v, and w as slack variables, we obtain the following sequence of tableaus:

	x	y	u	v	w	P	Constant	Ratio
	1	3	1	0	0	0	18	18
	3	2	0	1	0	0	19	$\frac{19}{3}$
Pivot row →	③	1	0	0	1	0	15	5
	−3	−2	0	0	0	1	0	

↑ Pivot column

$\xrightarrow{\frac{1}{3}R_3}$

x	y	u	v	w	P	Constant
1	3	1	0	0	0	18
3	2	0	1	0	0	19
①	$\frac{1}{3}$	0	0	$\frac{1}{3}$	0	5
−3	−2	0	0	0	1	0

$\xrightarrow[R_4+3R_3]{\substack{R_1-R_3 \\ R_2-3R_3}}$

	x	y	u	v	w	P	Constant	Ratio
	0	$\frac{8}{3}$	1	0	$-\frac{1}{3}$	0	13	$\frac{39}{8}$
Pivot row →	0	①	0	1	−1	0	4	4
	1	$\frac{1}{3}$	0	0	$\frac{1}{3}$	0	5	15
	0	−1	0	0	1	1	15	

↑ Pivot column

$\xrightarrow[R_4+R_2]{\substack{R_1-\frac{8}{3}R_2 \\ R_3-\frac{1}{3}R_2}}$

x	y	u	v	w	P	Constant
0	0	1	$-\frac{8}{3}$	$\frac{7}{3}$	0	$\frac{7}{3}$
0	①	0	1	−1	0	4
1	0	0	$-\frac{1}{3}$	$\frac{4}{3}$	0	$\frac{11}{3}$
0	0	0	1	0	1	19

We conclude that $x = \frac{11}{3}$, $y = 4$, $u = \frac{7}{3}$, $v = 0$, $w = 0$, and $P = 19$.

5. Using the simplex method to solve this regular linear programming problem, we calculate the following tableaus.

	x	y	z	u	v	P	Constant	Ratio
Pivot row →	1	2	③	1	0	0	12	4
	1	−3	2	0	1	0	10	5
	−2	−3	−5	0	0	1	0	

Pivot column (↑ under z)

$\xrightarrow{\frac{1}{3}R_1}$

x	y	z	u	v	P	Constant
$\frac{1}{3}$	$\frac{2}{3}$	①	$\frac{1}{3}$	0	0	4
1	−3	2	0	1	0	10
−2	−3	−5	0	0	1	0

$\xrightarrow[R_3+5R_1]{R_2-2R_1}$

	x	y	z	u	v	P	Constant	Ratio
	$\frac{1}{3}$	$\frac{2}{3}$	1	$\frac{1}{3}$	0	0	4	12
Pivot row →	$\left(\frac{1}{3}\right)$	$-\frac{13}{3}$	0	$-\frac{2}{3}$	1	0	2	6
	$-\frac{1}{3}$	$\frac{1}{3}$	0	$\frac{5}{3}$	0	1	20	

Pivot column (↑ under x)

$\xrightarrow{3R_2}$

x	y	z	u	v	P	Constant
$\frac{1}{3}$	$\frac{2}{3}$	1	$\frac{1}{3}$	0	0	4
①	−13	0	−2	3	0	6
$-\frac{1}{3}$	$\frac{1}{3}$	0	$\frac{5}{3}$	0	1	20

$\xrightarrow[R_3+\frac{1}{3}R_2]{R_1-\frac{1}{3}R_2}$

x	y	z	u	v	P	Constant
0	1	$\frac{1}{5}$	$\frac{1}{5}$	$-\frac{1}{5}$	0	$\frac{2}{5}$
1	−13	0	−2	3	0	6
0	−4	0	1	1	1	22

$\xrightarrow[R_3+4R_1]{R_2+13R_1}$

x	y	z	u	v	P	Constant
0	1	$\frac{1}{5}$	$\frac{1}{5}$	$-\frac{1}{5}$	0	$\frac{2}{5}$
1	0	$\frac{13}{5}$	$\frac{3}{5}$	$\frac{2}{5}$	0	$\frac{56}{5}$
0	0	$\frac{4}{5}$	$\frac{9}{5}$	$\frac{1}{5}$	1	$\frac{118}{5}$

We conclude that the P attains a maximum value of $\frac{118}{5}$ when $x = \frac{56}{5}$, $y = \frac{2}{5}$, $z = 0$, $u = 0$, and $v = 0$.

7. We wish to maximize $P = -C = 4x + 7y$ subject to the given constraints. Using the simplex method with u and v as slack variables, we obtain the following tableaus:

	x	y	u	v	P	Constant	Ratio
	3	1	1	0	0	8	8
Pivot row →	1	②	0	1	0	6	3
	−4	−7	0	0	1	0	

Pivot column (↑ under y)

$\xrightarrow{\frac{1}{2}R_2}$

x	y	u	v	P	Constant
3	1	0	0	0	8
$\frac{1}{2}$	①	0	$\frac{1}{2}$	0	3
−4	−7	0	0	1	0

$\xrightarrow[R_3+7R_2]{R_1-R_2}$

	x	y	u	v	P	Constant	Ratio
Pivot row →	$\left(\frac{5}{2}\right)$	0	1	$-\frac{1}{2}$	0	5	2
	$\frac{1}{2}$	1	0	$\frac{1}{2}$	0	3	6
	$-\frac{1}{2}$	0	0	$\frac{7}{2}$	1	21	

Pivot column (↑ under x)

$\xrightarrow{\frac{2}{5}R_1}$

x	y	u	v	P	Constant
①	0	$\frac{2}{5}$	$-\frac{1}{5}$	0	2
$\frac{1}{2}$	1	0	$\frac{1}{2}$	0	3
$-\frac{1}{2}$	0	0	$\frac{7}{2}$	1	21

$\xrightarrow[R_3+\frac{1}{2}R_1]{R_2-\frac{1}{2}R_1}$

x	y	u	v	P	Constant
1	0	$\frac{2}{5}$	$-\frac{1}{5}$	0	2
0	1	$-\frac{1}{5}$	$\frac{3}{5}$	0	2
0	0	$\frac{1}{5}$	$\frac{17}{5}$	1	22

We see that $x = 2$, $y = 2$, $u = 0$, $v = 0$, and $C = -P = -22$.

9. The solution to the primal problem is given by $x = 2$, $y = 1$, and $C = 9$. The solution to the dual problem is given by $u = \frac{3}{10}$, $v = \frac{11}{10}$, and $P = 9$.

11. We write a tableau for the primal problem, then interchange rows and columns to obtain a tableau from which we construct the dual problem:

x	y	Constant
2	3	6
2	1	4
3	2	

u	v	Constant
2	2	3
3	1	2
6	4	

Maximize $P = 6u + 4v$

subject to $2u + 2v \le 3$

$3u + v \le 2$

$u \ge 0, v \ge 0$

Using the simplex method with x and y as slack variables, we obtain the following tableaus:

	u	v	x	y	P	Constant	Ratio	
	2	2	1	0	0	3	$\frac{3}{2}$	$\xrightarrow{\frac{1}{3}R_2}$
Pivot row →	③	1	0	1	0	2	$\frac{2}{3}$	
	−6	−4	0	0	1	0		

↑ Pivot column

	u	v	x	y	P	Constant	
	2	2	1	0	0	3	$\xrightarrow[R_3 + 6R_2]{R_1 - 2R_2}$
	①	$\frac{1}{3}$	0	$\frac{1}{3}$	0	$\frac{2}{3}$	
	−6	−4	0	0	1	0	

	u	v	x	y	P	Constant	Ratio	
Pivot row →	0	$\frac{4}{3}$	1	$-\frac{2}{3}$	0	$\frac{5}{3}$	$\frac{5}{4}$	$\xrightarrow{\frac{3}{4}R_1}$
	1	$\frac{1}{3}$	0	$\frac{1}{3}$	0	$\frac{2}{3}$	2	
	0	−2	0	2	1	4		

↑ Pivot column

	u	v	x	y	P	Constant	
	0	①	$\frac{3}{4}$	$-\frac{1}{2}$	0	$\frac{5}{4}$	$\xrightarrow[R_3 + 2R_1]{R_2 - \frac{1}{3}R_1}$
	1	$\frac{1}{3}$	0	$\frac{1}{3}$	0	$\frac{2}{3}$	
	0	−2	0	2	1	4	

u	v	x	y	P	Constant
0	1	$\frac{3}{4}$	$-\frac{1}{2}$	0	$\frac{5}{4}$
1	0	$-\frac{1}{4}$	$\frac{1}{2}$	0	$\frac{1}{4}$
0	0	$\frac{3}{2}$	1	1	$\frac{13}{2}$

Therefore, C attains a minimum value of $\frac{13}{2}$ when $x = \frac{3}{2}$, $y = 1$, $u = 0$, and $v = 0$.

13. We write a tableau for the primal problem, then interchange rows and columns to obtain a tableau from which we construct the dual problem:

x	y	z	Constant
3	2	1	4
1	1	3	6
24	18	24	

u	v	Constant
3	1	24
2	1	18
1	3	24
4	6	

Maximize $P = 4u + 6v$

subject to $3u + v \le 24$

$2u + v \le 18$

$u + 3v \le 24$

$u \ge 0, v \ge 0$

Using the simplex method with x and y as slack variables, we obtain the following tableaus:

	u	v	x	y	z	P	Constant	Ratio
	3	1	1	0	0	0	24	36
	2	1	0	1	0	0	18	18
Pivot row $\rightarrow$	1	③	0	0	1	0	24	8
	-4	-6	0	0	0	1	0	

$\xrightarrow{\frac{1}{3}R_3}$

↑ Pivot column

	u	v	x	y	z	P	Constant
	3	1	1	0	0	0	24
	2	1	0	1	0	0	18
	$\frac{1}{3}$	①	0	0	$\frac{1}{3}$	0	8
	-4	-6	0	0	0	1	0

$\xrightarrow[\substack{R_2 - R_3 \\ R_4 + 6R_3}]{R_1 - R_3}$

	u	v	x	y	z	P	Constant	Ratio
Pivot row $\rightarrow$	⑧⁄₃	0	1	0	$-\frac{1}{3}$	0	16	6
	$\frac{5}{3}$	0	0	1	$-\frac{1}{3}$	0	10	6
	$\frac{1}{3}$	1	0	0	$\frac{1}{3}$	0	8	24
	-2	0	0	0	2	1	48	

$\xrightarrow{\frac{3}{8}R_1}$

↑ Pivot column

	u	v	x	y	z	P	Constant
	①	0	$\frac{3}{8}$	0	$-\frac{1}{8}$	0	6
	$\frac{5}{3}$	0	0	1	$-\frac{1}{3}$	0	10
	$\frac{1}{3}$	1	0	0	$\frac{1}{3}$	0	8
	-2	0	0	0	2	1	48

$\xrightarrow[\substack{R_3 - \frac{1}{3}R_1 \\ R_4 + 2R_1}]{R_2 - \frac{5}{3}R_1}$

u	v	x	y	z	P	Constant
1	0	$\frac{3}{8}$	0	$-\frac{1}{8}$	0	6
0	0	$-\frac{5}{8}$	1	$-\frac{1}{8}$	0	0
0	1	$-\frac{1}{8}$	0	$\frac{3}{8}$	0	6
0	0	$\frac{3}{4}$	0	$\frac{7}{4}$	1	60

We conclude that C attains a minimum value of 60 when $x = \frac{3}{4}$, $y = 0$, $z = \frac{7}{4}$, $u = 0$, and $v = 0$.

15. Rewriting the problem, we have the following:

$$\begin{aligned} \text{Maximize} \quad & P = 3x - 4y \\ \text{subject to} \quad & x + y \leq 45 \\ & -x + 2y \leq -10 \\ & x \geq 0, y \geq 0 \end{aligned}$$

Using the simplex method, we calculate the following tableaus.

	x	y	u	v	P	Constant	Ratio
Pivot row $\rightarrow$	1	1	1	0	0	45	45
	⊖1	2	0	1	0	-10	10
	-3	4	0	0	1	0	

$\xrightarrow{-R_2}$

↑ Pivot column

	x	y	u	v	P	Constant
	1	1	1	0	0	45
	①	-2	0	-1	0	10
	-3	4	0	0	1	0

$\xrightarrow[R_3 + 3R_2]{R_1 - R_2}$

x	y	u	v	P	Constant
0	3	1	1	0	35
1	-2	0	-1	0	10
0	-2	0	-3	1	30

$\xrightarrow[R_3 + 3R_1]{R_2 + R_1}$

x	y	u	v	P	Constant
0	3	1	1	0	35
1	1	1	0	0	45
0	7	3	0	1	135

We conclude that P attains a maximum value of 135 when $x = 45$, $y = 0$, $u = 0$, and $v = 35$.

17. We first rewrite the problem as follows:

$$\text{Maximize} \quad P = 2x + 3y$$
$$\text{subject to} \quad 2x + 5y \le 20$$
$$x - 5y \le -5$$
$$x \ge 0, y \ge 0$$

Using the simplex method, we calculate the following tableaus.

x	y	u	v	P	Constant
2	5	1	0	0	20
1	−5	0	1	0	−5
−2	−3	0	0	1	0

$\xrightarrow{\frac{1}{5}R_1}$

x	y	u	v	P	Constant
$\frac{2}{5}$	1	$\frac{1}{5}$	0	0	4
1	−5	0	1	0	−5
−2	−3	0	0	1	0

$\xrightarrow[R_3 + 3R_1]{R_2 + 5R_1}$

x	y	u	v	P	Constant
$\frac{2}{5}$	1	$\frac{1}{5}$	0	0	4
3	0	1	1	0	15
$-\frac{4}{5}$	0	$\frac{3}{5}$	0	1	12

$\xrightarrow{\frac{1}{3}R_2}$

x	y	u	v	P	Constant
$\frac{2}{5}$	1	$\frac{1}{5}$	0	0	4
1	0	$\frac{1}{3}$	$\frac{1}{3}$	0	5
$-\frac{4}{5}$	0	$\frac{3}{5}$	0	1	12

$\xrightarrow[R_3 + \frac{4}{5}R_2]{R_1 - \frac{2}{5}R_2}$

x	y	u	v	P	Constant
0	1	$\frac{1}{15}$	$-\frac{2}{15}$	0	2
1	0	$\frac{1}{3}$	$\frac{1}{3}$	0	5
0	0	$\frac{13}{15}$	$\frac{4}{15}$	1	16

We conclude that P attains a maximum value of 16 when $x = 5$, $y = 2$, $u = 0$, $v = 0$, and $P = 16$.

19. Refer to Exercise 3.2.11 on page 80 of this manual. The problem simplifies to the following:

$$\text{Minimize} \quad C = 14{,}000x + 16{,}000y$$
$$\text{subject to} \quad 2x + 3y \ge 26$$
$$3x + y \ge 18$$
$$x \ge 0, y \ge 0$$

We write a tableau for the primal problem, then interchange rows and columns to obtain a tableau from which we construct the dual problem:

x	y	Constant
2	3	26
3	1	18
14,000	16,000	

u	v	Constant
2	3	14,000
3	1	16,000
26	18	

$$\text{Maximize} \quad P = 26u + 18v$$
$$\text{subject to} \quad 2u + 3v \le 14{,}000$$
$$3u + v \le 16{,}000$$
$$u \ge 0, v \ge 0$$

Using the simplex method with x and y as slack variables, we obtain the following tableaus:

	u	v	x	y	P	Constant	Ratio
	2	3	1	0	0	14,000	7000
Pivot row →	③	1	0	1	0	16,000	$\frac{16{,}000}{3}$
	−26	−18	0	0	1	0	

↑ Pivot column

$\xrightarrow{\frac{1}{3}R_2}$

u	v	x	y	P	Constant
2	3	1	0	0	14,000
1	$\frac{1}{3}$	0	$\frac{1}{3}$	0	$\frac{16{,}000}{3}$
−26	−18	0	0	1	0

$\xrightarrow[R_3 + 26R_2]{R_1 - R_2}$

	u	v	x	y	P	Constant	Ratio
Pivot row →	0	$\left(\frac{7}{3}\right)$	1	$-\frac{2}{3}$	0	$\frac{10,000}{3}$	$\frac{10,000}{7}$
	1	$\frac{1}{3}$	0	$\frac{1}{3}$	0	$\frac{16,000}{3}$	16,000
	0	$-\frac{28}{3}$	0	$\frac{26}{3}$	1	$\frac{41,600}{3}$	

$\xrightarrow{\frac{3}{7}R_1}$

Pivot column

	u	v	x	y	P	Constant
	0	1	$\frac{3}{7}$	$-\frac{2}{7}$	0	$\frac{10,000}{7}$
	1	$\frac{1}{3}$	0	$\frac{1}{3}$	0	$\frac{16,000}{3}$
	0	$-\frac{28}{3}$	0	$\frac{26}{3}$	1	$\frac{41,600}{3}$

$\xrightarrow{\substack{R_2 - \frac{1}{3}R_1 \\ R_3 - \frac{28}{3}R_1}}$

Pivot column

u	v	x	y	P	Constant
0	1	$\frac{3}{7}$	$-\frac{2}{7}$	0	$\frac{10,000}{7}$
1	0	$-\frac{1}{7}$	$\frac{3}{7}$	0	$\frac{34,000}{7}$
0	0	4	6	1	152,000

The last tableau is in final form, and the Fundamental Theorem of Duality tells us that the solution to the primal problem is $x = 4$, $y = 6$, and $C = 152,000$. So, the Saddle Mine should be operated for 4 days and the Horseshoe Mine should be operated for 6 days at a minimum cost of \$152,000/day.

21. Let x and y denote the numbers of millions of gallons of water from the local reservoir and the pipeline. Then we have the following nonstandard linear programming problem:

$$\text{Minimize} \quad C = 300x + 500y$$
$$\text{subject to} \quad x + y \geq 10$$
$$x \leq 5$$
$$y \geq 6$$
$$y \leq 10$$
$$x \geq 0, y \geq 0$$

We rewrite the problem as a maximization problem with constraints bounded above:

$$\text{Maximize} \quad P = -C = -300x - 500y$$
$$\text{subject to} \quad -x - y \leq -10$$
$$x \leq 5$$
$$-y \leq -6$$
$$y \leq 10$$
$$x \geq 0, y \geq 0$$

Introducing slack variables u, v, w, and z, we calculate the following sequence of tableaus:

	x	y	u	v	w	z	P	Constant	Ratio
	-1	-1	1	0	0	0	0	-10	10
	-1	0	0	1	0	0	0	5	-
Pivot row →	0	$\left(-1\right)$	0	0	1	0	0	-6	-6
	0	1	0	0	0	1	0	10	10
	300	500	0	0	0	0	1	0	

$\xrightarrow{-R_3}$

Pivot column

x	y	u	v	w	z	P	Constant
-1	-1	1	0	0	0	0	-10
-1	0	0	1	0	0	0	5
0	$\left(1\right)$	0	0	-1	0	0	6
0	1	0	0	0	1	0	10
300	500	0	0	0	0	1	0

$\xrightarrow{\substack{R_1 + R_3 \\ R_4 - R_3 \\ R_5 - 500R_3}}$

	x	y	u	v	w	z	P	Constant	Ratio
Pivot row →	(-1)	0	1	0	-1	0	0	-4	4
	-1	0	0	1	0	0	0	5	-
	0	1	0	0	-1	0	0	6	-
	0	0	0	0	1	1	0	4	-
	300	0	0	0	500	0	1	-3000	

↑ Pivot column

$\xrightarrow{-R_1}$

x	y	u	v	w	z	P	Constant
(1)	0	-1	0	1	0	0	4
-1	0	0	1	0	0	0	5
0	1	0	0	-1	0	0	6
0	0	0	0	1	1	0	4
300	0	0	0	500	0	1	-3000

$\xrightarrow[R_5 - 300R_1]{R_2 + R_1}$

x	y	u	v	w	z	P	Constant
1	0	-1	0	1	0	0	4
0	0	-1	1	1	0	0	9
0	1	0	0	-1	0	0	6
0	0	0	0	1	1	0	4
0	0	300	0	200	0	1	-4200

We see that $x = 4$, $y = 6$, and $C = -P = 4200$, so 4 million gallons should be obtained from the reservoirs and 6 million gallons from the pipeline, at a minimum cost of $4200.

23. Let x, y, and z denote the number of units made of Products A, B, and C, respectively. Then the problem is:

$$\text{Maximize} \quad P = 4x + 6y + 8z$$
$$\text{subject to} \quad 9x + 12y + 18z \le 360$$
$$6x + 6y + 10z \le 240$$
$$x \ge 0, y \ge 0, z \ge 0$$

Using the simplex method, we obtain the following tableaus:

	x	y	z	u	v	P	Constant	Ratio
Pivot row →	9	12	(18)	1	0	0	360	20
	6	6	10	0	1	0	240	24
	-4	-6	-8	0	0	1	0	

↑ Pivot column

$\xrightarrow{\frac{1}{18}R_1}$

x	y	z	u	v	P	Constant
$\frac{1}{2}$	$\frac{2}{3}$	1	$\frac{1}{18}$	0	0	20
6	6	10	0	1	0	240
-4	-6	-8	0	0	1	0

$\xrightarrow[R_3 + 8R_1]{R_2 - 10R_1}$

	x	y	z	u	v	P	Constant	Ratio
Pivot row →	$\frac{1}{2}$	($\frac{2}{3}$)	1	$\frac{1}{18}$	0	0	20	30
	1	$-\frac{2}{3}$	0	$-\frac{5}{9}$	1	0	40	-
	0	$-\frac{2}{3}$	0	$\frac{4}{9}$	0	1	160	

↑ Pivot column

$\xrightarrow{\frac{3}{2}R_1}$

x	y	z	u	v	P	Constant
$\frac{3}{4}$	1	$\frac{3}{2}$	$\frac{1}{12}$	0	0	30
1	$-\frac{2}{3}$	0	$-\frac{5}{9}$	1	0	40
0	$-\frac{2}{3}$	0	$\frac{4}{9}$	0	1	160

$\xrightarrow[R_3 + \frac{2}{3}R_1]{R_2 + \frac{2}{3}R_1}$

x	y	z	u	v	P	Constant
$\frac{3}{4}$	1	$\frac{3}{2}$	$\frac{1}{12}$	0	0	30
$\frac{3}{2}$	0	1	$-\frac{1}{2}$	1	0	60
$\frac{1}{2}$	0	1	$\frac{1}{2}$	0	1	180

and conclude that the company should produce 30 units of Product B and none of Products A or C to realize a maximum profit of $180.

CHAPTER 4   Before Moving On...   page 278

1. We introduce slack variables u, v, and w.

	x	y	z	u	v	w	P	Constant	Ratio
Pivot row →	2	①	-1	1	0	0	0	3	3
	1	-2	3	0	1	0	0	1	–
	3	2	4	0	0	1	0	17	$\frac{17}{2}$
	-1	-2	3	0	0	0	1	0	

Pivot column (under y).

The pivot element is 1, as shown.

2. $x = 2$, $y = 0$, $z = 11$, $u = 2$, $v = w = 0$, and $P = 28$.

3. We introduce slack variables u and v.

	x	y	u	v	P	Constant	Ratio
	4	3	1	0	0	30	$\frac{15}{2}$
Pivot row →	②	-3	0	1	0	6	3
	-5	-2	0	0	1	0	

Pivot column (under x).

$\xrightarrow{\frac{1}{2}R_2}$

	x	y	u	v	P	Constant
	4	3	1	0	0	30
	1	$-\frac{3}{2}$	0	$\frac{1}{2}$	0	3
	-5	-2	0	0	1	0

$\xrightarrow{\begin{array}{c}R_1 - 4R_2 \\ R_3 + 5R_2\end{array}}$

	x	y	u	v	P	Constant	Ratio
Pivot row →	0	⑨	1	-2	0	18	2
	1	$-\frac{3}{2}$	0	$\frac{1}{2}$	0	3	–
	0	$-\frac{19}{2}$	0	$\frac{5}{2}$	1	15	

Pivot column (under y).

$\xrightarrow{\frac{1}{9}R_1}$

	x	y	u	v	P	Constant
	0	1	$\frac{1}{9}$	$-\frac{2}{9}$	0	2
	1	$-\frac{3}{2}$	0	$\frac{1}{2}$	0	3
	0	$-\frac{19}{2}$	0	$\frac{5}{2}$	1	15

$\xrightarrow{\begin{array}{c}R_2 + \frac{3}{2}R_1 \\ R_3 + \frac{19}{2}R_1\end{array}}$

x	y	u	v	P	Constant
0	1	$\frac{1}{9}$	$-\frac{2}{9}$	0	2
1	0	$\frac{1}{6}$	$\frac{1}{6}$	0	6
0	0	$\frac{19}{8}$	$\frac{7}{18}$	1	34

The optimal solution is $x = 6$, $y = 2$, $u = v = 0$, and $P = 34$.

4. We write a tableau for the primal problem, then interchange rows and columns to obtain a tableau from which we construct the dual problem:

x	y	Constant
1	1	3
2	3	6
1	2	

u	v	Constant
1	2	1
1	3	2
3	6	

Maximize $\quad P = -C = 3u + 6v$

subject to $\quad u + 2v \le 1$

$\qquad\qquad u + 3v \le 2$

$\qquad\qquad u \ge 0, v \ge 0$

Using the simplex method with x and y as slack variables, we obtain the following tableaus:

	u	v	x	y	P	Constant	Ratio
Pivot row $\to$	1	②	1	0	0	1	$\frac{1}{2}$
	1	3	0	1	0	2	$\frac{2}{3}$
	-3	-6	0	0	1	0	

$\xrightarrow{\ \frac{1}{2}R_1\ }$

$\uparrow$ Pivot column

u	v	x	y	P	Constant
$\frac{1}{2}$	1	$\frac{1}{2}$	0	0	$\frac{1}{2}$
1	3	0	1	0	2
-3	-6	0	0	1	0

$\xrightarrow[R_3 + 6R_1]{R_2 - 3R_1}$

u	v	x	y	P	Constant
$\frac{1}{2}$	1	$\frac{1}{2}$	0	0	$\frac{1}{2}$
$-\frac{1}{2}$	0	$-\frac{3}{2}$	1	0	$\frac{1}{2}$
0	0	3	0	1	3

Hence, the solution to the primal problem is $x = 3$, $y = 0$, $u = 0$, $v = \frac{1}{2}$, and $C = 3$.

5. We rewrite the problem:

Maximize $\quad P = 2x + y$

subject to $\quad 2x + 5y \le 20$

$\qquad\qquad -4x - 3y \le -16$

$\qquad\qquad x \ge 0, y \ge 0$

Using the simplex method with u and v as slack variables, we obtain the following tableaus:

	x	y	u	v	P	Constant	Ratio
	2	5	1	0	0	20	10
Pivot row $\to$	⊝4	-3	0	1	0	-11	$\frac{11}{4}$
	-2	-1	0	0	1	0	

$\xrightarrow{\ -\frac{1}{4}R_2\ }$

$\uparrow$ Pivot column

x	y	u	v	P	Constant
2	5	1	0	0	20
1	$\frac{3}{4}$	0	$-\frac{1}{4}$	0	$\frac{11}{4}$
-2	-1	0	0	1	0

$\xrightarrow[R_3 + 2R_2]{R_1 - 2R_2}$

	x	y	u	v	P	Constant	Ratio
Pivot row $\to$	0	$\frac{7}{2}$	1	①/②	0	$\frac{29}{2}$	29
	1	$\frac{3}{4}$	0	$-\frac{1}{4}$	0	$\frac{11}{4}$	—
	0	$\frac{1}{2}$	0	$-\frac{1}{2}$	1	$\frac{11}{2}$	

$\xrightarrow{\ 2R_1\ }$

$\uparrow$ Pivot column

x	y	u	v	P	Constant
0	7	2	1	0	29
1	$\frac{3}{4}$	0	$-\frac{1}{4}$	0	$\frac{11}{4}$
0	$\frac{1}{2}$	0	$-\frac{1}{2}$	1	$\frac{11}{2}$

$\xrightarrow[R_3 + \frac{1}{2}R_1]{R_2 + \frac{1}{4}R_1}$

x	y	u	v	P	Constant
0	7	2	1	0	29
1	$\frac{5}{2}$	$\frac{1}{2}$	0	0	10
0	4	1	0	1	20

Thus, the optimal solution is $x = 10$, $y = 0$, $u = 0$, $v = 29$, and $P = 20$.

5 MATHEMATICS OF FINANCE

5.1 Compound Interest

Problem-Solving Tips

In this section, you encountered several formulas for computing interest. As you work through the exercises that follow, first decide which formula you need to solve the problem. Then write out your solution. After doing this a few times, you should have the formulas memorized. The key here is to try not to look at the formula in the text, and to work the problem just as if you were taking a test. If you train yourself to work in this manner, test-taking will be a lot easier.

1. First decide if the problem involves *simple interest* or *compound interest*. This will be stated in the problem.

2. Determine whether the problem is asking for the *present value* or *future value* of an amount. For example, if you are asked to determine the value of an investment 5 years from now with interest compounded each year, then use a compound interest formula giving the accumulated amount. If you are asked to determine the current value of an investment that will have a value of $50,000 five years from now with interest compounded each year, then use a present value formula for compound interest. (If interest is compounded continuously, it will be stated in the problem.)

3. The effective rate of interest is the same as the APR rate that you see in advertisements involving loans. Because the interest for different loans may be compounded over different periods (daily, monthly, biannually, annually, or otherwise) it provides the consumer with a way to compare rates. It is the simple interest rate that would produce the same accumulated amount in 1 year as the nominal rate compounded m times per year.

Concept Questions   page 294

1. In simple interest, the interest is based on the original principal. In compound interest, interest earned is periodically added to the principal and thereafter earns interest at the same rate.

3. The effective rate of interest is the simple interest that would produce the same amount in 1 year as the nominal rate compounded m times per year.

Exercises   page 294

1. The interest is given by $I = (500)(2)(0.08) = 80$, or $80. The accumulated amount is $500 + 80$, or $580.

3. The interest is given by $I = (800)(0.06)(0.75) = 36$, or $36. The accumulated amount is $800 + 36$, or $836.

5. We are given that $A = 1160$, $t = 2$, and $r = 0.08$, and we are asked to find P. Because $A = P(1 + rt)$, we see that
$$P = \frac{A}{1 + rt} = \frac{1160}{1 + (0.08)(2)} = 1000, \text{ or } \$1000.$$

153

7. We use the formula $I = Prt$ and solve for t when $I = 20$, $P = 1000$, and $r = 0.025$. Thus,
   $20 = 1000\,(0.025)\left(\frac{t}{365}\right)$, and $t = \frac{365(20)}{25} = 292$, or 292 days.

9. We use the formula $A = P\,(1 + rt)$ with $A = 1075$, $P = 1000$, and $t = 0.75$, and solve for r. Thus,
   $1075 = 1000\,(1 + 0.75r)$, $75 = 750r$, and so $r = 0.10$. Therefore, the annual interest rate is 10%.

11. $A = 1000\,(1 + 0.04)^8 \approx 1368.57$, or $1368.57.   **13.** $A = 2500\left(1 + \frac{0.04}{2}\right)^{20} \approx 3714.87$, or $3714.87.

15. $A = 12{,}000\left(1 + \frac{0.05}{4}\right)^{42} \approx 20{,}219.60$, or $20,219.60.

17. $A = 150{,}000\left(1 + \frac{0.04}{12}\right)^{48} \approx 175{,}979.80$, or $175,979.80.

19. $A = 150{,}000\left(1 + \frac{0.09}{365}\right)^{1095} \approx 196{,}488.13$, or $196,488.13.

21. Using the formula $r_{\text{eff}} = \left(1 + \frac{r}{m}\right)^m - 1$ with $r = 0.06$ and $m = 2$, we have $r_{\text{eff}} = \left(1 + \frac{0.06}{2}\right)^2 - 1 = 0.0609$, or
    6.09% annually.

23. Using the formula $r_{\text{eff}} = \left(1 + \frac{r}{m}\right)^m - 1$ with $r = 0.04$ and $m = 12$, we have $r_{\text{eff}} = \left(1 + \frac{0.04}{12}\right)^{12} - 1 \approx 0.04074$,
    or 4.074% annually.

25. The present value is given by $P = 40{,}000\left(1 + \frac{0.04}{2}\right)^{-8} \approx 34{,}139.61$, or $34,139.61.

27. The present value is given by $P = 40{,}000\left(1 + \frac{0.03}{12}\right)^{-48} \approx 35{,}482.13$, or $35,482.13.

29. $A = 5000e^{0.06(4)} \approx 6356.25$, or approximately $6356.25.

31. Think of $300 as the principal and $306 as the accumulated amount at the end of 30 days. If r denotes the simple
    interest rate per annum, then we have $P = 300$, $A = 306$, and $t = \frac{1}{12}$, and we are required to find r. Using
    Equation 1(b), we have $306 = 300\left(1 + \frac{r}{12}\right) = 300 + r\left(\frac{300}{12}\right)$ and $r = \left(\frac{12}{300}\right)6 = 0.24$, or 24% annually.

33. The Abdullahs will owe $A = P\,(1 + rt) = 120{,}000\left[1 + (0.10)\left(\frac{3}{12}\right)\right] = 123{,}000$, or $123,000.

35. Here $P = 10{,}000$, $I = 3500$, and $t = 7$, and so from Formula 1(a), we have $3500 = 10{,}000\,(r)\,7$, and so
    $r = \frac{3500}{70{,}000} = 0.05$. Thus, the bond pays annual simple interest of 5%.

37. Using Equation 1(b) with $A = 15{,}000$, $P = 14{,}650$, and $t = \frac{52}{52} = 1$, we have $15{,}000 = 14{,}650\,(1 + r)$, so
    $r = \frac{15{,}000}{14{,}650} - 1 \approx 0.0239$. Thus, Maxwell's investment will earn simple interest at a rate of approximately 2.39%
    annually.

39. The value of Alan's stock portfolio after 1 year is $(1.2)\,P$ dollars, where P is the original amount invested.
    Its value after 2 years is $(1.1)\,(1.2)\,P$; after 3 years, it is $(0.9)\,(1.1)\,(1.2)\,P$; and finally after 4 years, it is
    $(0.8)\,(0.9)\,(1.1)\,(1.2)\,P$ or $0.9504P$ dollars. Thus, the value of Alan's stock portfolio after 4 years is less than its
    initial value.

41. Suppose Arabella's stock portfolio is worth $\$P$ initially. Then after 1 year, it is worth $(0.8) P$ dollars. Let r denote the annual rate (compounded annually) which the portfolio must earn in the second year in order to regain its original value at the end of the third year. Then $(1 + r)^2 (0.8) P = P$, so $(1 + r)^2 = \frac{1}{0.8}$, $1 + r = \sqrt{\frac{1}{0.8}} \approx 1.1180$, and $r \approx 0.1180$. The required rate is thus approximately 11.8% per year.

43. The rate that you would expect to pay is $A = 680 (1 + 0.08)^5 \approx 999.14$, or \$999.14 per day.

45. The amount that they can expect to pay is given by $A = 260,000 (1 + 0.05)^4 \approx 316,031.63$, or approximately \$316,032.

47. The investment will be worth $A = 1.5 \left(1 + \frac{0.025}{2}\right)^{20} \approx 1.92306$, or approximately \$1.92 million.

49. We use Formula 3 with $P = 15,000$, $r = 0.078$, $m = 12$, and $t = 4$, giving the worth of Jodie's account as $A = 15,000 \left(1 + \frac{0.078}{12}\right)^{(12)(4)} \approx 20,471.641$, or approximately \$20,471.64.

51. Using the formula $P = A \left(1 + \frac{r}{m}\right)^{-mt}$, we have $P = 40,000 \left(1 + \frac{0.035}{4}\right)^{-20} \approx 33,603.85$, or \$33,603.85.

53. a. They should set aside $P = 100,000 (1 + 0.045)^{-13} \approx 56,427.16$, or \$56,427.16.

   b. They should set aside $P = 100,000 \left(1 + \frac{0.045}{2}\right)^{-26} \approx 56,072.997$, or \$56,073.

   c. They should set aside $P = 100,000 \left(1 + \frac{0.045}{4}\right)^{-52} \approx 55,892.84$, or \$55,892.84.

55. The effective annual rate of interest for the Bendix Mutual Fund is $r_{\text{eff}} = \left(1 + \frac{0.064}{4}\right)^4 - 1 \approx 0.0656$, or 6.56%, whereas the rate for the Acme Mutual fund is $r_{\text{eff}} = \left(1 + \frac{0.065}{2}\right)^2 - 1 \approx 0.0661$, or 6.61%. We conclude that the Acme Mutual Fund has a better rate of return.

57. The present value of the \$8000 loan due in 3 years is given by $P = 8000 \left(1 + \frac{0.08}{2}\right)^{-6} \approx 6322.52$, or \$6322.52. The present value of the \$15,000 loan due in 6 years is given by $P = 15,000 \left(1 + \frac{0.08}{2}\right)^{-12} \approx 9368.96$, or \$9368.96. Therefore, the amount the proprietors of the inn will be required to pay at the end of 5 years is given by $A = 15,691.48 \left(1 + \frac{0.08}{2}\right)^{10} = 23,227.22$, or \$23,227.22.

59. Let $A = 10,000$, $r = 0.0525$, and $t = 10$. Using Formula 7, we have $P = 10,000 (1 + 0.0525)^{-10} \approx 5994.86$. Thus, Juan should pay \$5994.86 for the bond.

61. The projected online retail sales for 2009 were $1.243 (1.14) (1.305) (1.176) (1.105) (141.4) \approx 339.79$, or approximately \$339.79 billion.

63. Suppose \$1 is invested in each investment. For Investment A, the accumulated amount is $\left(1 + \frac{0.08}{2}\right)^8 \approx 1.36857$. For Investment B, the accumulated amount is $e^{0.078(4)} \approx 1.36615$. Thus, Investment A has a higher rate of return.

65. If they invest the money at 6.6% compounded quarterly, they should set aside

$$P = 120{,}000 \left(1 + \frac{0.066}{4}\right)^{-28} \approx 75{,}888.25, \text{ or } \$75{,}888.25.$$ If they invest the money at 6.6% compounded

continuously, they should set aside $P = 120{,}000e^{-7(0.066)} = 75{,}602.68$, or $\$75{,}602.68$.

67. $P(t) = V(t)e^{-rt} = 80{,}000e^{\sqrt{t}/2}e^{-rt} = 80{,}000e^{(\sqrt{t}/2 - 0.05t)}$, so $P(4) = 80{,}000e^{1-0.05(4)} \approx 178{,}043.27$, or approximately $\$178{,}043$.

69. By definition, $A = P(1 + r_{\text{eff}})^t$, so $(1 + r_{\text{eff}})^t = \frac{A}{P}$, $1 + r_{\text{eff}} = \left(\frac{A}{P}\right)^{1/t}$, and $r_{\text{eff}} = \left(\frac{A}{P}\right)^{1/t} - 1$.

71. Using the formula $r_{\text{eff}} = \left(\frac{A}{P}\right)^{1/t} - 1$ with $A = 366{,}000$, $P = 300{,}000$, and $t = 6$, we have

$$r_{\text{eff}} = \left(\frac{366{,}000}{300{,}000}\right)^{1/6} - 1 \approx 0.0337, \text{ or } 3.37\%.$$

73. Using the formula $r_{\text{eff}} = \left(\frac{A}{P}\right)^{1/t} - 1$ with $A = 10{,}000$, $P = 6{,}724.53$, and $t = 7$, we have

$$r_{\text{eff}} = \left(\frac{10{,}000}{6724.53}\right)^{1/7} - 1 \approx 0.0583, \text{ or } 5.83\%.$$

75. True. $A = P(1 + rt) = Prt$ is a linear function of t.

77. True. With $m = 1$, the effective rate is $r_{\text{eff}} = \left(1 + \frac{r}{1}\right)^1 - 1 = r$.

79. We use Formula 3 with $A = 6500$, $P = 5000$, $m = 12$, and $r = 0.06$. Thus, $6500 = 5000\left(1 + \frac{0.06}{12}\right)^{12t}$, so $(1.05)^{12t} = \frac{6500}{5000} = 1.3$, $12t \ln 1.05 = \ln 1.3$, and $t = \frac{\ln 1.3}{12 \ln 1.05} \approx 4.384$. Therefore, it will take approximately 4.4 years.

81. We use Formula 3 with $A = 4000$, $P = 2000$, $m = 12$, and $r = 0.05$. Thus, $4000 = 2000\left(1 + \frac{0.05}{12}\right)^{12t}$,

$\left(1 + \frac{0.05}{12}\right)^{12t} = 2$, $12t \ln\left(1 + \frac{0.05}{12}\right) = \ln 2$, and $t = \dfrac{\ln 2}{12 \ln\left(1 + \frac{0.05}{12}\right)} \approx 13.89$. Therefore, it will take

approximately 13.9 years.

83. We use Formula 5 with $A = 6000$, $P = 5000$, and $t = 3$. Thus, $6000 = 5000e^{3r}$, and so $e^{3r} = \frac{6000}{5000} = 1.2$. Next, taking logarithms of both sides of the equation, we have $3r = \ln 1.2$ (the natural logarithm of e^{3r} is $3r$), and so $r = \frac{\ln 1.2}{3} \approx 0.6077$. Therefore, the annual interest rate is 6.08%.

85. We use Formula 5 with $A = 7000$, $P = 6000$, and $r = 0.055$. Thus, $7000 = 6000e^{0.055t}$, giving $e^{0.055t} = \frac{7000}{6000} = \frac{7}{6}$. Taking logarithms of both sides and using the fact that the natural logarithm of $e^{0.055t}$ is $0.055t$, we have

$0.055t \ln e = \ln \frac{7}{6}$, so $t = \dfrac{\ln \frac{7}{6}}{0.055} \approx 2.803$. Therefore, it will take 2.8 years.

Technology Exercises   page 300

1. $5872.78   **3.** $475.49   **5.** 8.95%/yr   **7.** 10.20%/yr

9. $29,743.30   **11.** $53,303.25

5.2   Annuities

Problem-Solving Tips

1. Note the difference between annuities and the compound interest problems solved in Section 5.1. An annuity is a *sequence of payments* made at regular intervals. If we are asked to find the value of a sequence of payments at some future time, then we use the formula for the future value of an annuity: $S = R\left[\dfrac{(1+i)^n - 1}{i}\right]$. If we are asked to find the current value of a sequence of payments that will be made over a certain period of time, then we use the formula for the present value of an annuity: $P = R\left[\dfrac{1 - (1+i)^{-n}}{i}\right]$.

2. Note that the problems in this section deal with *ordinary annuities*—annuities in which the payments are made at the end of each payment period.

Concept Questions   page 308

1. In an ordinary annuity, the term is fixed, the periodic payments are of the same size, the payments are made at the end of the payment period and the payments coincide with the interest conversion periods.

3. The future value S of an annuity of n payments of R dollars each, paid at the end of each investment period into an account that earns interest at the rate of i per period, is $S = R\left[\dfrac{(1+i)^n - 1}{i}\right]$. One example is a retirement fund into which an employee makes a monthly deposit of a fixed amount for a certain period of time.

Exercises   page 308

1. $S = 1000\left[\dfrac{(1 + 0.05)^{10} - 1}{0.05}\right] \approx 12{,}577.89$, or \$12,577.89.

3. $S = 500\left[\dfrac{\left(1 + \frac{0.06}{2}\right)^{24} - 1}{\frac{0.06}{2}}\right] \approx 17{,}213.24$, or \$17,213.24.

5. $S = 600\left[\dfrac{\left(1 + \frac{0.05}{4}\right)^{36} - 1}{\frac{0.05}{4}}\right] \approx 27{,}069.30$, or \$27,069.30.

7. $S = 200\left[\dfrac{\left(1 + \frac{0.065}{12}\right)^{243} - 1}{\frac{0.065}{12}}\right] \approx 100{,}289.96$, or \$100,289.96.

9. $P = 5000\left[\dfrac{1 - (1 + 0.06)^{-8}}{0.06}\right] \approx 31{,}048.97$, or \$31,048.97.

11. $P = 1200\left[\dfrac{1 - \left(1 + \frac{0.05}{2}\right)^{-12}}{\frac{0.05}{2}}\right] \approx 12{,}309.32$, or \$12,309.32.

13. $P = 800 \left[\dfrac{1 - \left(1 + \frac{0.06}{4}\right)^{-28}}{\frac{0.06}{4}} \right] \approx 18{,}181.37$, or $18,181.37.

15. She will have $S = 1500 \left[\dfrac{(1 + 0.04)^{25} - 1}{0.04} \right] \approx 62{,}468.86$, or $62,468.86.

17. On October 31, Linda's account will be worth $S = 40 \left[\dfrac{\left(1 + \frac{0.025}{12}\right)^{11} - 1}{\frac{0.025}{12}} \right] \approx 444.61$, or $444.61. One month

later, this account will be worth $A \approx (444.61)\left(1 + \frac{0.025}{12}\right) \approx 445.54$, or $445.54.

19. To find how much Karen has at age 65, we use Formula 9 with $R = 150$, $i = \frac{r}{m} = \frac{0.04}{12}$, and

$n = mt = (12)(40) = 480$, giving $S = 150 \left[\dfrac{\left(1 + \frac{0.04}{12}\right)^{480} - 1}{\frac{0.04}{12}} \right] \approx 177{,}294.20$, or $177,294.20.

To find how much Matt will have upon attaining the age of 65, we use Formula 9 with $R = 250$, $i = \frac{r}{m} = \frac{0.04}{12}$, and

$n = mt = (12)(30) = 360$, giving $S = 250 \left[\dfrac{\left(1 + \frac{0.04}{12}\right)^{360} - 1}{\frac{0.04}{12}} \right] \approx 173{,}512.35$, or $173,512.35. Therefore,

Karen will have the bigger nest egg.

21. The sum of $150,000 will grow to $S_1 = 150{,}000 \left(1 + \frac{0.045}{4}\right)^{(4)(20)} \approx 367{,}091.25$, or $367,091.25. The annuity will

grow to $S_2 = R \left[\dfrac{(1 + i)^{mt} - 1}{i} \right] = 3000 \left[\dfrac{\left(1 + \frac{0.045}{4}\right)^{80} - 1}{\frac{0.045}{4}} \right] \approx 385{,}939.99$, or $385,939.99. So at the time of

his retirement, Luis will have $367{,}091.25 + 385{,}939.99 \approx 753{,}031.24$, or $753,031.24, in his retirement account.

23. At the end of the third year, the Piererras will have $S = 150 \left[\dfrac{\left(1 + \frac{0.03}{12}\right)^{36} - 1}{\frac{0.03}{12}} \right] \approx 5643.08$, or $5643.08.

25. We use the formula for the present value of an annuity, obtaining $P = 420 \left[\dfrac{1 - \left(1 + \frac{0.06}{12}\right)^{-36}}{\frac{0.06}{12}} \right] \approx 13{,}805.83$, or

$13,805.83. Because her down payment was $8000, the cash price of the car was $13{,}805.83 + 8000 = 21{,}805.83$, or $21,805.83.

27. With an $2400 monthly payment, the present value of their loan would be

$$P = 2400 \left[\frac{1 - \left(1 + \frac{0.055}{12}\right)^{-360}}{\frac{0.055}{12}} \right] \approx 422{,}692.23, \text{ or } \$422{,}692.23. \text{ With a } \$3000 \text{ monthly payment, the present}$$

value of their loan would be $P = 3000 \left[\dfrac{1 - \left(1 + \frac{0.055}{12}\right)^{-360}}{\frac{0.055}{12}} \right] \approx 528{,}365.29,$ or $528,365.29. Because they

intend to make a $40,000 down payment, they should consider homes priced from $462,692 to $568,365.29.

29. The lower limit of their house price range is $A = 2400 \left[\dfrac{1 - \left(1 + \frac{0.05}{12}\right)^{-180}}{\frac{0.05}{12}} \right] + 40{,}000 \approx 343{,}492.58,$

or approximately $343,493. The upper limit of their house price range is

$$A = 3000 \left[\frac{1 - \left(1 + \frac{0.05}{12}\right)^{-180}}{\frac{0.05}{12}} \right] + 40{,}000 \approx 419{,}365.73, \text{ or approximately } \$419{,}366. \text{ Therefore, they should}$$

consider homes priced from $343,493 to $419,366.

31. The deposits of $200/month into the bank account for a period of 2 years will grow to a sum of

$$A_1 = 200 \left[\frac{\left(1 + \frac{0.035}{12}\right)^{24} - 1}{\frac{0.035}{12}} \right] \approx 4964.50, \text{ or } \$4964.50. \text{ For the next 3 years, this amount will grow into a}$$

sum of $A_2 = A_1 \left(1 + \frac{0.035}{12}\right)^{36} = 4964.50 \left(1 + \frac{0.035}{12}\right)^{36} \approx 5513.28,$ or $5513.28. The deposits of $300/month

for a period of three years will grow to a sum of $A_3 = 300 \left[\dfrac{\left(1 + \frac{0.035}{12}\right)^{36} - 1}{\frac{0.035}{12}} \right] \approx 11{,}369.92,$ or $11,369.92.

Therefore, at the end of 5 years, he will have $A_2 + A_3 = 5513.28 + 11{,}369.92 \approx 16{,}883,$ or approximately $16,883.

33. Using Equation 9 with $R = 3000$, $r = 0.05$, $m = 1$, and $n = 10$, we see that the amount accumulated in Jacob's

account at the end of 10 years is $S = 3000 \left[\dfrac{(1 + 0.05)^{10} - 1}{0.05} \right] \approx 37{,}733.68,$ or $37,733.68. Then, using

Equation 3 with $P = 37{,}733.68$, $r = 0.05$, $m = 1$, and $n = 10$, we see that the amount accumulated in Jacob's

account at the end of 20 years is $S = 37{,}733.68 \left(1 + 0.05\right)^{10} \approx 61{,}464.19,$ or $61,464.19.

35. False. This statement is true only if the interest rate is zero.

Technology Exercises   page 312

1. $59,622.15   **3.** $8453.59   **5.** $35,607.23   **7.** $13,828.60

5.3 Amortization and Sinking Funds

Problem-Solving Tips

1. If a problem asks for the periodic payment that will amortize a loan over n periods, then use the amortization formula $R = \dfrac{Pi}{1 - (1 + i)^{-n}}$. For example, if you want to calculate the payment for a home mortgage, use this formula to find the payment.

2. If a problem asks for the periodic payment required to accumulate a certain sum of money over n periods, then use the formula $R = \dfrac{iS}{(1 + i)^n - 1}$. For example, if a businessman wants to set aside a certain sum of money through periodic payments for the purchase of new equipment, then use this formula to find the payment.

Concept Questions   page 320

1. $R = \dfrac{Pi}{1 - (1 + i)^{-n}}$.

   a. We rewrite $R = \dfrac{Pi}{1 - \frac{1}{(1+i)^n}}$. If n increases, then $(1 + i)^n$ increases and $\dfrac{1}{(1 + i)^n}$ decreases. Therefore $1 - \dfrac{1}{(1 + i)^n}$ increases, and so R decreases.

   b. If the principal and interest rate are fixed, and the number of payments is allowed to increase, then the size of each monthly payment decreases.

Exercises   page 320

1. The size of each installment is given by $R = \dfrac{100{,}000\,(0.06)}{1 - (1 + 0.06)^{-10}} \approx 13{,}586.80$, or \$13,586.80.

3. The size of each installment is given by $R = \dfrac{5000\,(0.01)}{1 - (1 + 0.01)^{-12}} \approx 444.24$, or \$444.24.

5. The size of each installment is given by $R = \dfrac{25{,}000\,(0.0075)}{1 - (1 + 0.0075)^{-48}} \approx 622.13$, or \$622.13.

7. The size of each installment is $R = \dfrac{80{,}000\left(\frac{0.055}{12}\right)}{1 - \left(1 + \frac{0.055}{12}\right)^{-360}} \approx 454.23$, or \$454.23.

9. The required periodic payment is $R = \dfrac{20{,}000\,(0.02)}{(1 + 0.02)^{12} - 1} \approx 1491.19$, or \$1491.19.

11. The required periodic payment is $R = \dfrac{100{,}000\,(0.0075)}{(1 + 0.0075)^{120} - 1} \approx 516.76$, or \$516.76.

13. The required periodic payment is $R = \dfrac{250{,}000\left(\frac{0.065}{12}\right)}{\left(1 + \frac{0.065}{12}\right)^{300} - 1} \approx 333.85$, or \$333.85.

15. The required periodic payment is $R = \dfrac{50{,}000\left(\frac{0.05}{4}\right)}{\left(1 + \frac{0.05}{4}\right)^{20} - 1} \approx 2216.02$, or \$2216.02.

17. The required periodic payment is $R = \dfrac{35{,}000\left(\frac{0.035}{2}\right)}{1 - \left(1 + \frac{0.035}{2}\right)^{-13}} \approx 3033.55$, or \$3033.55.

19. The size of each installment is given by $R = \dfrac{100{,}000\,(0.05)}{1 - (1 + 0.05)^{-10}} \approx 12{,}950.46$, or \$12,950.46.

21. The monthly payment in each case is given by $R = \dfrac{100{,}000\left(\frac{r}{12}\right)}{1 - \left(1 + \frac{r}{12}\right)^{-360}}$. Thus, if $r = 0.04$, then

$R = \dfrac{100{,}000\left(\frac{0.04}{12}\right)}{1 - \left(1 + \frac{0.04}{12}\right)^{-360}} \approx 477.42$, or \$477.42. If $r = 0.05$, then $R = \dfrac{100{,}000\left(\frac{0.05}{12}\right)}{1 - \left(1 + \frac{0.05}{12}\right)^{-360}} \approx 536.82$,

or \$536.82. If $r = 0.06$, then $R = \dfrac{100{,}000\left(\frac{0.06}{12}\right)}{1 - \left(1 + \frac{0.06}{12}\right)^{-360}} \approx 599.55$, or \$599.55. If $r = 0.07$, then

$R = \dfrac{100{,}000\left(\frac{0.07}{12}\right)}{1 - \left(1 + \frac{0.07}{12}\right)^{-360}} \approx 665.30$, or \$665.30.

a. The difference in monthly payments in the two loans is $665.30 - $421.60 = $243.70.

b. The monthly mortgage payment on a \$150,000 mortgage at 5% per year over 30 years would be 1.5 (536.82), or \$805.23. The monthly mortgage payment on a \$50,000 mortgage at 5% per year over 30 years would be 0.5 (536.82), or \$268.41.

23. a. The amount of the loan required is $20{,}000 - (0.25)\,(20{,}000)$ or \$15,000. If the car is financed over 36 months,

the monthly payments will be $R = \dfrac{15{,}000\left(\frac{0.06}{12}\right)}{1 - \left(1 + \frac{0.06}{12}\right)^{-36}} \approx 456.33$, or \$456.33. If the car is financed over

48 months, the monthly payments will be $R = \dfrac{15{,}000\left(\frac{0.06}{12}\right)}{1 - \left(1 + \frac{0.06}{12}\right)^{-48}} \approx 352.28$, or \$352.28.

b. The interest charges for the 36-month plan are $36\,(456.33) - 15{,}000 = 1427.88$, or \$1427.88. The interest charges for the 48-month plan are $48\,(352.28) - 15{,}000 = 1909.44$, or \$1909.44.

25. The amount borrowed is $270,000 - 30,000 = 240,000$, or $240,000. The size of the monthly installment

is $R = \dfrac{240,000 \left(\frac{0.06}{12} \right)}{1 - \left(1 + \frac{0.06}{12} \right)^{-360}} \approx 1438.92$, or $1438.92. To find their equity after five years, we compute

the present value of their remaining installments: $P = 1438.92 \left[\dfrac{1 - \left(1 + \frac{0.06}{12} \right)^{-300}}{\frac{0.06}{12}} \right] \approx 223,330.26$, or

$223,330.26, and so their equity is $270,000 - 223,330.26 = 46,669.74$, or $46,670. To find their equity

after ten years, we compute $P = 1438.92 \left[\dfrac{1 - \left(1 + \frac{0.06}{12} \right)^{-240}}{\frac{0.06}{12}} \right] \approx 200,845.56$, or $200,846, so their

equity is $270,000 - 200,846 = 69,154.44$, or $69,154. To find their equity after twenty years, we compute

$P = 1438.92 \left[\dfrac{1 - \left(1 + \frac{0.06}{12} \right)^{-120}}{\frac{0.06}{12}} \right] \approx 129,608.49$, or $129,608.49, and their equity is $270,000 - 129,608.49$, or

$140,392.

27. The amount that must be deposited annually into this fund is given by $R = \dfrac{(0.04)\,(2.5)}{(1 + 0.04)^{20} - 1} = 0.08395438$ million,

or approximately $83,954.38.

29. The amount that must be deposited quarterly into this fund is $R = \dfrac{\left(\frac{0.09}{4} \right) 200,000}{\left(1 + \frac{0.09}{4} \right)^{40} - 1} \approx 3,135.48$, or $3,135.48.

31. The size of each monthly installment is given by $R = \dfrac{\left(\frac{0.045}{12} \right) 250,000}{\left(1 + \frac{0.045}{12} \right)^{300} - 1} \approx 452.08$, or $452.08.

33. Here $S = 450,000$, $i = \frac{0.06}{12}$, and $n = mt = (12)\,(30) = 360$. Thus, Formula 9 gives

$450,000 = R \left[\dfrac{\left(1 + \frac{0.06}{12} \right)^{360} - 1}{\frac{0.06}{12}} \right]$, and so $R = \dfrac{450,000 \left(\frac{0.06}{12} \right)}{\left(1 + \frac{0.06}{12} \right)^{360} - 1} \approx 447.98$. Her monthly payment is $447.98.

35. The value of the IRA account after 20 years is $S = 375 \left[\dfrac{\left(1 + \frac{0.04}{4}\right)^{80} - 1}{\frac{0.04}{4}} \right] \approx 45{,}626.82$, or

$45,626.82. The payment he would receive at the end of each quarter for the next 15 years is given by

$R = \dfrac{\left(\frac{0.04}{4}\right) 45{,}626.82}{1 - \left(1 + \frac{0.04}{4}\right)^{-60}} \approx 1014.94$, or $1014.94. If he continues working and makes quarterly payments

until age 65, the value of the IRA account would be $S = 375 \left[\dfrac{\left(1 + \frac{0.04}{4}\right)^{100} - 1}{\frac{0.04}{4}} \right] \approx 63{,}930.52$, or

$63,930.52. The payment he would receive at the end of each quarter for the next 10 years is given by

$R = \dfrac{\left(\frac{0.04}{4}\right) 63{,}930.52}{1 - \left(1 + \frac{0.04}{4}\right)^{-40}} \approx 1947.04$, or $1947.04.

37. The amount of the sinking fund Jason needs to accumulate is found using Equation 10 with $R = 8000$,

$i = \dfrac{r}{m} = \dfrac{0.06}{12}$, and $n = 12 \cdot 20 = 240$. Thus, $S = \dfrac{8000 \left[1 - \left(1 + \frac{0.06}{12}\right)^{-240} \right]}{\frac{0.06}{12}} \approx 1{,}116{,}646.173$,

or $1,116,646.17. Next, using Equation 14 with $S = 1{,}116{,}646.17$, $i = \frac{0.06}{12}$, and $n = 360$, we obtain

$R = \dfrac{1{,}116{,}646.17 \left(\frac{0.06}{12}\right)}{\left(1 + \frac{0.06}{12}\right)^{360} - 1} \approx 1111.63$. Thus, he must contribute $1111.63 per month.

39. Using Equation 13 with $R = 400$, $i = \frac{0.042}{12}$, and $n = 48$, we have $P = 400 \left[\dfrac{1 - \left(1 + \frac{0.042}{12}\right)^{-48}}{\frac{0.042}{12}} \right] \approx 17{,}645.51$,

or $17,645.51. Because he can get $8000 for the trade-in, Dan can afford a car that costs no more than $25,645.51.

41. Kim's monthly payment is found using Formula 12 with $P = 180{,}000$, $i = \dfrac{r}{m} = \dfrac{0.045}{12}$, and $n = (12)(30) = 360$.

Thus, $R = 180{,}000 \left[\dfrac{\frac{0.045}{12}}{1 - \left(1 + \frac{0.045}{12}\right)^{-360}} \right] \approx 912.0336$. After 8 years, he has made $8 \cdot 12 = 96$ payments. His

outstanding principal is given by the sum of the remaining $360 - 96 = 264$ installments. Using Formula 10, we find

$P = 912.0336 \left[\dfrac{1 - \left(1 + \frac{0.045}{12}\right)^{-264}}{\frac{0.045}{12}} \right] \approx 152{,}670.69$, so his outstanding principal is $152,670.69.

43. As of now, Paul owes his sister $A = Pe^{rt} = 10{,}000e^{(0.04)(2)} \approx 10{,}832.87$, or $10,832.87. To repay the loan,

Paul's monthly payment will be $R = \dfrac{Pi}{1 - (1 + i)^{-n}} = \dfrac{10{,}832.87 \left(\frac{0.03}{12}\right)}{1 - \left(1 + \frac{0.03}{12}\right)^{-60}} \approx 194.65$, or approximately

$194.65 per month.

45. To find Emilio's monthly payment, we use Equation 12 with $P = 280,000$, $r = 0.045$, $m = 12$, and $t = 30$,

obtaining $R = \dfrac{280,000 \left(\frac{0.045}{12} \right)}{1 - \left(1 + \frac{0.045}{12} \right)^{-360}} \approx 1418.72$, or $1418.72 per month. After $7 \cdot 12 = 84$ payments have been

made, there are 276 payments remaining. The present value of an annuity with $n = 276$, $R = 1418.72$, and

$i = \frac{0.045}{12}$ is $P = 1418.72 \left[\dfrac{1 - \left(1 + \frac{0.045}{12} \right)^{-276}}{\frac{0.045}{12}} \right] \approx 243,673.79$, so Emilio's balloon payment is \$243,673.79.

47. a. Here $P = 200,000$, $i = \dfrac{r}{m} = \dfrac{0.065}{12}$, and $n = mt = 12 \cdot 30 = 360$. Therefore,

$R = \dfrac{200,000 \left(\frac{0.065}{12} \right)}{1 - \left(1 + \frac{0.065}{12} \right)^{-360}} \approx 1264.1360$, and so her monthly payment is \$1264.14.

b. After $4 \cdot 12 = 48$ monthly payments have been made, her outstanding principal is given by the sum of
the present values of the remaining $360 - 48 = 312$ installments. Using Formula 10, we find it to be

$P = 1264.136 \left[\dfrac{1 - \left(1 + \frac{0.065}{12} \right)^{-312}}{\frac{0.065}{12}} \right] \approx 190,119.14$, or approximately \$190,119.14.

c. Here $P = 190,119.14$, $i = \dfrac{r}{m} = \dfrac{0.0475}{12}$, and $n = 12 \cdot 30 = 360$. Thus, $R = \dfrac{190,119.14 \left(\frac{0.0475}{12} \right)}{1 - \left(1 + \frac{0.0475}{12} \right)^{-360}} \approx 991.75$,

and so her new monthly payment is \$991.75.

d. Emily will save $1264.14 - 991.75$, or \$272.39 per month.

49. The monthly payment the Sandersons are required to make under the terms of their original loan is given by

$R = \dfrac{100,000 \left(\frac{0.05}{12} \right)}{1 - \left(1 + \frac{0.05}{12} \right)^{-240}} \approx 659.96$, or \$659.96. Their monthly payment under the terms of their new loan is given

by $R = \dfrac{100,000 \left(\frac{0.042}{12} \right)}{1 - \left(1 + \frac{0.042}{12} \right)^{-240}} \approx 616.57$, or \$616.57. The amount of money that the Sandersons can expect to save

over the life of the loan by refinancing is given by $240 (659.96 - 616.57) = 10,413.60$, or \$10,413.60.

51. First we find Samantha's monthly payment on the original loan amount. Here $P = 150{,}000$, $i = \dfrac{r}{m} = \dfrac{0.055}{12}$, and

$n = mt = 12 \cdot 30 = 360$. Therefore, $R = \dfrac{150{,}000\left(\frac{0.055}{12}\right)}{1 - \left(1 + \frac{0.055}{12}\right)^{-360}} \approx 851.6835$. Next, to find her current outstanding

principal, observe that this is just the sum of the present values of the $360 - 36 = 324$ remaining payments. Using

Formula 10, we have $P = 851.6835 \left[\dfrac{1 - \left(1 + \frac{0.055}{12}\right)^{-324}}{\frac{0.055}{12}} \right] \approx 143{,}589.7288$. Finally, using Formula 13 with

$P = 143{,}589.7288$, $i = \dfrac{r}{m} = \dfrac{0.04}{12}$, and $n = mt = 12 \cdot 27 = 324$, we find $R = \dfrac{143{,}589.7288\left(\frac{0.04}{12}\right)}{1 - \left(1 + \frac{0.04}{12}\right)^{-324}} \approx 725.43$,

and so Samantha's new monthly payment will be $725.43 per month.

53. The amount of the loan the Meyers need to secure is $280,000. Using the bank's financing, the monthly payment

would be $R = \dfrac{280{,}000\left(\frac{0.55}{12}\right)}{1 - \left(1 + \frac{0.55}{12}\right)^{-300}} \approx 1719.44$, or \$1719.44. Using the seller's financing, the monthly payment

would be $R = \dfrac{280{,}000\left(\frac{0.049}{12}\right)}{1 - \left(1 + \frac{0.049}{12}\right)^{-300}} \approx 1620.58$, or \$1620.58. By choosing the seller's financing rather than the

bank's, the Meyers would save $(1719.44 - 1620.58)(300) = 29{,}658$, or \$29,658 in interest.

55. To find the Foleys' monthly payment for a 30-year, $400,000 mortgage at 4% per year compounded monthly, we

compute $R = \dfrac{400{,}000\left(\frac{0.04}{12}\right)}{1 - \left(1 + \frac{0.04}{12}\right)^{-360}} \approx 1909.66$, or \$1909.66. Since the property tax is $\frac{1}{12}(6000)$, or \$500 per month,

their monthly payment would be $2409.66. To qualify for the mortgage, their monthly payment must be less than $\frac{1}{12}(72{,}000) \cdot 0.43 = 2580$, or \$2580. Thus, the Foleys will qualify for the mortgage, provided they do not have other significant debts.

57. a. Here $i = \dfrac{r}{m} = \dfrac{0.044}{12}$, $n = 12 \cdot 5 = 60$, and $R = 3000$. Using Equation 11, we find

$P = \dfrac{3000\left[1 - \left(1 + \frac{0.044}{12}\right)^{-360}\right]}{\frac{0.044}{12}} \approx 599{,}088.30$, or \$599,088.30. Thus, if the Carlsons choose the 5/1 ARM,

they can borrow at most $599,088.30.

b. If P denotes the maximum amount that the Carlsons can borrow, then we have $12 \cdot 3000 = 0.0462P$, so

$P = \dfrac{12\,(3000)}{0.0462} \approx 779{,}220.78$. Thus, if the Carlsons choose the interest-only loan, they can borrow at most \$779,220.78.

1. $628.02   **3.** $1379.28   **5.** $1988.41   **7.** $894.12

9. The annual payment is $15,165.46. The amortization schedule follows.

End of Period	Interest Charged	Repayment Made	Payment toward Principal	Outstanding Principal
0				$120,000
1	$5400	$15,165.46	$ 9765.46	110,234.54
2	4960.55	15,165.46	10,204.91	100,029.63
3	4501.33	15,165.46	10,664.13	89,365.50
4	4021.45	15,165.46	11,144.01	78,221.49
5	3519.97	15,165.46	11,645.49	66,576.00
6	2995.92	15,165.46	12,169.54	54,406.46
7	2448.29	15,165.46	12,717.17	41,689.29
8	1876.02	15,165.46	13,289.44	28,399.85
9	1277.99	15,165.46	13,877.47	14,512.38
10	653.06	15,165.44	14,512.38	0

5.4   Arithmetic and Geometric Progressions

Problem-Solving Tips

Note the difference between an arithmetic progression and a geometric progression:

An *arithmetic progression* is a sequence of numbers in which each term after the first is obtained by *adding a constant d to the preceding term.*

A *geometric progression* is a sequence of numbers in which each term after the first is obtained by *multiplying the preceding term by a constant r*.

Concept Questions   page 333

1. a. $a_n = a + (n - 1)d$ **b.** $S_n = \dfrac{n}{2}[2a + (n - 1)d]$

Exercises   page 333

1. $a_9 = 6 + (9 - 1)3 = 30$ **3.** $a_8 = -15 + (8 - 1)\left(\dfrac{3}{2}\right) = -\dfrac{9}{2} = -4.5$

5. $a_{11} - a_4 = (a_1 + 10d) - (a_1 + 3d) = 7d$. Also, $a_{11} - a_4 = 107 - 30 = 77$. Therefore, $7d = 77$ and $d = 11$. Next, $a_4 = a + 3d = a + 3(11) = a + 33 = 30$ and $a = -3$. Therefore, the first five terms are $-3, 8, 19, 30, 41$.

7. Here $a = x$, $n = 7$, and $d = y$. Therefore, the required term is $a_7 = x + (7 - 1)y = x + 6y$.

9. Using the formula for the sum of the terms of an arithmetic progression with $a = 4$, $d = 7$ and $n = 15$, we have $S_n = \dfrac{n}{2}[2a + (n - 1)d]$, so $S_{15} = \dfrac{15}{2}[2(4) + (15 - 1)7] = \dfrac{15}{2}(106) = 795$.

11. The common difference is $d = 2$ and the first term is $a = 15$. Using the formula for the nth term $a_n = a + (n - 1)d$, we have $57 = 15 + (n - 1)(2) = 13 + 2n$, so $2n = 44$ and $n = 22$. Using the formula for the sum of the terms of an arithmetic progression with $a = 15$, $d = 2$ and $n = 22$, we have $S_n = \dfrac{n}{2}[2a + (n - 1)d]$, so $S_{22} = \dfrac{22}{2}[2(15) + (22 - 1)2] = 11(72) = 792$.

13. $f(1) + f(2) + f(3) + \cdots + f(22) = [3(1) - 4] + [3(2) - 4] + [3(3) - 4] + \cdots + [3(22) - 4]$

$$= 3(1 + 2 + 3 + \cdots + 22) + 22(-4) = 3\left(\frac{22}{2}\right)[2(1) + (22 - 1)1] - 88 = 671$$

15. $S_n = \frac{n}{2}[2a_1 + (n - 1)d] = \frac{n}{2}[a_1 + a_1 + (n - 1)d] = \frac{n}{2}(a_1 + a_n)$

a. $S_{11} = \frac{11}{2}(3 + 47) = 275$ **b.** $S_{20} = \frac{20}{2}[5 + (-33)] = -280$

17. Let n be the number of weeks until she reaches 10 miles. Then $a_n = 1 + (n - 1)\frac{1}{4} = 1 + \frac{1}{4}n - \frac{1}{4} = \frac{1}{4}n + \frac{3}{4} = 10$.
Therefore, $n + 3 = 40$, so $n = 37$; that is, at the beginning of the 37th week.

19. To compute Kunwoo's fare by taxi, take $a = 2$, $d = 1.20$, and $n = 25$. Then the required fare is given by
$a_{25} = 2 + (25 - 1)1.20 = 30.8$, or \$30.80. Therefore, by taking the airport limousine, Kunwoo will save
$30.80 - 15.00 = 15.80$, or \$15.80.

21. a. Using the formula for the sum of an arithmetic progression, we have
$$S_n = \frac{n}{2}[2a + (n - 1)d] = \frac{N}{2}[2(1) + (N - 1)(1)] = \frac{N}{2}(N + 1).$$

b. $S_{10} = \frac{10}{2}(10 + 1) = 5(11) = 55$, so
$$D_3 = (C - S)\frac{N - (n - 1)}{S_N} = (6000 - 500)\frac{10 - (3 - 1)}{55} = 5500\left(\frac{8}{55}\right) = 800, \text{ or } \$800.$$

23. This is a geometric progression with $a = 4$ and $r = 2$. Next, $a_7 = 4(2)^6 = 256$ and $S_7 = \frac{4(1 - 2^7)}{1 - 2} = 508$.

25. If we compute the ratios $\dfrac{a_2}{a_1} = \dfrac{-\frac{3}{8}}{\frac{1}{2}} = -\dfrac{3}{4}$ and $\dfrac{a_3}{a_2} = \dfrac{\frac{1}{4}}{-\frac{3}{8}} = -\dfrac{2}{3}$, we see that the given sequence is not geometric
because the ratios are not equal.

27. This is a geometric progression with $a = 243$, and $r = \frac{1}{3}$. Thus, $a_7 = 243\left(\frac{1}{3}\right)^6 = \frac{1}{3}$ and

$$S_7 = \frac{243\left[1 - \left(\frac{1}{3}\right)^7\right]}{1 - \frac{1}{3}} = \frac{1093}{3}.$$

29. First, we compute $r = \dfrac{a_2}{a_1} = \dfrac{3}{-3} = -1$. Next, $a_{20} = -3(-1)^{19} = 3$, and so $S_{20} = \dfrac{-3\left[1 - (-1)^{20}\right]}{1 - (-1)} = 0$.

31. The population in five years is expected to be $200{,}000(1.08)^{6-1} = 200{,}000(1.08)^5 \approx 293{,}866$.

33. The salary of a union member whose salary was \$42,000 six years ago is given by the 7th term of a geometric
progression whose first term is 42,000 and whose common ratio is 1.05. Thus, $a_7 = (42{,}000)(1.05)^6 \approx 56{,}284.02$,
or \$56,284.

35. With 8% raises per year, the employee would make $S_4 = 48{,}000\left[\dfrac{1 - (1.08)^4}{1 - 1.08}\right] \approx 216{,}293.38$,
or \$216,293.38 over the next four years. With \$4000 raises per year, the employee would make
$S_4 = \frac{4}{2}[2(48{,}000) + (4 - 1)4000] = 216{,}000$ or \$216,000 over the next four years. We conclude that the
employee should choose annual raises of 8% per year.

37. a. During the sixth year, she will receive $a_6 = 10,000\,(1.15)^5 \approx 20,113.57$, or $20,113.57.

b. The total amount of the six payments will be given by $S_6 = \dfrac{10,000[1 - (1.15)^6]}{1 - 1.15} \approx 87,537.38$, or $87,537.38.

39. The book value of the office equipment at the end of the eighth year is given by

$$V\,(8) = 150,000\left(1 - \tfrac{2}{10}\right)^8 \approx 25,165.82, \text{ or } \$25,165.82.$$

41. The book value of the restaurant equipment at the end of six years is given by $V\,(6) = 150,000\,(0.8)^6 = 39,321.60$, or $39,321.60. By the end of the sixth year, the equipment will have depreciated by
$D\,(6) = 150,000 - 39,321.60 = 110,678.40$, or $110,678.40.

43. True. Suppose d is the common difference of $a_1, a_2, ..., a_n$ and e is the common difference of $b_1, b_2, ..., b_n$. Then $d + e$ is the common difference of $a_1 + b_1, a_2 + b_2, ..., a_n + b_n$, and we see that the latter is indeed an arithmetic progression.

CHAPTER 5   Concept Review Questions   page 336

1. a. original, $P\,(1 + rt)$
    b. interest, $P\,(1 + i)^n$, $A\,(1 + i)^{-n}$

3. annuity, ordinary annuity, simple annuity

5. $\dfrac{Pi}{1 - (1 + i)^{-n}}$

7. constant d, $a + (n - 1)\,d$, $\tfrac{n}{2}\,[2a + (n - 1)\,d]$

CHAPTER 5   Review Exercises   page 337

1. a. Here $P = 5000$, $r = 0.05$, and $m = 1$. Thus, $i = r = 0.05$ and $n = 4$, so $A = 5000\,(1.05)^4 = 6077.53$, or $6077.53.

b. Here $m = 2$, so $i = \tfrac{0.05}{2} = 0.025$ and $n = 4 \cdot 2 = 8$. Thus, $A = 5000\,(1.025)^8 \approx 6092.01$, or $6092.01.

c. Here $m = 4$, so $i = \tfrac{0.05}{4} = 0.0125$ and $n = 4 \cdot 4 = 16$. Thus, $A = 5000\,(1.0125)^{16} \approx 6099.45$, or $6099.45.

d. Here $m = 12$, so that $i = \tfrac{0.05}{12}$ and $n = 4 \cdot 12 = 48$. Thus, $A = 5000\left(1 + \tfrac{0.05}{12}\right)^{48} \approx 6104.48$, or $6104.48.

3. a. The effective rate of interest is given by $r_{\text{eff}} = \left(1 + \tfrac{r}{m}\right)^m - 1 = (1 + 0.06) - 1 = 0.06$, or 6%.

b. The effective rate of interest is given by $r_{\text{eff}} = \left(1 + \tfrac{r}{m}\right)^m - 1 = \left(1 + \tfrac{0.06}{2}\right)^2 - 1 = 0.0609$, or 6.09%.

c. The effective rate of interest is given by $r_{\text{eff}} = \left(1 + \tfrac{r}{m}\right)^m - 1 = \left(1 + \tfrac{0.06}{4}\right)^4 - 1 \approx 0.061363$, or 6.1363%.

d. The effective rate of interest is given by $r_{\text{eff}} = \left(1 + \tfrac{r}{m}\right)^m - 1 = \left(1 + \tfrac{0.06}{12}\right)^{12} - 1 \approx 0.06168$, or 6.168%.

5. The present value is given by $P = 41,413\left(1 + \tfrac{0.045}{4}\right)^{-20} \approx 33,110.52$, or approximately $33,110.52.

7. $S = 150 \left[\dfrac{\left(1 + \frac{0.05}{4}\right)^{28} - 1}{\frac{0.05}{4}} \right] \approx 4991.91$, or \$4991.91.

9. Using the formula for the present value of an annuity with $R = 250$, $n = 36$, and $i = \frac{0.045}{12} = 0.00375$, we have

$$P = 250 \left[\dfrac{1 - (1.00375)^{-36}}{0.00375} \right] \approx 8404.23, \text{ or } \$8404.23.$$

11. Using the amortization formula with $P = 22{,}000$, $n = 36$, and $i = \frac{0.035}{12}$, we find

$$R = \dfrac{22{,}000 \left(\frac{0.035}{12}\right)}{1 - \left(1 + \frac{0.035}{12}\right)^{-36}} \approx 644.65, \text{ or } \$644.65.$$

13. Using the sinking fund formula with $S = 18{,}000$, $n = 48$, and $i = \frac{0.03}{12}$, we have $R = \dfrac{\left(\frac{0.03}{12}\right) 18{,}000}{\left(1 + \frac{0.03}{12}\right)^{48} - 1} \approx 353.42$,

or \$353.42.

15. The effective rate of interest is given by $r_{\text{eff}} = \left(1 + \frac{r}{m}\right)^m - 1 = \left(1 + \frac{0.036}{12}\right)^{12} - 1 \approx 0.03660$, or 3.660%.

17. $P = 119{,}346 e^{-(0.05)4} \approx 97{,}712.24$, or \$97,712.24.

19. At the end of five years, the investment will be worth $A = P\left(1 + \frac{r}{m}\right)^{mt} = 4.2\left(1 + \frac{0.054}{4}\right)^{4(5)} = 5.491922$, or \$5,491,922.

21. Using the present value formula for compound interest, we have

$$P = A\left(1 + \frac{r}{m}\right)^{-mt} = 19{,}440.31 \left(1 + \frac{0.035}{12}\right)^{-12(4)} \approx 16{,}904.04, \text{ or } \$16{,}904.04.$$

23. The future value of his investment is given by $A = P\left(1 + r_{\text{eff}}\right)^t$. Therefore, $34{,}616 = 24{,}000\left(1 + r_{\text{eff}}\right)^5$,
$1 + r_{\text{eff}} = (1.4423)^{1/5}$, and $r_{\text{eff}} = (1.4423)^{1/5} - 1 = 0.075997149$, or approximately 7.6%.

25. Using the formula for the future value of an annuity with $R = 200 + 200 = 400$, $n = 120$, and $i = \frac{0.05}{12}$, we have

$$S = 400 \left[\dfrac{\left(1 + \frac{0.05}{12}\right)^{120} - 1}{\frac{0.05}{12}} \right] \approx 62{,}112.91, \text{ or } \$62{,}112.91.$$

27. Using the formula for the present value of an annuity, we see that the purchase price of the furniture is

$$400 + P = 400 + 75.32 \left[\dfrac{1 - \left(1 + \frac{0.06}{12}\right)^{-24}}{\frac{0.06}{12}} \right] \approx 400 + 1699.44, \text{ or approximately } \$2099.44.$$

29. a. The monthly payment is given by $P = \dfrac{(120{,}000)(0.00375)}{1 - (1 + 0.00375)^{-180}} \approx 917.99$, or \$917.99.

b. We can find the total interest payment by computing $180(917.99) - 120{,}000 = 45{,}238.20$, or \$45,238.20.

c. We first compute the present value of their remaining payments:

$$P = 917.99 \left[\frac{1 - (1 + 0.00375)^{-60}}{0.00375} \right] \approx 49{,}240.41, \text{ or } \$49{,}240.41. \text{ Then their equity is } 150{,}000 - 49{,}240.41, \text{ or}$$

approximately $100,760.

31. Using the sinking fund formula with $S = 120{,}000$, $n = 24$, and $i = \frac{0.058}{12}$, we find that the amount of each

installment should be $R = \dfrac{\left(\frac{0.058}{12}\right) 120{,}000}{\left(1 + \frac{0.058}{12}\right)^{24} - 1} \approx 4727.67$, or $4727.67.

33. We use Formula 9 with $R = 250$, $r = 0.05$, $m = 12$, and $n = 12$, obtaining

$$P = 250 \left[\frac{1 - \left(1 + \frac{0.05}{12}\right)^{-9}}{\frac{0.05}{12}} \right] \approx 2203.83, \text{ or } \$2203.83. \text{ Thus, Matt's parents need to deposit } \$2203.83.$$

CHAPTER 5   Before Moving On...   page 339

1. Here $P = 2000$, $r = 0.08$, $t = 3$, and $m = 12$, so $i = \frac{0.08}{12}$. Therefore, $A = 2000 \left(1 + \frac{0.08}{12}\right)^{(12)(3)} \approx 2540.47$, or
$2540.47.

2. Here $r = 0.06$ and $m = 365$, so $r_{\text{eff}} = \left(1 + \frac{0.06}{365}\right)^{365} - 1 \approx 0.0618$, or approximately 6.18% per year.

3. Here $R = 800$, $n = 10 \cdot 52 = 520$, $r = 0.06$, and $m = 52$, so $S = \dfrac{800 \left[\left(1 + \frac{0.06}{52}\right)^{520} - 1 \right]}{\frac{0.06}{52}} \approx 569{,}565.47$, or

$569,565.47.

4. Here $P = 100{,}000$, $t = 10$, $r = 0.08$, $m = 12$, so $R = \dfrac{100{,}000 \left(\frac{0.08}{12}\right)}{1 - \left(1 + \frac{0.08}{12}\right)^{-120}} \approx 1213.276$, or approximately

$1213.28.

5. Here $S = 15{,}000$, $t = 6$, $r = 0.1$, and $m = 52$, so $R = \dfrac{\left(\frac{0.1}{52}\right) 15{,}000}{\left(1 + \frac{0.1}{52}\right)^{(6)(52)} - 1} \approx 35.132$, or $35.13.

6. a. Here $a_1 = a = 3$ and $d = 4$, so with $n = 10$, $S_{10} = \frac{10}{2} \left[2(3) + 9(4) \right] = 210$.

   b. Here $a = 2$ and $r = 2$, so with $n = 8$, $S_8 = \dfrac{\frac{1}{2} \left[1 - (2)^8 \right]}{1 - 2} = 127.5$.

6 SETS AND COUNTING

6.1 Sets and Set Operations

Problem-Solving Tips

It's often easier to remember a formula if you learn to express the formula in words. For example, DeMorgan's Laws can be expressed as follows:

$(A \cup B)^c = A^c \cap B^c$ says that the complement of the union of two sets is equal to the intersection of their complements.

$(A \cap B)^c = A^c \cup B^c$ says that the complement of the intersection of two sets is equal to the union of their complements.

Concept Questions   page 348

1. a. A set is a well-defined collection of objects. As an example, consider the set of all freshmen at a college.

b. Two sets A and B are equal if they have exactly the same elements. $\{a, c\} = \{c, a\}$. (Examples will vary.)

c. The empty set is the set that contains no element.

3. a. If $A \subset B$, then $B^c \subset A^c$.   **b.** If $A^c = \varnothing$, then $A = U$, the universal set.

Exercises   page 348

1. $\{x \mid x$ is a gold medalist in the 2014 Winter Olympic Games$\}$

3. $\{x \mid x$ is an integer greater than 2 and less than 8$\}$

5. $\{2, 3, 4, 5, 6\}$

7. $\{-2\}$

9. a. True. The order in which the elements are listed is not important.

b. False. No finite set contains itself.

11. a. False. The empty set has no element.

b. False. 0 is not a set.

13. True.

15. a. True. 2 belongs to A.   **b.** False. For example, 5 belongs to A but $5 \notin \{2, 4, 6\}$.

17. a and b.

19. a. $\varnothing, \{1\}, \{2\}, \{1, 2\}$

b. $\varnothing, \{1\}, \{2\}, \{3\}, \{1, 2\}, \{1, 3\}, \{2, 3\}, \{1, 2, 3\}$

c. $\varnothing, \{1\}, \{2\}, \{3\}, \{4\}, \{1, 2\}, \{1, 3\}, \{1, 4\}, \{2, 3\}, \{2, 4\}, \{3, 4\}, \{1, 2, 3\}, \{1, 2, 4\}, \{1, 3, 4\}, \{2, 3, 4\}, \{1, 2, 3, 4\}$

21. {1, 2, 3, 4, 6, 8, 10}.

23. {Jill, John, Jack, Susan, Sharon}.

25. a. 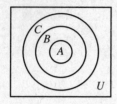   **b.** 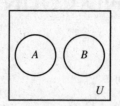   **c.**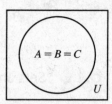

27. a. The required set consists of all elements that are in A but not in B, along with all elements that are in B but not in A. So the required expression is $(A \cap B^c) \cup (A^c \cap B)$

b. The required set consists of all elements that are not in both A and B, so the required expression is $(A \cap B)^c$ or $A^c \cup B^c$.

29. a. $A \cap B^c$ **b.** $A^c \cap B$ **31. a.** $A \cup B \cup C$ **b.** $A \cap B \cap C$

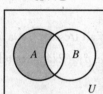   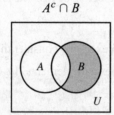   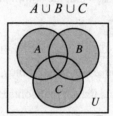   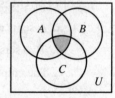

33. a. $A^c \cap B^c \cap C^c$ **b.** $(A \cup B)^c \cap C$

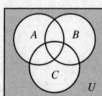   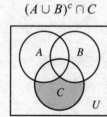

35. a. $A^c = \{2, 4, 6, 8, 10\}$

b. $B \cup C = \{2, 4, 6, 8, 10\} \cup \{1, 2, 4, 5, 8, 9\} = \{1, 2, 4, 5, 6, 8, 9, 10\}$

c. $C \cup C^c = U = \{1, 2, 3, 4, 5, 6, 7, 8, 9, 10\}$

37. a. $(A \cap B) \cup C = C = \{1, 2, 4, 5, 8, 9\}$

b. $(A \cup B \cup C)^c = \varnothing$

c. $(A \cap B \cap C)^c = U = \{1, 2, 3, 4, 5, 6, 7, 8, 9, 10\}$

39. a. The sets are not disjoint because 4 is an element of both sets.

b. The sets are disjoint as they have no common element.

41. a. The set of all employees at the Universal Life Insurance Company who do not drink tea.

b. The set of all employees at the Universal Life Insurance Company who do not drink coffee.

43. a. The set of all employees at the Universal Life Insurance Company who drink tea but not coffee.

b. The set of all employees at the Universal Life Insurance Company who drink coffee but not tea.

45. a. The set of all employees at the hospital who are not doctors.

  b. The set of all employees at the hospital who are not nurses.

47. a. The set of all employees at the hospital who are female doctors.

  b. The set of all employees at the hospital who are both doctors and administrators.

49. a. $D \cap F$ **b.** $R \cap F^c \cap L^c$ **51. a.** B^c **b.** $A \cap B$ **c.** $A \cap B \cap C^c$

53. a. $A = \{$New York, Chicago, Boston$\}$, $B = \{$Chicago, Boston$\}$, and $C = \{$Las Vegas, San Francisco$\}$.

  b. $A \cup B = \{$New York, Chicago, Boston$\}$.

  c. $A \cap B = \{$Chicago, Boston$\}$.

  d. $A^c \cap B = \{$Las Vegas, San Francisco$\} \cap \{$Chicago, Boston$\} = \varnothing$.

  e. $A \cap B^c = \{$New York, Chicago, Boston$\} \cap \{$New York, Las Vegas, San Francisco$\} = \{$New York$\}$.

  f. Using the result of part (a), $(A \cup B)^c = \{$Las Vegas, San Francisco$\}$.

55. a. Region 1: $A \cap B \cap C$ is the set of tourists who used all three modes of transportation over a one-week period in London.

  b. Regions 1 and 4: $A \cap C$ is the set of tourists who have taken the underground and a bus over a one-week period in London.

  c. Regions 4, 5, 7, and 8: B^c is the set of tourists who have not taken a cab over a one-week period in London.

57. $A \subseteq A \cup B$ $B \subseteq A \cup B$ **59.** $A \cup (B \cup C)$ $=$ $(A \cup B) \cup C$

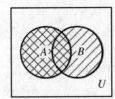   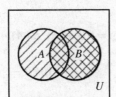   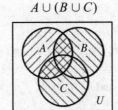   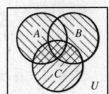

61. $A \cap (B \cup C)$ $=$ $(A \cap B) \cup (A \cap C)$

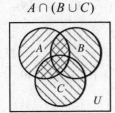   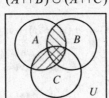

63. a. $A \cup (B \cup C) = \{1, 3, 5, 7, 9\} \cup (\{1, 2, 4, 7, 8\} \cup \{2, 4, 6, 8\}) = \{1, 3, 5, 7, 9\} \cup \{1, 2, 4, 6, 7, 8\}$

  $= \{1, 2, 3, 4, 5, 6, 7, 8, 9\}$

  $(A \cup B) \cup C = (\{1, 3, 5, 7, 9\} \cup (\{1, 2, 4, 7, 8\}) \cup \{2, 4, 6, 8\} = \{1, 2, 3, 4, 5, 7, 8, 9\} \cup \{2, 4, 6, 8\}$

  $= \{1, 2, 3, 4, 5, 6, 7, 8, 9\}$

  b. $A \cap (B \cap C) = \{1, 3, 5, 7, 9\} \cap (\{1, 2, 4, 7, 8\} \cap \{2, 4, 6, 8\}) = \{1, 3, 5, 7, 9\} \cap (\{2, 4, 8\}) = \varnothing$

  $(A \cap B) \cap C = (\{1, 3, 5, 7, 9\} \cap \{1, 2, 4, 7, 8\}) \cap \{2, 4, 6, 8\} = \{1, 7\} \cap \{2, 4, 6, 8\} = \varnothing$

65. a. $r, u, v, w, x\ y$ **b.** v, r **67. a.** t, y, s **b.** t, s, w, x, z

69. $A \subset C$

71. False. Because every element in a set A belongs to A, A is a subset of itself.

73. True. If at least one of the sets A or B is nonempty, then $A \cup B \neq \varnothing$.

75. True. $(A \cup A^c)^c = U^c = \varnothing$.

77. True. Because $A \subseteq B$, all of the elements in A are also in B, so $A \cup B = B$.

79. True. Because A is a proper subset of B, all of the elements in A are also in B and there is at least one element in B that is not in A. Therefore, there is at least one element in A^c that is not in B^c, and so $B^c \subset A^c$.

6.2   The Number of Elements in a Finite Set

Problem-Solving Tips

In the problems that follow, it is often helpful to draw a Venn diagram.

Concept Questions   page 357

1. a. If A and B are sets with $A \cap B = \varnothing$, then $n(A) + n(B) = n(A \cup B)$.

  b. If $n(A \cup B) \neq n(A) + n(B)$, then $A \cap B \neq \varnothing$.

Exercises   page 357

1. $A \cup B = \{a, e, g, h, i, k, l, m, o, u\}$ and so $n(A \cup B) = 10$. Next, $n(A) + n(B) = 5 + 5 = 10$.

3. a. $A = \{2, 4, 6, 8\}$ and $n(A) = 4$.   **b.** $B = \{6, 7, 8, 9, 10\}$ and $n(B) = 5$.

  c. $A \cup B = \{2, 4, 6, 7, 8, 9, 10\}$ and $n(A \cup B) = 7$.   **d.** $A \cap B = \{6, 8\}$ and $n(A \cap B) = 2$.

5. Using the results of Exercise 3, we see that $n(A \cup B) = 7$ and $n(A) + n(B) - n(A \cap B) = 4 + 5 - 2 = 7$.

7. Because $n(A \cup B) = n(A) + n(B) - n(A \cap B)$, $n(B) = n(A \cup B) + n(A \cap B) - n(A) = 30 + 5 - 15 = 20$.

9. a. $n(A) = 12 + 3 = 15$ **b.** $n(A \cup B) = 12 + 3 + 15 = 30$

  c. $n(A^c \cap B) = 15$ **d.** $n(A \cap B^c) = 12$

  e. $n(U) = 50$ **f.** $n[(A \cup B)^c] = 20$

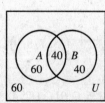

11. Refer to the Venn diagram at left.
  a. $n(A \cup B) = 60 + 40 + 40 = 140$.
  b. $n(A^c) = 40 + 60 = 100$.
  c. $n(A \cap B^c) = 60$.

13. $n(A \cup B) = n(A) + n(B) - n(A \cap B) = 6 + 10 - 3 = 13$.

15. $n(A \cup B) = n(A) + n(B) - n(A \cap B)$, so $n(A \cap B) = n(A) + n(B) - n(A \cup B) = 4 + 5 - 9 = 0$.

17. $n(A \cap B \cap C) = n(A \cup B \cup C) - n(A) - n(B) - n(C) + n(A \cap B) + n(A \cap C) + n(B \cap C)$, so

$n(C) = n(A \cup B \cup C) - n(A \cap B \cap C) - n(A) - n(B) + n(A \cap B) + n(A \cap C) + n(B \cap C)$

$= 25 - 2 - 12 - 12 + 5 + 5 + 4 = 13$.

19. Let A denote the set of prisoners in the Wilton County Jail who were accused of a felony and B the set of prisoners in that jail who were accused of a misdemeanor. Then we are given that $n(A \cup B) = 190$. Referring to the Venn diagram, the number of prisoners who were accused of both a felony and a misdemeanor is given by $(A \cap B) = n(A) + n(B) - n(A \cup B) = 130 + 121 - 190 = 61$.

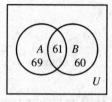

21. Let U denote the set of all customers surveyed, $A = \{x \in U \mid x$ buys brand $A\}$, and $B = \{x \in U \mid x$ buys brand $B\}$. Then $n(U) = 120$, $n(A) = 80$, $n(B) = 68$, and $n(A \cap B) = 42$.

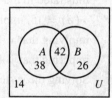

a. The number of customers who buy at least one of these brands is
$n(A \cup B) = 80 + 68 - 42 = 106$.

b. The number who buy exactly one of these brands is $n(A \cap B^c) + n(A^c \cap B) = 38 + 26 = 64$.

c. The number who buy only brand A is $n(A \cap B^c) = 38$.

d. The number who buy none of these brands is $n[(A \cup B)^c] = 120 - 106 = 14$.

23. Let U denote the set of 200 investors, $A = \{x \in U \mid x$ uses a discount broker$\}$, and $B = \{x \in U \mid x$ uses a full-service broker$\}$.

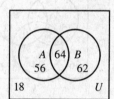

a. The number of investors who use at least one kind of broker is
$n(A \cup B) = n(A) + n(B) - n(A \cap B) = 120 + 126 - 64 = 182$.

b. The number of investors who use exactly one kind of broker is
$n(A \cap B^c) + n(A^c \cap B) = 56 + 62 = 118$.

c. The number of investors who use only discount brokers is $n(A \cap B^c) = 56$.

d. The number of investors who do not use a broker is $n(A \cup B)^c = n(U) - n(A \cup B) = 200 - 182 = 18$.

25. Let U denote the set of 200 households in the survey,
$A = \{x \in U \mid x$ owns a desktop computer$\}$, and
$B = \{x \in U \mid x$ owns a laptop computer$\}$. Referring to the figure, we see that the number of households that own both desktop and laptop computers is
$n(A \cap B) = 200 - 120 - 10 - 40 = 30$.

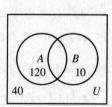

27. a. $n(A) = 7 + 2 + 3 + 4 = 16$ **b.** $n(A \cup B) = 7 + 2 + 3 + 5 + 10 + 4 = 31$

c. $n(A \cap B \cap C^c) = 4$ **d.** $n[(A \cup B) \cap C^c] = 7 + 4 + 10 = 21$

e. $n[(A \cup B \cup C)^c] = 11$

29. The first five pieces of information are illustrated in the first figure. Using the remaining information, we have $n(A) = 16$, implying that $x + y + 3 + 8 = 16$, so $x + y = 5$; $n(B) = 17$, implying that $y + 10 + z + 3 = 17$, so $y + z = 4$; and $n(U) = 92$, implying that $x + y + 10 + z + 5 + 8 + 3 + 60 = 92$, so $x + y + z = 6$.
Solving the system

$$
\begin{aligned}
x + y \quad\;\; &= 5 \\
y + z &= 4 \\
x + y + z &= 6
\end{aligned}
$$

we obtain $x = 2$, $y = 3$, and $z = 1$, as shown in the second figure.

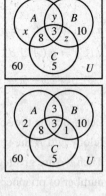

31. The first four pieces of information, along with the last piece, are illustrated in the first figure. Using the remaining information, we have $n(A) = 15$, implying that $x + y + 3 + 4 = 15$, so $x + y = 8$; $n(B) = 22$, implying that $x + y + z + 12 = 22$, so $x + y + z = 10$; and $n(C) = 24$, implying that $y + z + 14 + 3 = 24$, so $y + z = 7$.
Solving the system

$$
\begin{aligned}
x + y \quad\;\; &= 8 \\
x + y + z &= 10 \\
y + z &= 7
\end{aligned}
$$

we obtain $x = 3$, $y = 5$, and $z = 2$, as shown in the second figure.

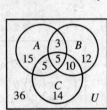

33. **a.** $n(A \cup B \cup C) = 64$
  b. $n(A^c \cap B \cap C) = 10$

35. **a.** $n(A^c \cap B^c \cap C^c) = n\left[(A \cup B \cup C)^c\right] = 36$
  b. $n\left[A^c \cap (B \cup C)\right] = 36$

37. $A = \{R_1, R_2, R_3\}$, $B = \{R_2, R_3\}$, and $C = \{R_3, R_4, R_5, R_6\}$.

  a. $n(A) = 3$, $N(B) = 2$, and $N(C) = 4$.

  b. $n(A \cup B) = n(\{R_1, R_2, R_3\}) = 3$.

  c. $n(A \cap B^c) = n(\{R_1, R_2, R_3\} \cap \{R_1, R_4, R_5, R_6\}) = n(\{R_1\}) = 1$.

  d. $n(B \cap C^c) = n(\{R_2, R_3\} \cap \{R_1, R_2\}) = n(\{R_2\}) = 1$.

  e. $n(A \cap B^c \cap C) = n(\{R_1, R_2, R_3\} \cap \{R_1, R_4, R_5, R_6\} \cap \{R_3, R_4, R_5, R_6\}) = n(\varnothing) = 0$.

39. $A = \{F_1, F_2, F_3, F_4, F_5\}$, $B = \{F_4, F_5, F_6\}$, and $C = \{F_4, F_5, F_6, F_7, F_8\}$.

  a. $n(A) = 5$, $N(B) = 3$, and $N(C) = 5$.

  b. $n(A \cap B) = n(\{F_4, F_5\}) = 2$.

  c. $n(A^c \cap C) = n(\{F_6, F_7, F_8\} \cap \{F_4, F_5, F_6, F_7, F_8\}) = n(\{F_6, F_7, F_8\}) = 3$.

  d. $n(A \cap B^c) = n(\{F_1, F_2, F_3, F_4, F_5\} \cap \{F_1, F_2, F_3, F_7, F_8\}) = n(\{F_1, F_2, F_3\}) = 3$.

  e. $n(A^c \cap C^c) = n(\{F_6, F_7, F_8\} \cap \{F_1, F_2, F_3\}) = n(\varnothing) = 0$.

  f. $n((A \cup B) \cap C) = n(\{F_1, F_2, F_3, F_4, F_5, F_6\} \cap \{F_4, F_5, F_6, F_7, F_8\}) = n(\{F_4, F_5, F_6\}) = 3$.

41. Let U denote the set of all economists surveyed,

$A = \{x \in U \mid x$ had lowered his or her estimate of the consumer inflation rate$\}$,

and $B = \{x \in U \mid x$ had raised his or her estimate of the GNP growth rate$\}$. Then

$n(U) = 10$, $n(A) = 7$, $n(B) = 8$, and $n(A \cap B^c) = 2$, so the number of

economists who had both lowered their estimate of the consumer inflation rate and

raised their estimate of the GNP growth rate is given by $n(A \cap B) = 5$.

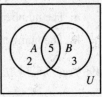

43. Let U denote the set of 100 college students who were surveyed,

$A = \{x \in U \mid x$ reads *Time*$\}$, $B = \{x \in U \mid x$ reads *The New Yorker*$\}$, and

$C = \{x \in U \mid x$ reads *Vanity Fair*$\}$. Then $n(A) = 40$, $n(B) = 30$, $n(C) = 25$,

$n(A \cap B) = 15$, $n(A \cap C) = 12$, $n(B \cap C) = 10$, and $n(A \cap B \cap C) = 4$.

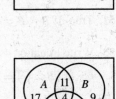

a. The number of students surveyed who read at least one magazine is

$n(A \cup B \cup C) = 17 + 11 + 4 + 8 + 6 + 7 + 9 = 62$.

b. The number of students surveyed who read exactly one magazine is

$n(A \cap B^c \cap C^c) + n(A^c \cap B \cap C^c) + n(A^c \cap B^c \cap C) = 17 + 9 + 7 = 33$.

c. The number of students surveyed who read exactly two magazines is

$n(A \cap B \cap C^c) + n(A^c \cap B \cap C) + n(A \cap B^c \cap C) = 11 + 6 + 8 = 25$.

d. The number of students surveyed who did not read any of these magazines is $n(A \cup B \cup C)^c = 100 - 62 = 38$.

45. Let U denote the set of all customers surveyed, $A = \{x \in U \mid x$ buys brand $A\}$,

$B = \{x \in U \mid x$ buys brand $B\}$, and $C = \{x \in U \mid x$ buys brand $C\}$. Then

$n(U) = 120$, $n(A \cap B \cap C^c) = 15$, $n(A^c \cap B \cap C^c) = 25$,

$n(A^c \cap B^c \cap C) = 26$, $n(A \cap B \cap C^c) = 15$, $n(A \cap B^c \cap C) = 10$,

$n(A^c \cap B \cap C) = 12$, and $n(A \cap B \cap C) = 8$.

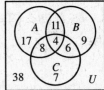

a. The number of customers who buy at least one of these brands is

$n(A \cup B \cup C) = 12 + 15 + 25 + 12 + 8 + 10 + 26 = 108$.

b. The number who buy brands A and B but not C is $n(A \cap B \cap C^c) = 15$.

c. The number who buy brand A is $n(A) = 12 + 10 + 15 + 8 = 45$.

d. The number who buy none of these brands is $n\left[(A \cup B \cup C)^c\right] = 120 - 108 = 12$.

47. Let U denote the set of 200 employees surveyed,

$A = \{x \in U \mid x$ had investments in stock funds$\}$,

$B = \{x \in U \mid x$ had investments in bond funds$\}$, and

$C = \{x \in U \mid x$ had investments in money market funds$\}$. Then

$n(U) = 200$, $n(A) = 141$, $n(B) = 91$, $n(C) = 60$, $n(A \cap B) = 47$,

$n(A \cap C) = 36$, $n(B \cap C) = 36$, and

$n(A^c \cap B^c \cap C^c) = n\left[(A \cup B \cup C)^c\right] = 5$.

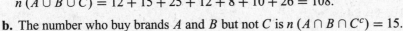

Letting $n(A \cap B \cap C) = z$, we arrive at the Venn diagram shown.

Next, using the facts that $n(A) = 141$, $n(B) = 91$, and $n(C) = 60$, we obtain

$a + (36 - z) + (47 - z) + z = 141$, $b + (47 - z) + (36 - z) + z = 91$, $c + (36 - z) + (36 - z) + z = 60$, and

$a + b + c + (36 - z) + (47 - z) + (36 - z) + z + 5 = 200$. These equations simplify to $a - z = 58$, $b - z = 8$,

$c - z = -12$, and $a + b + c - 2z = 76$. Solving, we find $a = 80$, $b = 30$, $c = 10$, and $z = 22$.

a. The number of employees surveyed who had invested in all three investments is $n(A \cap B \cap C) = z = 22$.

b. The number who had invested in stock funds only is given by $n(A \cap B^c \cap C^c) = a = 80$.

49. True. $n(A \cup B) = n(A) + n(B) - n(A \cap B)$.

51. True. If $A \cap B \neq \varnothing$, then $n(A \cup B) = n(A) + n(B) - n(A \cap B)$.

53. Write Equation 4 as $n(D \cup E) = n(D) + n(E) - n(D \cap E)$ and let $D = A \cup B$ and $E = C$. Then

$$n(A \cup B \cup C) = n(A \cup B) + n(C) - n[(A \cup B) \cap C] = n(A) + n(B) - n(A \cap B) + n(C) - n[(A \cup B) \cap C]$$

$$= n(A) + n(B) - n(A \cap B) + n(C) - n[(A \cap C) \cup (B \cap C)]$$

$$= n(A) + n(B) - n(A \cap B) + n(C) - [n(A \cap C) + n(B \cap C) - n(A \cap C \cap B \cap C)]$$

$$= n(A) + n(B) + n(C) - n(A \cap B) - n(A \cap C) - n(B \cap C) + n(A \cap B \cap C).$$

6.3   The Multiplication Principle

Concept Questions   page 365

1. If task T_1 can be performed in N_1 ways, task T_2 can be performed in N_2 ways, ..., and task T_n can be performed in N_n ways, then the number of ways of performing the tasks $T_1, T_2, \ldots, T_n$ in succession is given by $N_1 N_2 \cdots N_n$.

Exercises   page 365

1. By the multiplication principle, the number of rates is given by $4 \times 3 = 12$.

3. By the multiplication principle, the number of ways that a blackjack hand can be dealt is $4 \times 16 = 64$.

5. By the multiplication principle, she can create $2 \cdot 4 \cdot 3 = 24$ different ensembles.

7. The number of paths is $2 \times 4 \times 3 = 24$.

9. By the multiplication principle, we see that the number of ways a health-care plan can be selected is $(10)(3)(2) = 60$.

11. $10^9 = 1,000,000,000$

13. Jeanne has 9 choices for the first digit, 10 choices for each of the second, third, and fourth digits, and 5 choices for the last digit, so the total number of PINs available is $(9)(10)(10)(10)(5) = 45,000$.

15. There are $2^8 = 256$ binary strings possible.

17. There are $(3)(5)(4) + (3)(5)(4) = 120$ different three-course meals available.

19. The number of different responses is $\underbrace{(5)(5) \cdots (5)}_{50 \text{ terms}} = 5^{50}$.

21. The number of selections is given by $(6)(2)(4)(6)(2) = 576$.

23. a. The number of license plate numbers that may be formed is $(26)(26)(26)(10)(10)(10) = 17,576,000$.

b. The number of license plate numbers that may be formed is $(10)(10)(10)(26)(26)(26) = 17{,}576{,}000$.

25. $(26)(10)(10)(10)(26)(26)(26) = 456{,}976{,}000$

27. The number of possible identification numbers is $(26)(9)(10)(10)(10)(10) = 2{,}340{,}000$.

29. The number of ways the first, second and third prizes can be awarded is $(15)(14)(13) = 2730$.

31. a. There are $10^6 = 1{,}000{,}000$ possible combinations.

    b. There are $(10)(10)(5)(10)(10)(5) = 250{,}000$ possible combinations.

33. The number of ways in which the nine symbols on the wheels can appear in the window slot is $(9)(9)(9) = 729$. The number of ways in which the eight symbols other than the "lucky dollar" can appear in the window slot is $(8)(8)(8) = 512$. Therefore, the number of ways in which the "lucky dollars" can appear in the window slot is $729 - 512 = 217$.

35. True. There are four choices for the digit in the hundreds position, four choices in the tens position, and two choices in the units position, for a total of $4 \cdot 4 \cdot 2 = 32$ such numbers.

6.4   Permutations and Combinations

Problem-Solving Tips

1. Note the difference between a permutation and a combination. A permutation is an arrangement of a set of distinct objects in a *definite order* whereas a combination is an arrangement of a set of distinct objects without regard to order. In a permutation of two distinct objects A and B, we distinguish between the selections AB and BA, whereas in a combination these selections would be considered the same.

2. Sometimes the solution of an applied problem involves the multiplication principle and a permutation and/or a combination. (See Example 12 on page 376 of the text.)

Concept Questions   page 378

1. a. Given a set of distinct objects, a permutation of the set is an arrangement of these objects in a *definite order*.

    b. $P(n, r) = \dfrac{n!}{(n-r)!}$, so $P(5, 3) = \dfrac{5!}{(5-3)!} = \dfrac{5 \cdot 4}{2 \cdot 1} = 10$.

3. a. $C(n, r) = \dfrac{n!}{r!\,(n-r)!}$

    b. $C(6, 3) = \dfrac{6!}{3!\,(6-3)!} = \dfrac{6 \cdot 5 \cdot 4}{3 \cdot 2} = 20$

Exercises   page 378

1. $3(5!) = 3(5)(4)(3)(2)(1) = 360$

3. $\dfrac{5!}{2!\,3!} = 5(2) = 10$

5. $P(5, 5) = \dfrac{5!}{(5-5)!} = \dfrac{5!}{0!} = 120$

7. $P(5, 2) = \dfrac{5!}{(5-2)!} = \dfrac{5!}{3!} = (5)(4) = 20$

9. $P(n, 1) = \dfrac{n!}{(n-1)!} = n$

11. $C(6, 6) = \dfrac{6!}{6!\,0!} = 1$

13. $C(7, 4) = \dfrac{7!}{4!\,3!} = \dfrac{7 \cdot 6 \cdot 5}{3 \cdot 2} = 35$ **15.** $C(5, 0) = \dfrac{5!}{5!\,0!} = 1$

17. $C(9, 6) = \dfrac{9!}{3!\,6!} = \dfrac{9 \cdot 8 \cdot 7}{3 \cdot 2} = 84$ **19.** $C(n, 2) = \dfrac{n!}{(n-2)!\,2!} = \dfrac{n(n-1)}{2}$

21. $P(n, n-2) = \dfrac{n!}{[n-(n-2)]!} = \dfrac{n!}{(n-n+2)!} = \dfrac{n!}{2}$

23. Order is important here because the word *GLACIER* is different from the word *REICALG*. Therefore, this is a permutation.

25. Order is not important here. Therefore, we are dealing with a combination. If we consider a sample of three cellphones of which one is defective, it does not matter whether the first, second, or third member of our sample is defective. The net result is a sample of three cellphones, of which one is defective.

27. The order is important here. Therefore, we are dealing with a permutation. Consider, for example, nine books on a library shelf. Each of the nine books has a call number, and the books are filed in order of their call numbers; that is, a call number of 902 comes before a call number of 910.

29. The order is not important here, and consequently we are dealing with a combination. There is no difference between the hand Q Q Q 5 5 and the hand 5 5 Q Q Q. In each case the hand consists of three queens and a pair of fives.

31. The number of four-letter permutations is $P(4, 4) = \dfrac{4!}{0!} = 4 \cdot 3 \cdot 2 \cdot 1 = 24$.

33. The number of seating arrangements is $P(4, 4) = \dfrac{4!}{0!} = 24$.

35. The number of different batting orders is $P(9, 9) = \dfrac{9!}{0!} = 362{,}880$.

37. The number of different ways the three candidates can be selected is $C(12, 3) = \dfrac{12!}{9!\,3!} = \dfrac{12 \cdot 11 \cdot 10}{3 \cdot 2 \cdot 1} = 220$.

39. There are ten letters in the word *ANTARCTICA*, including 3 *A*s, 2 *C*s, 1 *I*, 1 *N*, 1 *R*, and 2 *T*s. Therefore, we use the formula for the permutation of n objects, not all distinct: $\dfrac{n!}{n_1!\,n_2! \cdots n_r!} = \dfrac{10!}{3!\,2!\,2!} = 151{,}200$.

41. The vowels cannot be permuted among themselves and may be considered as identical, so we can view the problem as that of finding the number of permutations of seven letters, taken all together, where two of the letters are identical. Thus, the result is $\dfrac{7!}{2!\,(1!)^5} = (7)(6)(5)(4)(3) = 2520$.

43. Here we use Formula 7. The number of distinct numbers is given by $\dfrac{5!}{3!\,1!\,1!} = 20$.

45. The number of ways the three sites can be selected is $C(12, 3) = \dfrac{12!}{9!\,3!} = \dfrac{12 \cdot 11 \cdot 10}{3 \cdot 2 \cdot 1} = 220$.

47. The number of ways in which the sample of three microprocessors can be selected is
$C(100, 3) = \dfrac{100!}{97!\,3!} = \dfrac{100 \cdot 99 \cdot 98}{3 \cdot 2 \cdot 1} = 161{,}700$.

49. In this case order is important, as it makes a difference whether a commercial is shown first, last, or in between. The number of ways that the director can schedule the commercials is given by $P(6, 6) = 6! = 720$.

51. Jenny has $C(3, 1) + C(4, 1) = 3 + 4 = 7$ choices.

53. The inquiries can be directed in $P(12, 6) = \dfrac{12!}{6!} = 12 \cdot 11 \cdot 10 \cdot 9 \cdot 8 \cdot 7 = 665,280$ ways.

55. There are
$$C(8, 4) \cdot C(7, 4) + C(8, 4) \cdot C(6, 3) \quad = C(8, 4)\,[C(7, 4) + C(6, 3)] = \frac{8!}{4!\,4!}\left(\frac{7!}{3!\,4!} + \frac{6!}{3!\,3!}\right)$$
$$= 70\,(35 + 20) = 3850 \text{ possible ensembles.}$$

57. a. The ten books can be arranged in $P(10, 10) = 10! = 3,628,800$ ways.

    b. The books on the same subject are grouped together, so we multiply the number of ways the mathematics books can be arranged, the number of ways the social science books can be arranged, the number of ways the biology books can be arranged, and the number of ways the three sets of books can be arranged. Thus, there are $P(3, 3) \times P(4, 4) \times P(3, 3) \times P(3, 3) = 5184$ ways.

59. Notice that order is certainly important here.

    a. The number of ways that the 20 featured items can be arranged is given by $P(20, 20) = 20! = 2.43 \times 10^{18}$.

    b. If items from the same department must appear in the same row, then the number of ways they can be arranged on the page is
$$\begin{pmatrix}\text{number of ways} \\ \text{of arranging the rows}\end{pmatrix} \cdot \begin{pmatrix}\text{number of ways of arranging} \\ \text{the items in each of the five rows}\end{pmatrix}$$
$$= P(5, 5) \cdot [P(4, 4) \times P(4, 4) \times P(4, 4) \times P(4, 4) \times P(4, 4)] = 5! \cdot (4!)^5 = 955,514,880.$$

61. a. $P(12, 9) = \dfrac{12!}{3!} = 79,833,600.$ **b.** $C(12, 9) = \dfrac{12!}{3!\,9!} = 220$ **c.** $C(12, 9) \cdot C(3, 2) = 220 \cdot 3 = 660$

63. The number of ways is given by
$$(\text{number of players}) \left[\begin{pmatrix}\text{number of ways to win} \\ \text{in exactly two sets}\end{pmatrix} + \begin{pmatrix}\text{number of ways to win} \\ \text{in exactly three sets}\end{pmatrix}\right]$$
$$= 2\,\{C(2, 2) + [C(3, 2) - C(2, 2)]\} = 2\,[1 + (3 - 1)] = 2 \cdot 3 = 6.$$

65. The number of ways the measure can be passed is $C(3, 3)\,[C(8, 6) + C(8, 7) + C(8, 8)] = 37$. Here three of the three permanent members must vote for passage of the bill and this can be done in $C(3, 3) = 1$ way. Of the eight nonpermanent members who are voting, six, seven, or eight can vote for passage of the bill. Therefore, there are $C(8, 6) + C(8, 7) + C(8, 8) = 28 + 8 + 1 = 37$ ways that the nonpermanent members can vote to ensure passage of the measure. This gives $1 \times 37 = 37$ ways that the members can vote so that the bill is passed.

67. a. If no preference is given to any student, then the number of ways of awarding the three teaching assistantships is
$$C(12, 3) = \frac{12!}{3!\,9!} = 220.$$

    b. If it is stipulated that one particular student receive one of the assistantships, then the remaining two assistantships must be awarded to two of the remaining 11 students. Thus, the number of ways is
$$C(11, 2) = \frac{11!}{2!\,9!} = 55.$$

c. If at least one woman is to be awarded one of the assistantships, and the group of students consists of seven men and five women, then the number of ways the assistantships can be awarded is

$$C(5,1) \times C(7,2) + C(5,2) \times C(7,1) + C(5,3) = \frac{5!}{4!\,1!} \cdot \frac{7!}{5!\,2!} + \frac{5!}{3!\,2!} \cdot \frac{7!}{6!\,1!} + \frac{5!}{3!\,2!} = 105 + 70 + 10 = 185.$$

69. The number of ways of awarding the three contracts to seven different firms is given by $P(7,3) = \frac{7!}{4!} = 210$. The number of ways of awarding the three contracts to two different firms (so one firm gets two contracts) from a choice of seven different firms is $C(7,2) \times P(3,2) = 126$ (first pick the two firms, and then award the 3 contracts). Therefore, the number of ways the contracts can be awarded if no firm is to receive more than two contracts is $210 + 126 = 336$.

71. The number of different curricula that are available for the student's consideration is given by

$$C(5,1) \times C(3,1) \times C(6,2) \times [C(4,1) + C(3,1)] = \frac{5!}{4!\,1!} \cdot \frac{3!}{2!\,1!} \cdot \frac{6!}{4!\,2!} \cdot \left(\frac{4!}{3!\,1!} + \frac{3!}{2!\,1!} \right)$$

$$= (5)(3)(15)(4) + (5)(3)(15)(3) = 900 + 675 = 1575.$$

73. The number of ways of dealing a straight flush (five cards in sequence in the same suit) is given by

$$\binom{\text{number of ways of selecting five cards}}{\text{in sequence in the same suit}} \cdot \binom{\text{number of ways}}{\text{of selecting a suit}} = 10 \cdot 4 = 40.$$

75. The number of ways of dealing a flush (five cards in one suit that are not all in sequence) is given by

$$\binom{\text{number of ways of selecting}}{\text{five cards in the same suit}} - \binom{\text{number of}}{\text{straight flushes}} = 4C(13,5) - 40 = 5148 - 40 = 5108.$$

77. The number of ways of dealing a full house (three of a kind and a pair) is given by

$$\binom{\text{number of}}{\text{different ranks}} \cdot \binom{\text{number of ways of picking}}{\text{three of a kind of that rank}} \cdot \binom{\text{number of ways of picking a pair}}{\text{from the 12 remaining ranks}}$$

$$= 13 \cdot C(4,3) \cdot 12C(4,2) = 13 \cdot 4 \cdot 12(6) = 3744.$$

79. The bus will travel a total of 6 blocks. Each route must include 2 blocks running north and south and 4 blocks running east and west. To compute the total number of possible routes, it suffices to compute the number of ways the 2 blocks running north and south can be selected from the 6 blocks. Thus, the number of possible routes is

$$C(6,2) = \frac{6!}{2!\,4!} = 15.$$

81. The number of ways that the quorum can be formed is given by

$$C(12,6) + C(12,7) + C(12,8) + C(12,9) + C(12,10) + C(12,11) + C(12,12)$$

$$= \frac{12!}{6!\,6!} + \frac{12!}{7!\,5!} + \frac{12!}{8!\,4!} + \frac{12!}{9!\,3!} + \frac{12!}{10!\,2!} + \frac{12!}{11!\,1!} + \frac{12!}{12!\,0!} = 924 + 792 + 495 + 220 + 66 + 12 + 1 = 2510.$$

83. Using the formula from Exercise 78, we see that the number of ways of seating the five commentators at a round table is $(5 - 1)! = 4! = 24$.

85. The number of possible corner points is $C(8,3) = \frac{8!}{5!\,3!} = 56$.

87. True.

89. True. $C(n,r) = \dfrac{n!}{(n-r)!\,r!}$ and $C(n,n-r) = \dfrac{n!}{[n-(n-r)]!\,(n-r)!} = \dfrac{n!}{r!\,(n-r)!}$, so

$C(n,r) = C(n,n-r)$.

Technology Exercises   page 382

1. $1.307674368 \times 10^{12}$ **3.** $2.56094948229 \times 10^{16}$ **5.** 674,274,182,400   **7.** 133,784,560

9. 4,656,960

11. Using the multiplication principle, the number of 10-question exams she can set is given by

$C(25,3) \times C(40,5) \times C(30,2) = 658{,}337{,}004{,}000.$

CHAPTER 6   Concept Review Questions   page 383

1. set, elements, set

3. subset

5. a. union   **b.** intersection

7. $A^c \cap B^c \cap C^c$

CHAPTER 6   Review Exercises   page 384

1. {3}. The set consists of all solutions to the equation $3x - 2 = 7$.

3. {4, 6, 8, 10}

5. Yes

7. Yes

9. $A \cup (B \cap C)$

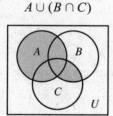

11. $A^c \cap B^c \cap C^c$

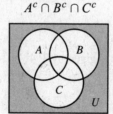

13. $A \cup (B \cup C) = \{a,b\} \cup [\{b,c,d\} \cup \{a,d,e\}] = \{a,b\} \cup \{a,b,c,d,e\} = \{a,b,c,d,e\}$, while

$(A \cup B) \cup C = [\{a,b\} \cup \{b,c,d\}] \cup \{a,d,e\} = \{a,b,c,d\} \cup \{a,d,e\} = \{a,b,c,d,e\}$.

15. $A \cap (B \cup C) = \{a,b\} \cap [\{b,c,d\} \cup \{a,d,e\}] = \{a,b\} \cap \{a,b,c,d,e\} = \{a,b\}$, while

$(A \cap B) \cup (A \cap C) = [\{a,b\} \cap \{b,c,d\}] \cup [\{a,b\} \cap \{a,d,e\}] = \{b\} \cup \{a\} = \{a,b\}$.

17. The set of all participants in a consumer behavior survey who both avoided buying a product because it is not recyclable and boycotted a company's products because of its record on the environment.

19. The set of all participants in a consumer behavior survey who both did not use cloth diapers rather than disposable diapers and voluntarily recycled their garbage.

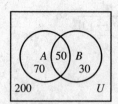

21. $n(A \cup B) = n(A) + n(B) - n(A \cap B) = 120 + 80 - 50 = 150.$

23. $n(B^c) = n(U) - n(B) = 350 - 80 = 270.$

25. $n(A \cap B^c) = n(A) - n(A \cap B) = 120 - 50 = 70.$

27. $C(20, 18) = \dfrac{20!}{18! \, 2!} = 190.$

29. $C(5, 3) \cdot P(4, 2) = \dfrac{5!}{3! \, 2!} \cdot \dfrac{4!}{2!} = 10 \cdot 12 = 120.$

31. Let U denote the set of 5 major cards, $A = \{x \in U \mid x \text{ offered cash advances}\}$,
$B = \{x \in U \mid x \text{ offered extended payments for all goods and services purchased}\}$,
and $C = \{x \in U \mid x \text{ required an annual fee that was less than \$35}\}$. Thus,
$n(A) = 3$, $n(B) = 3$, $n(C) = 2$, $n(A \cap B) = 2$, $n(B \cap C) = 1$, and
$n(A \cap B \cap C) = 0$. From the diagram, we have $x + y + 2 = 3$ and $y + 2 = 2$.
Solving, we find $x = 1$, and $y = 0$. Therefore, the number of cards that offer cash
advances and have an annual fee that is less than \$35 is given by
$n(A \cap C) = y = 0.$

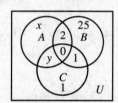

33. The number of ways the DVDs can be arranged on a shelf is $P(6, 6) = 6! = 720.$

35. The number of ways is given by $3! \, 4! = 144.$

37. a. Because there is repetition of the letters C, I, and N, we use the formula for the permutation of n objects (not all distinct) with $n = 10$, $n_1 = 2$, $n_2 = 3$, and $n_3 = 3$. Then the number of permutations that can be formed is given by $\dfrac{10!}{2! \, 3! \, 3!} = 50{,}400.$

b. Again we use the formula for the permutation of n objects (not all distinct), this time with $n = 8$, $n_1 = 2$, $n_2 = 2$, and $n_3 = 2$. Then the number of permutations is given by $\dfrac{8!}{2! \, 2! \, 2!} = 5040.$

39. The number of selections a customer can make is $(3)(3)(4)(3) = 108.$

41. Let U denote the set comprising Halina's clients, $A = \{x \in U \mid x \text{ owns stocks}\}$, $B = \{x \in U \mid x \text{ owns bonds}\}$, and $C = \{x \in U \mid x \text{ owns mutual funds}\}$. Then $n(A) = 300$, $n(B) = 180$, $n(C) = 160$, $n(A \cap B) = 110$, $n(A \cap C) = 120$, and $n(B \cap C) = 90$. With $n(A \cap B \cap C) = z$, we have the Venn diagram shown. Using the facts that $n(A) = 300$, $n(B) = 180$, and $n(C) = 160$, we have the system

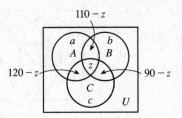

$$\begin{cases} a + (110 - z) + (120 - z) + z = 300 \\ b + (110 - z) + (90 - z) + z = 180 \\ c + (120 - z) + (90 - z) + z = 160 \\ a + b + c + (110 - z) + (120 - z) + (90 - z) + z = 400 \end{cases}$$

which simplifies to $a - z = 70$, $b - z = -20$, $c - z = -50$, and $a + b + c - 2z = 80$. Solving, we find $a = 150$, $b = 60$, $c = 30$, and $z = 80$. Therefore, the number who own stocks, bonds and mutual funds is $n(A \cap B \cap C) = z = 80$.

43. The number of possible outcomes is $(6)(4)(5)(6) = 720$.

45. **a.** If order matters, the number of possibilities is given by $52 \cdot 52 = 2704$.

    b. If order does not matter, the number of possibilities is given by $52 \cdot 51 = 2652$.

47. **a.** The number of different ways the sample can be selected is given by $C(60, 4) = \dfrac{60!}{56! \, 4!} = 487{,}635$.

    b. The number of samples that contain three defective CPUs is given by
    $$C(5, 3) \cdot C(55, 1) = \frac{5!}{3! \, 2!} \cdot \frac{55!}{54! \, 1!} = 10(55) = 550.$$

    c. The number of samples that do not contain a defective CPU is given by $C(55, 4) = \dfrac{55!}{51! \, 4!} = 341{,}055$.

49. **a.** If there are no restrictions, then the number of ways is given by $6! = 720$.

    b. If the women and men must alternate then the number of ways is given by
    $$\begin{pmatrix} \text{number of ways of} \\ \text{seating the men} \end{pmatrix} \cdot \begin{pmatrix} \text{number of ways of} \\ \text{seating the women} \end{pmatrix} \cdot \begin{pmatrix} \text{number of ways the first seat} \\ \text{is filled (man or woman)} \end{pmatrix} = 3! \cdot 3! \cdot 2 = 72.$$
    Once again, we can also think of the number of possible seating arrangements as the number of ways of filling in six blanks. Then the number of possibilities is given by $\underline{6} \cdot \underline{3} \cdot \underline{2} \cdot \underline{2} \cdot \underline{1} \cdot \underline{1} = 72$.

    c. Following the same reasoning as above, we find that the number of ways three married couples can be seated in a row if each married couple sits together is $3! \cdot 2 \cdot 2 \cdot 2 = 48$.

CHAPTER 6   Before Moving On...   page 386

1. **a.** $B \cup C = \{b, c, d, e, f, g\}$, so $A \cap (B \cup C) = \{d, f, g\}$.

   b. $A \cap C = \{f\}$, and so $(A \cap C) \cup (B \cup C) = \{b, c, d, e, f, g\}$.

   c. $A^c = \{b, c, e\}$.

2.

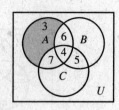

The shaded area is $A \cap (B \cup C)^c$ and

$$n\left[A \cap (B \cup C)^c\right] = 20 - (7 + 4 + 6) = 3.$$

3. The number of possibilities is $C(6, 4) = \dfrac{6!}{4!\,2!} = 15$.

4. The number of ways of obtaining the three deuces is $C(4, 3) = \dfrac{4!}{3!\,1!} = 4$. The number of ways of obtaining the two face cards is $C(12, 2) = \dfrac{12!}{2!\,10!} = 66$. Therefore, the number of hands with three deuces and two face cards is $4\,(66) = 264$.

5. There are $C(6, 3) = \dfrac{6!}{3!\,3!} = 20$ ways of picking the three seniors and $C(5, 2) = \dfrac{5!}{2!\,3!} = 10$ ways of picking the two juniors. Therefore, there are $20 \cdot 10 = 200$ possible teams.

7 PROBABILITY

7.1 Experiments, Sample Spaces, and Events

Problem-Solving Tips

1. The **union of two events** A **and** B, written $A \cup B$, is the set of outcomes in A and/or B. The **intersection of two events** A **and** B, written $A \cap B$, is the set of outcomes in both A and B. **The complement of an event** A, written A^c, is the set of outcomes in the sample space S that are not in A.

2. Two events A and B are mutually exclusive if $A \cap B = \varnothing$. In other words, the two events cannot occur at the same time.

Concept Questions   page 393

1. An experiment is an activity with observable results. Examples vary.

Exercises   page 393

1. $E \cup F = \{a, b, d, f\}$, $E \cap F = \{a\}$.

3. $F^c = \{b, c, e\}$, $E \cap G^c = \{a, b\} \cap \{a, d, f\} = \{a\}$.

5. Because $E \cap F = \{a\}$ is not a null set, we conclude that E and F are not mutually exclusive.

7. $E \cup F \cup G = \{2, 4, 6\} \cup \{1, 3, 5\} \cup \{5, 6\} = \{1, 2, 3, 4, 5, 6\}$.

9. $(E \cup F \cup G)^c = \{1, 2, 3, 4, 5, 6\}^c = \varnothing$.

11. Yes, $E \cap F = \varnothing$, that is, E and F do not contain any common elements.

13. $E^c = \{2, 4, 6\}^c = \{1, 3, 5\} = F$ and so E and F are complementary.

15. $E \cup F$ 　　　　　　　　　　　17. G^c 　　　　　　　　　　　19. $(E \cup F \cup G)^c$

21. **a.** Refer to Example 4 on page 390 of the text.
    $E = \{(2, 1), (3, 1), (4, 1), (5, 1), (6, 1), (3, 2), (4, 2), (5, 2), (6, 2), (4, 3), (5, 3), (6, 3), (5, 4), (6, 4), (6, 5)\}$.
    b. $E = \{(1, 2), (2, 4), (3, 6)\}$.

23. $\varnothing, \{a\}, \{b\}, \{c\}, \{a, b\}, \{a, c\}, \{b, c\}, \{a, b, c\}$.

25. **a.** $S = \{R, B\}$ 　　　　　　**b.** $\varnothing, \{B\}, \{R\}, \{B, R\}$

27. **a.** $S = \{(H, 1), (H, 2), (H, 3), (H, 4), (H, 5), (H, 6), (T, 1), (T, 2), (T, 3), (T, 4), (T, 5), (T, 6)\}$
    b. $E = \{(H, 2), (H, 4), (H, 6)\}$

29. Here $S = \{1, 2, 3, 4, 5, 6\}$, $E = \{2\}$, and $F = \{2, 4, 6\}$.

   a. Because $E \cap F = \{2\} \neq \varnothing$, we conclude that E and F are not mutually exclusive.

   b. $E^c = \{1, 3, 4, 5, 6\} \neq F$, and so E and F are not complementary.

31. $S = \{ddd, ddn, dnd, ndd, dnn, ndn, nnd, nnn\}$

33. a.

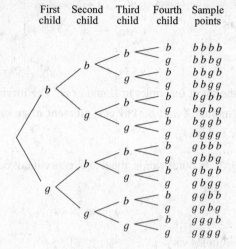

From the tree diagram, we see that the required sample space is

$S = \{bbbb, bbbg, bbgb, bbgg, bgbb, bgbg, bggb, bggg,$
$\quad gbbb, gbbg, gbgb, gbgg, ggbb, ggbg, gggb, gggg\}$.

   b. $E = \{bbbg, bbgb, bgbb, gbbb\}$

   c. $F = \{bbbg, bbgg, bgbg, bggg, gbbg, gbgg, ggbg, gggg\}$

   d. $G = \{gbbg, gbgg, ggbg, gggg\}$

35. a. $\{ABC, ABD, ABE, ACD, ACE, ADE, BCD, BCE, BDE, CDE\}$.   **b.** 6   **c.** 3   **d.** 6

37. a. E^c **b.** $E^c \cap F^c$ **c.** $E \cup F$ **d.** $(E \cap F^c) \cup (E^c \cap F)$

39. a. $S = \{t \mid t > 0\}$ **b.** $E = \{t \mid 0 < t \leq 2\}$ **c.** $F = \{t \mid t > 2\}$

41. a. $S = \{0, 1, 2, 3, \ldots, 10\}$ **b.** $E = \{0, 1, 2, 3\}$ **c.** $F = \{5, 6, 7, 8, 9, 10\}$

43. a. $S = \{0, 1, 2, \ldots, 20\}$ **b.** $E = \{0, 1, 2, \ldots, 9\}$ **c.** $F = \{20\}$

45. Let S denote the sample space of the experiment that is the set of 52 cards. Then $E = \{x \in S \mid x \text{ is an ace}\}$, $F = \{x \in S \mid x \text{ is a spade}\}$, and $E \cap F = \{x \in S \mid x \text{ is the ace of spades}\}$. Now $n(E) = 4$, $n(F) = 13$, and $n(E \cap F) = 1$. Also, $E \cup F = \{x \in S \mid x \text{ is an ace or a spade}\}$ and $n(E \cup F) = 16$, so $n(E) + n(F) - n(E \cap F) = 4 + 13 - 1 = 16 = n(E \cup F)$.

47. $E^c \cap F^c = (E \cup F)^c$ by DeMorgan's Law. Because $(E \cup F) \cap (E \cup F)^c = \varnothing$, they are mutually exclusive.

49. False. Let $E = \{1, 2, 3\}$, $F = \{4, 5, 6\}$, and $G = \{4, 5\}$. Then $E \cap F = \varnothing$ and $E \cap G = \varnothing$, but $F \cap G = \{4, 5\} \neq \varnothing$.

7.2   Definition of Probability

Problem-Solving Tips

1. **Uniform sample spaces** are sample spaces in which the outcomes are equally likely.

2. Events consisting of a single outcome are called **simple events**.

3. To find the probability of an event E, follow these steps:

a. Determine an appropriate space S associated with the experiment.

b. Assign probabilities to the simple events of the experiment. Then $P(E) = P(s_1) + P(s_2) + P(s_3) + \cdots + P(s_n)$, where $E = \{s_1, s_2, s_3, \ldots, s_n\}$ and $\{s_1\}, \{s_2\}, \{s_3\}, \ldots, \{s_n\}$ are the simple events of S.

Concept Questions   page 401

1. a. By assigning probabilities to each simple event of an experiment, we obtain a *probability distribution* that gives the probability of each simple event. Examples vary.

b. The function P that assigns a probability to each of the simple events is called a *probability function*. Examples vary.

3. $P(E) = P(s_1) + P(s_2) + \cdots + P(s_n)$; $P(\varnothing) = 0$.

Exercises   page 401

1. $\{(H, H)\}, \{(H, T)\}, \{(T, H)\}, \{(T, T)\}$.

3. $\{(D, m)\}, \{(D, f)\}, \{(R, m)\}, \{(R, f)\}, \{(I, m)\}, \{(I, f)\}$

5. $\{(1, i)\}, \{(1, d)\}, \{(1, s)\}, \{(2, i)\}, \{(2, d)\}, \{(2, s)\}, \ldots, \{(5, i)\}, \{(5, d)\}, \{(5, s)\}$

7. $\{(A, Rh^+)\}, \{(A, Rh^-)\}, \{(B, Rh^+)\}, \{(B, Rh^-)\}, \{(AB, Rh^+)\}, \{(AB, Rh^-)\}, \{(O, Rh^+)\}, \{(O, Rh^-)\}$

9. a.

Answer	1–2	3–4	5 or more
Probability	.65	.20	.15

b. The probability is .20.

11. a.

Answer	Low	Middle	Extreme	No response
Probability	.34	.44	.20	.02

b. The probability is .20.

13.

Grade	A	B	C	D	F
Probability	.10	.25	.45	.15	.05

15. a.

Answer	New processes	Getting to know...	New technology	Fitting in	Other
Probability	.44	.20	.17	.12	.07

b. The probability is .12.

17.

Rating	A	B	C	D	E
Probability	.026	.199	.570	.193	.012

19.

Number of days	0	1	2	3	4	5	6	7
Probability	.05	.06	.09	.15	.11	.20	.17	.17

21. a. $S = \{(0 < x \le 200), (200 < x \le 400), (400 < x \le 600), (600 < x \le 800), (800 < x \le 1000), (x > 1000)\}$

b.

Number of cars (x)	Probability
$0 < x \le 200$	.075
$200 < x \le 400$	.1
$400 < x \le 600$	.175
$600 < x \le 800$	.35
$800 < x \le 1000$	.225
$x > 1000$	.075

23. The probability is $\frac{84,000,000}{179,000,000} \approx .469$.

25. a. The probability that a person killed by lightning is a male is $\frac{376}{439} \approx .856$.

b. The probability that a person killed by lightning is a female is $\frac{439 - 376}{439} = \frac{63}{439} \approx .144$.

27. The probability that the retailer uses electronic tags as antitheft devices is $\frac{81}{176} \approx .460$.

29. a. $P(D) = \frac{13}{52} = \frac{1}{4}$ **b.** $P(B) = \frac{26}{52} = \frac{1}{2}$ **c.** $P(A) = \frac{4}{52} = \frac{1}{13}$

31. The probability of arriving at the traffic light when it is red is $\frac{30}{30 + 5 + 45} = \frac{30}{80} = \frac{3}{8}$.

33. No. Because the die is loaded, the outcomes are not equally likely.

35. Yes, the outcomes are equally likely.

37. a. $P(A) = P(s_1) + P(s_2) + P(s_4) = \frac{1}{14} + \frac{3}{14} + \frac{2}{14} = \frac{3}{7}$.

b. $P(B) = P(s_1) + P(s_5) = \frac{1}{14} + \frac{2}{14} = \frac{3}{14}$.

c. $P(C) = P(S) = 1$.

39. a. The event E that the interviewees include applicant a is given by $E = \{ab, ac, ad, ae, ba, ca, da, ea\}$. The required probability is $\frac{n(E)}{n(S)} = \frac{8}{20} = .4$.

b. The event F that the interviewees include applicants a and c is given by $F = \{ac, ca\}$. The required probability is $\frac{n(F)}{n(S)} = \frac{2}{20} = .1$.

c. The event G that the interviewees include applicants d and e is $G = \{de, ed\}$. The required probability is $\frac{n(G)}{n(S)} = \frac{2}{20} = .1$.

41. a. $P(E) = \frac{62}{9 + 62 + 27} = \frac{62}{98} \approx .633$ **b.** $P(E) = \frac{27}{98} \approx .276$

43. a. The probability that a registered voter favors the proposition is .35.

b. The probability that a registered voter is undecided about the proposition is $1 - .35 - .32 = .33$.

45. The required probability is $\frac{281 + 251}{382 + 281 + 251 + 90} \approx .530$.

47. a. The required probability is $\dfrac{25 + 15}{37 + 14 + 25 + 15 + 9} \approx .4$.

  b. The required probability is $\dfrac{14 + 9}{37 + 14 + 25 + 15 + 9} \approx .23$.

49. a. The required probability is $\dfrac{448}{448 + 169 + 155 + 100 + 22 + 106} = .448$.

  b. The required probability is $\dfrac{155 + 100}{448 + 169 + 155 + 100 + 22 + 106} = .255$.

51. The probability that the primary cause of the crash was due to pilot error or bad weather is given by

$$\dfrac{327 + 22}{327 + 49 + 14 + 22 + 19 + 15} = \dfrac{349}{446} \approx .783.$$

53. True

7.3   Rules of Probability

Problem-Solving Tips

If S is a sample space of an experiment and E and F are events of the experiment, then the following are always true:

a. $P(E) \geq 0$ for any E.

b. $P(S) = 1$.

c. If E and F are mutually exclusive, then $P(E \cup F) = P(E) + P(F)$. More generally, if E and F are any two events of an experiment, then $P(E \cup F) = P(E) + P(F) - P(E \cap F)$.

d. $P(E^c) = 1 - P(E)$.

Concept Questions   page 411

1. a. The event E cannot occur.

  b. There is a 50% chance that the event F will occur.

  c. The event S will certainly occur.

  d. The probability of the event $E \cup F$ occurring is given by the sum of the probabilities of E and F minus the probability of $E \cap F$.

Exercises   page 411

1. Refer to Example 4 on page 390 of the text. Let E denote the event of interest. Then $P(E) = \frac{18}{36} = \frac{1}{2}$

3. Refer to Example 4 on page 390 of the text. The event of interest is $E = \{1, 1\}$, and $P(E) = \frac{1}{36}$.

5. Let E denote the event of interest. Then $E = \{(6, 2), (6, 1), (1, 6), (2, 6)\}$ and $P(E) = \frac{4}{36} = \frac{1}{9}$.

7. Let E denote the event that the card drawn is a king and F the event that the card drawn is a diamond. Then the required probability is $P(E \cap F) = \frac{1}{52}$.

9. Let E denote the event that a face card is drawn. Then $P(E) = \frac{12}{52} = \frac{3}{13}$.

11. Let E denote the event that an ace is drawn. Then $P(E) = \frac{1}{13}$. Then E^c is the event that an ace is not drawn and $P(E^c) = 1 - P(E) = \frac{12}{13}$.

13. Let E denote the event that a ticketholder will win first prize. Then $P(E) = \frac{1}{500} = .002$, and the probability of the event that a ticketholder will not win first prize is $P(E^c) = 1 - .002 = .998$.

15. Property 2 of the laws of probability is violated. The sum of the probabilities must add up to 1. In this case $P(S) = 1.1$, which is not possible.

17. The five events are not mutually exclusive: the probability of winning at least one purse is
$$1 - (\text{probability of losing all five times}) = 1 - \frac{9^5}{10^5} \approx 1 - .5905 = .4095.$$

19. The two events are not mutually exclusive; hence, the probability of the given event is $\frac{1}{6} + \frac{1}{6} - \frac{1}{36} = \frac{11}{36}$.

21. $E^c \cap F^c = \{c, d, e\} \cap \{a, b, e\} = \{e\} \neq \varnothing$.

23. **a.** $P(E \cap F) = 0$ because E and F are mutually exclusive.
    b. $P(E \cup F) = P(E) + P(F) - P(E \cap F) = .2 + .5 = .7$.
    c. $P(E^c) = 1 - P(E) = 1 - .2 = .8$.
    d. $P(E^c \cap F^c) = P\big[(E \cup F)^c\big] = 1 - P(E \cup F) = 1 - .7 = .3$.

25. **a.** $P(A) = P(s_1) + P(s_2) = \frac{1}{8} + \frac{3}{8} = \frac{1}{2}$; $P(B) = P(s_1) + P(s_3) = \frac{1}{8} + \frac{1}{4} = \frac{3}{8}$.
    b. $P(A^c) = 1 - P(A) = 1 - \frac{1}{2} = \frac{1}{2}$ and $P(B^c) = 1 - P(B) = 1 - \frac{3}{8} = \frac{5}{8}$.
    c. $P(A \cap B) = P(s_1) = \frac{1}{8}$.
    d. $P(A \cup B) = P(A) + P(B) - P(A \cap B) = \frac{1}{2} + \frac{3}{8} - \frac{1}{8} = \frac{3}{4}$.
    e. $P(A^c \cap B^c) = P\big((A \cup B)^c\big) = 1 - P(A \cup B) = 1 - \frac{3}{4} = \frac{1}{4}$.
    f. $P(A^c \cup B^c) = P\big((A \cap B)^c\big) = 1 - P(A \cap B) = 1 - \frac{1}{8} = \frac{7}{8}$.

27. **a.** Let E denote the event that the moviegoer is between 12 and 24 years old. Then $P(E) = .30$.
    b. Let F denote the event that the moviegoer is between 25 and 64 years old. Then $P(F) = .36 + .28 = .64$.
    c. Let G denote the event that the moviegoer is between 12 and 24 years old or between 65 and 74 years old. Then $P(G) = .30 + .06 = .36$.

29. **a.** The probability is .06.
    b. The probability is $.25 + .14 = .39$.
    c. The probability is $1 - .55 = .45$.

31. **a.** The probability is $.29 + .24 = .53$.
    b. The probability is $.24 + .27 = .51$.
    c. The probability is $.29 + .27 = .56$.

33. **a.** The probability is $.063 + .015 + .006 + .066 = .15$.
    b. The probability is $.374 + .066 = .44$.

35. a. The sum of the numbers $45.1 + 16.5 + 6.9 + 6.1 + 4.2 + 3.8 + 2.5 + 14.9$ is 100, and so the table does give a probability distribution.

  b. The probability is $45.1 + 6.9 = 52$ percent, or .52.

  c. The probability is $1 - (.061 + .042 + .038) = .859$.

37. Let U denote the set of cars in the experiment,
  $A = \{x \in U \mid x \text{ failed the tread test}\}$, and
  $B = \{x \in U \mid x \text{ failed the pressure test }\}$.

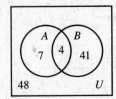

  a. The probability is $P(B \cap A^c) = \frac{41}{100} = .41$.

  b. The probability is $P\left[(A \cup B)^c\right] = \frac{48}{100} = .48$.

39.

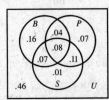

  B: Probability of buying a blouse

  P: Probability of buying pants

  S: Probability of buying a skirt

From the given information and the diagram, we have $P(B) = .35$, $P(P) = .30$, $P(S) = .27$, $P(B \cap S) = .15$, $P(S \cap P) = .19$, $P(B \cap P) = .12$, and $P(B \cap P \cap S) = .08$.

  a. The probability of exactly one item is
$$P(B \cap S^c \cap P^c) + P(P \cap B^c \cap S^c) + P(S \cap B^c \cap P^c)$$
$$= .16 + .07 + .01 = .24.$$

  b. The probability of buying no item is
$$1 - P(A \cup B \cup C)$$
$$= 1 - (.16 + .07 + .01 + .08 + .04 + .11 + .07) = .46.$$

41.

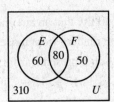

Let E and F denote the events that the person surveyed learned of the products from *Good Housekeeping* and *Ladies' Home Journal*, respectively. Then $P(E) = \frac{140}{500} = \frac{7}{25}$, $P(F) = \frac{130}{500} = \frac{13}{50}$, and $P(E \cap F) = \frac{80}{500} = .16$.

  a. $P(E \cap F) = \frac{80}{500} = .16$.

  b. $P(E \cup F) = \frac{14}{50} + \frac{13}{50} - \frac{8}{50} = \frac{19}{50} = .38$.

  c. $P(E \cap F^c) + P(E^c \cap F) = \frac{60}{500} + \frac{50}{500} = \frac{110}{500} = .22$.

43. Let A denote the event that the person has a cell phone and B the event that the person has a land line. From the Venn diagram, we see that

  a. $P(A \cap B^c) = .33$ and   **b.** $P(A^c \cap B) = .09$.

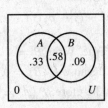

45.

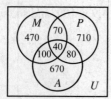

From the diagram, we see that the probability that a teacher selected at random from this group said that lack of parental support is the only problem hampering a student's schooling is $\frac{710}{2140} \approx .332$.

P: Lack of parental support

M: Malnutrition

A: Abused or neglected

47.

	Cars	Light trucks	Large trucks	Buses	Other
	$\frac{2748}{8242}$	$\frac{4814}{8242}$	$\frac{443}{8242}$	$\frac{10}{8242}$	$\frac{227}{8242}$

a. The probability is $\frac{2748}{8242} \approx .333$.

b. The probability is $\frac{4814}{8242} \approx .584$.

c. The probability is $\frac{10}{8242} + \frac{443}{8242} \approx .055$.

49. a. The probability is $\frac{253+304+81}{1012} \approx .63$.   **b.** The probability is $\frac{253+304}{1012} \approx .55$.

c. The probability is $\frac{334}{1012} \approx .33$.

51. a. The probability that the plane crashed while taxiing on the ground or while en route is given by $\frac{4+5}{100} = .09$.

b. The probability that the plane crashed during takeoff or landing is given by $\frac{10+31}{100} = .41$.
It is at the highest risk during descent and approach and during landing.

53.

Number of years car is kept (x)	Probability
$0 \leq x < 1$	.03
$1 \leq x < 3$	.22
$3 \leq x < 5$	.18
$5 \leq x < 7$	.17
$7 \leq x < 10$	.12
$x \geq 10$	.28

a. The probability that an automobile owner selected at random from those surveyed plans to keep his or her present car for less than five years is $.18 + .22 + .03 = .43$.

b. The probability that an automobile owner selected at random from those surveyed plans to keep his or her present car for three or more years is $.18 + .17 + .12 + .28 = .75$.

55. The probability that Bill will fail to solve the problem is $1 - p_1$ and the probability that Mike will fail to solve the problem is $1 - p_2$. Therefore, the probability that both Bill and Mike will fail to solve the problem is $(1 - p_1)(1 - p_2)$. So, the probability that at least one of them will solve the problem is $1 - (1 - p_1)(1 - p_2) = 1 - (1 - p_2 - p_1 + p_1 p_2) = p_1 + p_2 - p_1 p_2$.

57. True. Write $B = A \cup (B - A)$. Because A and $B - A$ are mutually exclusive, we have $P(B) = P(A) + P(B - A)$. Because $P(B) = 0$, we have $P(A) + P(B - A) = 0$. If $P(A) > 0$, then $P(B - A) < 0$ and this is not possible. Therefore, $P(A) = 0$.

59. False. Take $E_1 = \{1, 2\}$ and $E_2 = \{2, 3\}$, where $S = \{1, 2, 3\}$. Then $P(E_1) = \frac{2}{3}$ and $P(E_2) = \frac{2}{3}$, but $P(E_1 \cup E_2) = P(S) = 1$.

61. True. Suppose that $E \subseteq F$, but $E \neq F$. Then $E^c \cap F \neq \varnothing$, and furthermore $F = E \cup (E^c \cap F)$, where E and $E^c \cap F$ are mutually exclusive. Therefore, $P(F) = P(E) + P(E^c \cap F)$, and since $P(E^c \cap F) \neq 0$, we see that $P(F) < P(E)$, contradicting the given condition that $P(E) = P(F)$. Thus we must have $E = F$.

7.4   Use of Counting Techniques in Probability

Problem-Solving Tips

If S is a uniform sample space and E is any event in S, then $P(E) = \dfrac{\text{Number of favorable outcomes in } E}{\text{Number of possible outcomes in } S} = \dfrac{n(E)}{n(S)}$.

Concept Questions   page 422

1. If E is an event in a uniform sample space S, then the probability of E occurring is
$$P(E) = \frac{\text{Number of outcomes in } E}{\text{Number of outcomes in } S} = \frac{n(E)}{n(S)}.$$

Exercises   page 422

1. Let E denote the event that the coin lands heads all five times. Then $P(E) = \dfrac{1}{2^5} = \dfrac{1}{32}$.

3. Let E denote the event that the coin lands tails all five times. Then E^c is the event that the coin lands heads at least once, and $P(E^c) = 1 - P(E) = 1 - \frac{1}{32} = \frac{31}{32}$.

5. Let E denote the event that a pair is drawn. Then $P(E) = \dfrac{13 \cdot C(4, 2)}{C(52, 2)} = \dfrac{78}{1326} \approx .059$.

7. Let E denote the event that two black cards are drawn. Then $P(E) = \dfrac{C(26, 2)}{C(52, 2)} = \dfrac{325}{1326} \approx .245$.

9. The probability of the event that two of the balls will be white and two will be blue is
$$P(E) = \frac{n(E)}{n(S)} = \frac{C(3, 2) \cdot C(5, 2)}{C(8, 4)} = \frac{3 \cdot 10}{70} = \frac{3}{7}.$$

11. The probability of the event that exactly three of the balls are blue is $P(E) = \dfrac{n(E)}{n(S)} = \dfrac{C(5, 3) \, C(3, 1)}{C(8, 4)} = \dfrac{30}{70} = \dfrac{3}{7}$.

13. The probability of the event that a family will have one boy and two girls is $P(E) = \dfrac{C(3, 2)}{8} = \dfrac{3}{8}$.

15. The probability of the event that a family with three children has three boys is $P(E) = \dfrac{C(3, 3)}{8} = \dfrac{1}{8}$.

17. The number of elements in the sample space is 2^{10}. There are $C(10, 6) = \dfrac{10!}{6! \, 4!} = 210$ ways of answering exactly six questions correctly. Therefore, the required probability is $\dfrac{210}{2^{10}} = \dfrac{210}{1024} \approx .205$.

19. a. Let E denote the event that both of the bulbs are defective. Then

$$P(E) = \frac{C(4,2)}{C(24,2)} = \frac{\frac{4!}{2!\,2!}}{\frac{24!}{22!\,2!}} = \frac{4 \cdot 3}{24 \cdot 23} = \frac{1}{46} \approx .022.$$

b. Let F denote the event that none of the bulbs is defective. Then

$$P(F) = \frac{C(20,2)}{C(24,2)} = \frac{20!}{18!\,2!} \cdot \frac{22!\,2!}{24!} = \frac{20}{24} \cdot \frac{19}{23} \approx .6884.$$ Therefore, the probability that at least one of the light

bulbs is defective is $1 - P(F) \approx 1 - .6884 = .3116.$

21. a. The probability that both of the cartridges are defective is $P(E) = \dfrac{C(6,2)}{C(80,2)} = \dfrac{15}{3160} \approx .005.$

b. Let F denote the event that none of the cartridges is defective. Then $P(F) = \dfrac{C(74,2)}{C(80,2)} = \dfrac{2701}{3160} \approx .855,$ and so

$P(F^c) = 1 - P(F) \approx 1 - .855 = .145$ is the probability that at least one of the cartridges is defective.

23. a. The probability that Mary's name will be selected is $P(E) = \frac{12}{100} = .12.$ The probability that both Mary's and

John's names will be selected is $P(F) = \dfrac{C(98,10)}{C(100,12)} = \dfrac{\frac{98!}{88!\,10!}}{\frac{100!}{88!\,12!}} = \dfrac{12 \cdot 11}{100 \cdot 99} \approx .013.$

b. The probability that Mary's name will be selected is $P(M) = \frac{6}{40} = .15.$ The probability that both Mary's and

John's names will be selected is $P(M) \cdot P(J) = \frac{6}{60} \cdot \frac{6}{40} = \frac{36}{2400} = .015.$

25. The probability is $\dfrac{C(12,8) \cdot C(8,2)}{C(20,10)} + \dfrac{C(12,9)\,C(8,1)}{C(20,10)} + \dfrac{C(12,10)}{C(20,10)} = \dfrac{(28)(495) + (220)(8) + 66}{184{,}756} \approx .085.$

27. a. The probability that he will select brand B is

$$\left(\frac{\text{the number of selections that include brand } B}{\text{the number of possible selections}} \right) = \frac{C(4,2)}{C(5,3)} = \frac{6}{10} = \frac{3}{5}.$$

b. The probability that he will select brands B and C is $\dfrac{C(3,1)}{C(5,3)} = .3.$

c. The probability that he will select at least one of the two brands B and C is

$$(1 - \text{probability that he selects neither of brands } B \text{ and } C) = 1 - \frac{C(3,3)}{C(5,3)} = .9.$$

29. The probability that the three "Lucky Dollar" symbols will appear in the window of the slot machine is

$$P(E) = \frac{n(E)}{n(S)} = \frac{(1)(1)(1)}{C(9,1)\,C(9,1)\,C(9,1)} = \frac{1}{729}.$$

31. The probability of a ticket holder having all four digits in exact order is

$$\frac{1}{C(10,1) \cdot C(10,1) \cdot C(10,1) \cdot C(10,1)} = \frac{1}{10{,}000} = .0001.$$

33. The probability of a ticket holder having one specified digit is $\dfrac{C(1,1)\,C(10,1)\,C(10,1)\,C(10,1)}{10^4} = .1.$

35. The number of ways of selecting a five-card hand from 52 cards is $C(52,5) = 2{,}598{,}960.$ The number of straight
flushes that can be dealt in each suit is 10, so there are $4 \cdot 10 = 40$ possible straight flushes. Therefore, the

probability of being dealt a straight flush is $\dfrac{4(10)}{C(52,5)} = \dfrac{40}{2{,}598{,}960} \approx .0000154.$

37. The number of ways of being dealt a flush in one suit is $C(13, 5)$. Because there are four suits, the number of ways of being dealt a flush is $4 \cdot C(13, 5)$. Because we wish to exclude the hands that are straight flushes we subtract the number of possible straight flushes from $4 \cdot C(13, 5)$. Therefore, the probability of being drawn a flush (but not a straight flush) is $\dfrac{4 \cdot C(13, 5) - 40}{C(52, 5)} = \dfrac{5108}{2{,}598{,}960} \approx .00197$.

39. The total number of ways to select three cards of one rank is $13 \cdot C(4, 3)$. The remaining two cards must form a pair of another rank, and there are $12 \cdot C(4, 2)$ ways of selecting the pair. Thus, the total number of ways to be dealt a full house is $13 \cdot C(4, 3) \cdot 12 \cdot C(4, 2) = 3744$. Hence, the probability of being dealt a full house is $\dfrac{13 \cdot C(4, 3) \cdot 12 \cdot C(4, 2)}{C(52, 5)} = \dfrac{3{,}744}{2{,}598{,}960} \approx .00144$.

41. **a.** Let E denote the event that in a group of five, no two have the same sign. Then $P(E) = \dfrac{12 \cdot 11 \cdot 10 \cdot 9 \cdot 8}{12^5} \approx .3819$. Therefore, the probability that at least two will have the same sign is given by $1 - P(E) = 1 - .3819 \approx .618$.

   b. $P(\text{no Aries}) = \dfrac{11 \cdot 11 \cdot 11 \cdot 11 \cdot 11}{12^5} \approx .647$ and $P(\text{one Aries}) = \dfrac{C(5, 1) \cdot (1)(11)(11)(11)(11)}{12^5} \approx .294$.
   Therefore, the probability that at least two will have the sign Aries is given by
   $1 - [P(\text{no Aries}) + P(\text{one Aries})] = 1 - .941 \approx .059$.

43. Referring to the table on page 421 of the text, we see that in a group of 50 people, the probability that no two have the same birthday is approximately $1 - .970 = .030$.

45. Using the formula on page 421 of the text, we have $P(E) = 1 - \dfrac{(365)(364)(363) \cdots (365 - 12 + 1)}{365^{12}} \approx .167$.

7.5   Conditional Probability and Independent Events

Problem-Solving Tips

1. The probability that the event B will occur given that the event A has already occurred is called the **conditional probability of B given A**, written $P(B \mid A)$.

2. If $P(A) \neq 0$, then $P(B \mid A) = \dfrac{P(A \cap B)}{P(A)}$.

3. The **Product Rule** states that $P(A \cap B) = P(A) P(B \mid A)$.

4. Two events are **independent** if the outcome of one does not affect the outcome of the other; that is, if $P(A \mid B) = P(A)$ and $P(B \mid A) = P(B)$. Do not confuse independent events with mutually exclusive events. (The latter cannot occur at the same time.)

5. Two events are independent if and only if $P(A \mid B) = P(A) \cdot P(B)$.

Concept Questions   page 435

1. The conditional probability of an event is the probability of an event occurring given that another event has already occurred. Examples will vary.

3. $P(A \cap B) = P(A) P(B \mid A)$.

Exercises   page 436

1. a. $P(A \mid B) = \dfrac{P(A \cap B)}{P(B)} = \dfrac{.2}{.5} = \dfrac{2}{5}$.   **b.** $P(B \mid A) = \dfrac{P(A \cap B)}{P(A)} = \dfrac{.2}{.6} = \dfrac{1}{3}$.

3. $P(A \cap B) = P(A) P(B \mid A) = (.6)(.5) = .3$.

5. $P(A) \cdot P(B) = (.3)(.6) = .18 = P(A \cap B)$. Therefore, the events are independent.

7. $P(A \cap B) = P(A) + P(B) - P(A \cup B) = .5 + .7 - .85 = .35 = P(A) \cdot P(B)$, so A and B are independent events.

9. a. $P(A \cap B) = P(A) P(B) = (.4)(.6) = .24$.

b. $P(A \cup B) = P(A) + P(B) - P(A \cap B) = .4 + .6 - .24 = .76$.

c. $P(A \mid B) = P(A) = .4$ because A and B are independent.

d. $P(A^c \cup B^c) = P\left[(A \cap B)^c\right] = 1 - P(A \cap B) = 1 - .24 = .76$.

11. a. $P(A) = .5$.

b. $P(E \mid A) = .4$.

c. $P(A \cap E) = P(A) P(E \mid A) = (.5)(.4) = .2$.

d. $P(E) = (.5)(.4) + (.5)(.3) = .35$.

e. No. $P(A \cap E) \neq P(A) \cdot P(E) = (.5)(.35)$.

f. A and E are not independent events.

13.

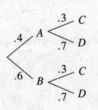

a. $P(A) = .4$.

b. $P(C \mid A) = .3$.

c. $P(A \cap C) = P(A) P(C \mid A) = (.4)(.3) = .12$.

d. $P(C) = (.4)(.3) + (.6)(.3) = .30$.

e. Yes. $P(A \cap C) = .12 = P(A) P(C)$.

f. Yes.

15. a. Refer to Figure 15 on page 426 of the text. Here $E = \{(5, 1), (5, 2), (5, 3), (5, 4), (5, 5), (5, 6)\}$ and $F = \{(6, 4), (5, 5), (4, 6)\}$, so $P(F) = \dfrac{3}{36} = \dfrac{1}{12}$.

b. $P(E \cap F) = \dfrac{1}{36}$ because $E \cap F = \{(5, 5)\}$.

c. $P(F \mid E) = \dfrac{1}{6}$.

d. $P(E) = \dfrac{6}{36} = \dfrac{1}{6}$.

e. $P(F \mid E) = \dfrac{P(E \cap F)}{P(E)} = \dfrac{\frac{1}{36}}{\frac{1}{6}} = \dfrac{1}{6} \neq P(F) = \dfrac{1}{12}$, and so the events are not independent.

17. Let A denote the event that the sum of the numbers is less than 9 and B the event that at least one of the numbers is a 6. Then, $P(A \mid B) = \dfrac{P(A \cap B)}{P(B)} = \dfrac{\frac{4}{36}}{\frac{11}{36}} = \dfrac{4}{11}$.

19. Refer to Figure 15 on page 426 of the text. Here $E = \{(3,1), (3,2), (3,3), (3,4), (3,5), (3,6)\}$ and $F = \{(1,6), (6,1), (2,5), (5,2), (3,4), (4,3)\}$, so $E \cap F = \{(3,4)\}$. Now $P(E \cap F) = \frac{1}{36}$, and this is equal to $P(E) \cdot P(F) = \frac{6}{36} \cdot \frac{6}{36} = \frac{1}{36}$, so E and F are independent events.

21. $P(E \cap F) = \frac{13}{24} = \frac{1}{4}$, $P(E) = \frac{26}{52} = \frac{1}{2}$, and $P(F) = \frac{13}{52} = \frac{1}{4}$. Now $P(E) \cdot P(F) = \frac{1}{2} \cdot \frac{1}{4} = \frac{1}{8} \neq P(E \cap F) = \frac{1}{4}$, so E and F are not independent events. The fact that the card drawn is black increases the probability that it is a spade.

23. Let A denote the event that the battery lasts 10 or more hours and let B denote the event that the battery lasts 15 or more hours. Then $P(A) = .8$, $P(B) = .15$, and $P(A \cap B) = .15$. Therefore, the probability that the battery will last 15 hours or more is $P(B \mid A) = \dfrac{P(A \cap B)}{P(A)} = \dfrac{.15}{.8} = \dfrac{3}{16} = .1875$.

25. **a.** Let A denote the event that a person has type Rh^- blood and B the event that a person has type A blood. Then
$$P(A \mid B) = \dfrac{P(A \cap B)}{P(B)} = \dfrac{.063}{.15} = .42.$$

b. Let C denote the event that a person has type Rh^+ blood and D the event that a person has type B blood. Then
$$P(C \mid D) = \dfrac{P(C \cap D)}{P(D)} = \dfrac{.085}{.10} = .85.$$

27. Let A denote the event that a potential buyer will read the ad and B the event that a reader will buy Jack's car. Then $P(A) = .3$ and $P(B \mid A) = .2$, so the probability that the person who reads the ad will buy Jack's car is $P(A \cap B) = P(A) P(B \mid A) = (.3)(.2) = .06$.

29.

a. The probability that a student selected at random from this medical school is black is $\left(\frac{1}{7}\right)\left(\frac{1}{3}\right) = \frac{1}{21}$.

b. The probability that a student selected at random from this medical school is black if it is known that the student is a member of a minority group is $P(B \mid M) = \frac{1}{3}$.

31. Let A denote the event that Sandy takes Olivia to the supermarket on Friday and B the event that Sandy buys Olivia a popsicle. Then $P(A) = .6$ and $P(B \mid A) = .8$, so the probability that Sandy takes Olivia to the supermarket on Friday and buys her a popsicle is $P(A \cap B) = P(A) P(B \mid A) = (.6)(.8) = .48$.

33. Let C denote the event that a person in the survey was a heavy coffee drinker and Q the event that a person in the survey had cancer of the pancreas. Then $P(C) = \frac{3200}{10,000} = .32$, $P(Q) = \frac{160}{10,000} = .016$, $P(C \cap Q) = \frac{132}{160} = .825$, and $P(C) \cdot P(Q) = .00512 \neq P(C \cap Q)$. Therefore, the events are not independent.

35. a. $P(A) = \dfrac{8120}{10,730} \approx .757$, $P(B) = \dfrac{6101}{10,730} \approx .569$, $P(A \cap B) = \dfrac{4222}{10,730} \approx .393$,

$$P(B \mid A) = \frac{P(A \cap B)}{P(A)} = \frac{n(A \cap B)}{n(A)} = \frac{4222}{8120} \approx .520, \text{ and}$$

$$P(B \mid A^c) = \frac{P(A^c \cap B)}{P(A^c)} = \frac{n(A^c \cap B)}{n(A^c)} = \frac{1879}{2610} \approx .720$$

b. $P(B \mid A) \neq P(B)$, so A and B are not independent events.

37. The sample space for a three-child family is $S = \{GGG, GGB, GBG, GBB, BGG, BGB, BBG, BBB\}$. Because we know that there is at least one girl in the three-child family we are dealing with a reduced sample space $S_1 = \{GGG, GGB, GBG, GBB, BGG, BGB, BBG\}$ in which there are 7 outcomes. Then the probability that all three children are girls is $P(E) = \dfrac{n(E)}{n(S)} = \dfrac{1}{7}$.

39. Let A denote the event that the first pilot is successful and B the event that the second is successful. Then we have

$$P(A \cup B) = P(A) + P(B) - P(A)P(B) \quad (A \text{ and } B \text{ are independent})$$
$$= .9 + .8 - (.9)(.8) = .98.$$

41. Let A denote the event that a broadband user had switched his or her service and let B, C, and D denote the events that a user was very satisfied, somewhat satisfied, and not satisfied with the service, respectively. Referring to the figure, we see that

a. the required probability is

$P(A^c \cap B) = P(A^c)P(B \mid A^c) = (.625)(.48) = .3$, and

b. the required probability is

$P(A \cap D) + P(A^c \cap D) = P(A)P(D \mid A) + P(A^c)P(D \mid A^c) = (.375)(.10) + (.625)(.09) = .09375.$

43. $P(D) = P(A \cap D) + P(B \cap D) + P(C \cap D) = (.45)(.01) + (.25)(.02) + (.30)(.015) = .014.$

45.

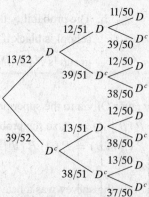

Let D denote the event that the card drawn is a diamond. Then the required probability is

$\dfrac{13}{52} \cdot \dfrac{12}{51} \cdot \dfrac{11}{50} + \dfrac{13}{52} \cdot \dfrac{39}{51} \cdot \dfrac{12}{50} + \dfrac{39}{52} \cdot \dfrac{13}{51} \cdot \dfrac{12}{50} + \dfrac{39}{52} \cdot \dfrac{38}{51} \cdot \dfrac{13}{50} = .25.$

47.

Let D denote the event that a light is defective and D^c the event that a light is nondefective. The probability that both defective lights will be found after three trials is

$$P(D \cap D) + P(D \cap D^c \cap D) + P(D^c \cap D \cap D)$$

$$= \frac{2}{10} \cdot \frac{1}{9} + \frac{2}{10} \cdot \frac{8}{9} \cdot \frac{1}{8} + \frac{8}{10} \cdot \frac{2}{9} \cdot \frac{1}{8} = \frac{16 + 16 + 16}{720} = \frac{1}{15}.$$

49. Let A denote the event that a prospective home buyer shows up for the open house after seeing the ad, B the event that he or she returns for a second showing, and C the event that he or she makes an offer to buy the house. The required probability is $P(A \cap B \cap C) = P(A) P(B \mid A) P(C \mid B) = (.24)(.30)(.75) = .054$.

51. Let A, B, C, and D denote the events that Dad, Mom, Janet and Bob strike acceptable poses. Since the events are independent, we see that the required probability is
$$P(A \cap B \cap C \cap D) = P(A) P(B) P(C) P(D) = (.9)(.9)(.7)(.3) = .1701.$$

53. The probability that the first test will fail is .03, the probability that the second test will fail is .015, and the probability that the third test will fail is .015. Because these are independent events, the probability that all three tests will fail is $(.03)(.015)(.015) = .0000068$.

55. a. The probability that none of the dozen eggs is broken is $(.992)^{12} \approx .908$. Therefore, the probability that at least one egg is broken is $1 - .908 = .092$.

b. Using the results of part (a), we see that the required probability is $(.092)(.092)(.908) \approx .008$.

57. Let A denote the event that at least one of the floodlights remain functional over the one-year period. Then $P(A) = .99999$ and $P(A^c) = 1 - P(A) = .00001$. Letting n represent the minimum number of floodlights needed, we have $(.01)^n = .00001$, so $n \log(.01) = -5$, $n(-2) = -5$, and $n = \frac{5}{2} = 2.5$. Therefore, the minimum number of floodlights needed is 3.

59. a. No, because $E \cap S = E \neq \varnothing$ unless $E = \varnothing$.

b. Yes, because $E \cap \varnothing = \varnothing$.

61. $P(E \mid F) = \dfrac{P(E \cap F)}{P(F)}$. Because E and F are mutually exclusive, $P(E \cap F) = 0$ and so $P(E \mid F) = 0$.
Interpretation: Because E and F are mutually exclusive, the occurrence of F implies that E cannot occur. Therefore, the probability that E occurs given that F has occurred is 0.

63. Using Formula 2, we find $P(A \mid A \cup B) = \dfrac{P[A \cap (A \cup B)]}{P(A \cup B)}$. Because A and B are mutually exclusive,
$A \cap (A \cup B) = A$ and $P(A \cup B) = P(A) + P(B)$. Therefore, $P(A \mid A \cup B) = \dfrac{P(A)}{P(A) + P(B)}$.

65. True. Because A and B are mutually exclusive, $A \cap B = \varnothing$ and $P(A \mid B) = \dfrac{P(A \cap B)}{P(B)} = \dfrac{P(\varnothing)}{P(B)} = 0$.

67. True. This follows from Formula 3: $P(A \cap B) = P(A) P(B \mid A)$.

7.6   Bayes' Theorem

If $A_1, A_2, \ldots A_n$ is a partition of a sample space S and E is an event of the experiment such that $P(E) \neq 0$, then

$$P(A_i \mid E) = \frac{P(A_i) P(E \mid A_i)}{P(A_1) P(E \mid A_1) + P(A_2) P(E \mid A_2) + \cdots + P(A_n) P(E \mid A_n)}$$

If you draw a tree diagram to represent the experiment, then this formula can also be remembered by noting that

$$P(A_i \mid E) = \frac{\text{The product of the probabilities along the limb through } A_i}{\text{The sum of the products of the probabilities along each limb terminating at } E}$$

Concept Questions   page 444

1. An *a priori probability* gives the likelihood that an event *will* occur and an *a posteriori probability* gives the probability that an event *did occur* after the outcomes of an experiment have been observed. Examples will vary.

3. It represents the a posteriori probability that the component having the property described by E was produced in factory A.

Exercises   page 445

1.

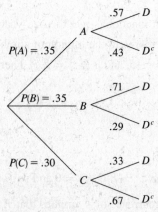

3. **a.** $P(D^c) = \dfrac{15 + 10 + 20}{35 + 35 + 30} = .45$

   b. $P(B \mid D^c) = \dfrac{10}{15 + 10 + 20} \approx .22$

5. **a.** $P(D) = \dfrac{25 + 20 + 15}{50 + 40 + 35} = .48$

   b. $P(B \mid D) = \dfrac{20}{25 + 20 + 15} \approx .33$

7. **a.** $P(A) \cdot P(D \mid A) = (.4)(.2) = .08$

   b. $P(B) \cdot P(D \mid B) = (.6)(.25) = .15$

   c. $P(A \mid D) = \dfrac{P(A) \cdot P(D \mid A)}{P(A) \cdot P(D \mid A) + P(B) \cdot P(D \mid B)} = \dfrac{.4(.2)}{.08 + .15} \approx .348$

9. **a.** $P(A) \cdot P(D \mid A) = \dfrac{1}{3} \cdot \dfrac{1}{4} = \dfrac{1}{12}$

   b. $P(B) \cdot P(D \mid B) = \dfrac{1}{2} \cdot \dfrac{1}{2} = \dfrac{1}{4}$

   c. $P(C) \cdot P(D \mid C) = \dfrac{1}{6} \cdot \dfrac{1}{3} = \dfrac{1}{18}$

   d. $P(A \mid D) = \dfrac{P(A) \cdot P(D \mid A)}{P(A) \cdot P(D \mid A) + P(B) \cdot P(D \mid B) + P(C) \cdot P(C \mid B)} = \dfrac{\frac{1}{12}}{\frac{1}{12} + \frac{1}{4} + \frac{1}{18}} = \dfrac{1}{12} \cdot \dfrac{36}{14} = \dfrac{3}{14}$

11. a. $P(B) = P(A)P(B \mid A) + P(A^c)P(B \mid A^c) = (.3)(.2)(.7)(.3) = .27$

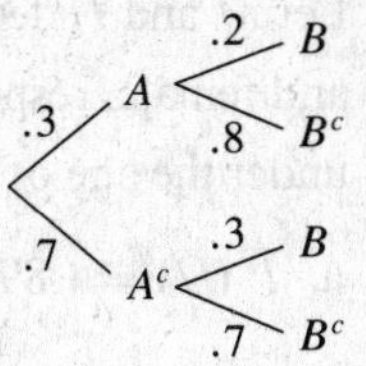

b. $P(A \mid B) = \dfrac{P(A \cap B)}{P(B)} = \dfrac{(.3)(.2)}{.27} \approx .22$

c. $P(B^c) = P(A)P(B^c \mid A) + P(A^c)P(B^c \mid A)$

$\qquad = (.3)(.8) + (.7)(.7) = .73$

d. $P(A \mid B^c) = \dfrac{P(A \cap B^c)}{P(B^c)} = \dfrac{.3(.8)}{.73} \approx .33$

(Tree diagram at top right:)

- $.3$ → A: $.2$ → B, $.8$ → B^c
- $.7$ → A^c: $.3$ → B, $.7$ → B^c

13. Let A denote the event that the first card drawn is a heart and B the event that the second card drawn is a heart.

Then $P(A \mid B) = \dfrac{P(A) \cdot P(B \mid A)}{P(A) \cdot P(B \mid A) + P(A^c)P(B \mid A^c)} = \dfrac{\frac{1}{4} \cdot \frac{12}{51}}{\frac{1}{4} \cdot \frac{12}{51} + \frac{3}{4} \cdot \frac{13}{51}} = \dfrac{4}{17}.$

15.

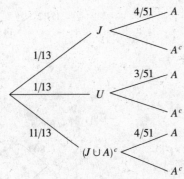

$P(J \mid A) = \dfrac{\frac{1}{13} \cdot \frac{4}{51}}{\frac{1}{13} \cdot \frac{4}{51} + \frac{1}{13} \cdot \frac{3}{51} + \frac{11}{13} \cdot \frac{4}{51}} = \dfrac{\frac{4}{13 \cdot 51}}{\frac{51}{13 \cdot 51}}$

$\qquad\quad = \dfrac{4}{51} \approx .0784.$

17.

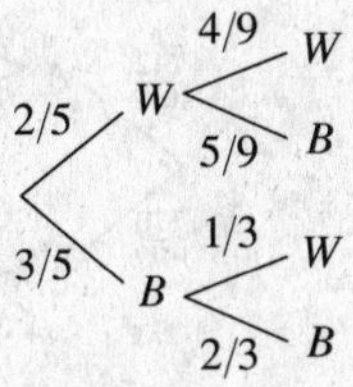

(Tree diagram:)

- $2/5$ → W: $4/9$ → W, $5/9$ → B
- $3/5$ → B: $1/3$ → W, $2/3$ → B

19. Referring to the tree diagram in the solution to Exercise 17, we see that the probability that the transferred ball was black given that the second ball was white is $P(B \mid W) = \dfrac{\frac{3}{5} \cdot \frac{1}{3}}{\frac{2}{5} \cdot \frac{4}{9} + \frac{3}{5} \cdot \frac{1}{3}} = \dfrac{9}{17}.$

21. a. $P(I \mid S) = \dfrac{P(I) \cdot P(S \mid I)}{P(I)P(S \mid I) + P(II) \cdot P(S \mid II)} = \dfrac{(.64)(.002)}{(.64)(.002) + (.36)(.005)} \approx .416.$

b. $P(II \mid S) = \dfrac{P(II) \cdot P(S \mid II)}{P(I)P(S \mid I) + P(II) \cdot P(S \mid II)} = \dfrac{(.36)(.005)}{(.64)(.002) + (.36)(.005)} \approx .584.$

23. Let D denote the event that the person tested has the disease and E the event that the test result is positive. Then the required probability is

$P(D \mid E) = \dfrac{P(D)P(E \mid D)}{P(D)P(E \mid D) + P(D^c)P(E \mid D^c)} = \dfrac{(.003)(.95)}{(.003)(.95) + (.997)(.02)}$

$\qquad \approx .125.$

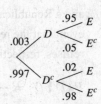

(Tree diagram:)

- $.003$ → D: $.95$ → E, $.05$ → E^c
- $.997$ → D^c: $.02$ → E, $.98$ → E^c

25. Let M and F denote the events that a person arrested for crime was male and female, respectively; and let U denote the event that the person was under the age of 18.

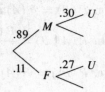

    a. $P(U) = (.89)(.30) + (.11)(.27) = .297.$

    b. $P(F \mid U) = \dfrac{(.11)(.27)}{(.89)(.30) + (.11)(.27)} \approx .100.$

27. Let E, F, and G denote the events that the child selected at random is 12, 13, and 14 years old, respectively; and let C^c denote the event that the child does not have a cavity. Then

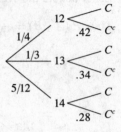

$$P(G \mid C^c) = \frac{\frac{5}{12}(.28)}{\frac{1}{4}(.42) + \frac{1}{3}(.34) + \frac{5}{12}(.28)} \approx .348.$$

29. $P(S_2 \mid S) = \dfrac{(.8)(.95)}{(.8)(.95) + (.2)(.3)} \approx .927.$

31. Let A denote the event that a respondent is between the ages of 18 and 49, and let H denote the event that a respondent experiences hardship due to gas prices. Using the tree diagram, or otherwise, we find

$$P(A \mid H^c) = \frac{P(A)P(H^c \mid A)}{P(A)P(H^c \mid A) + P(A^c)P(H^c \mid A^c)}$$

$$= \frac{\left(\frac{742}{1012}\right)\left(\frac{40}{100}\right)}{\left(\frac{742}{1012}\right)\left(\frac{40}{100}\right) + \left(\frac{270}{1012}\right)\left(\frac{45}{100}\right)} \approx .710.$$

33. Let D and R denote the events that the respondent is a Democratic voter and a Republican voter, respectively. Next, let S, O, and X denote the event that the respondent supports, opposes, or either doesn't know or refuses, respectively. The required probability is then

$$P(D \mid O) = \frac{\frac{4}{7}(.14)}{\frac{4}{7}(.14) + \frac{3}{7}(.31)} \approx .3758.$$

35. Let C denote the event that a study was corporate-sponsored, F the event that a participant found the products favorable, N the event that a participant was neutral, and U the event that a participant found the products unfavorable.

a. The probability is

$$P(C)P(F) + P(C^c)P(F) = (.625)(.63) + (.375)(.47) = .57.$$

b. The probability is

$$\frac{P(C)P(F)}{P(C)P(F) + P(C^c)P(F)} = \frac{(.625)(.63)}{(.57)} \approx .691.$$

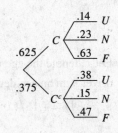

37. Let x denote the age of an adult selected at random from the population, and let R denote the event that the adult is a renter.

a. $P(R) = (.51)(.58) + (.31)(.45) + (.18)(.60) \approx .543.$

b. $P(21 \le x \le 44 \mid R) = \dfrac{(.51)(.58)}{.543} \approx .545.$

c. $P(E) = 1 - P(21 \le x \le 44 \mid R) = 1 - .545 = .455.$

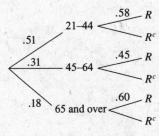

39. Let H_2 denote the event that the coin tossed is the two-headed coin, H_B the event that the coin is the biased coin, and H_F the event that the coin is the fair coin.

a. $P(H) = \frac{1}{3} \cdot 1 + \frac{1}{3} \cdot \frac{3}{4} + \frac{1}{3} \cdot \frac{1}{2} = \frac{1}{3} + \frac{1}{4} + \frac{1}{6} = \frac{9}{12} = \frac{3}{4}.$

b. $P(H_F \mid H) = \dfrac{\frac{1}{3} \cdot \frac{1}{2}}{\frac{3}{4}} = \frac{2}{9}$

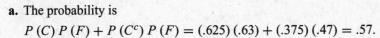

41. a.

$$P(D \mid +) = \frac{(.04)(.95)}{(.04)(.95) + (.96)(.04)}$$

$$\approx .497.$$

b. From part (a), we know that 49.7% of those who test positive have the disease, whereas 50.3% of those who test positive do not have the disease. Using this information, we construct another tree diagram.

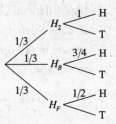

Thus, $P(D \mid ++) = \dfrac{(.497)(.95)}{(.497)(.95) + (.503)(.04)} \approx .959.$

43. a. $P(D) = P(\text{I}) \cdot P(D \mid \text{I}) + P(\text{II}) + P(D \mid \text{II}) + P(\text{III}) \cdot (D \mid \text{III}) + P(\text{IV}) \cdot (D \mid \text{IV})$

$$= (.15)(.04) + (.3)(.02) + (.35)(.02) + (.2)(.03) = .025.$$

b. $P(\text{I} \mid D) = \dfrac{P(\text{I})P(D \mid \text{I})}{P(D)} = \dfrac{(.15)(.04)}{.025} = .24.$

c. $P(\text{II} \mid D) = \dfrac{P(\text{II})P(D \mid \text{II})}{P(D)} = \dfrac{(.3)(.02)}{.025} = .24.$

45. $P(\text{III} \mid D) = \dfrac{(.30)(.02)}{(.35)(.015) + (.35)(.01) + (.30)(.02)} = \dfrac{.006}{.01475} \approx .407$

47. Let x denote the age of an insured driver and A the event that an insured driver is in an accident.

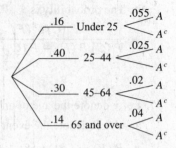

a. $P(A) = (.16)(.055) + (.4)(.025) + (.3)(.02) + (.14)(.04) \approx .03$.

b. $P(x < 25 \mid A) = \dfrac{(.16)(.055)}{.03} \approx .29$.

49. Let A, B, C, and D denote the events that a person in the survey belongs to the Millennial Generation, Generation X, the Baby Boomer generation, and the Silent Generation, respectively, and let S denote the event that a person in the survey has slept with a cell phone nearby. The required probability is

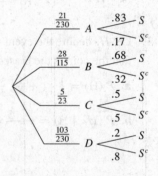

$$P(S^c \mid A) = \dfrac{P(A)\,P(S^c \mid A)}{\left[\begin{array}{l} P(A)\,P(S^c \mid A) + P(B)\,P(S^c \mid B) \\ \qquad + P(C)\,P(S^c \mid C) + P(D)\,P(S^c \mid D) \end{array}\right]}$$

$$= \dfrac{\frac{21}{230}(.17)}{\frac{21}{230}(.17) + \frac{28}{115}(.32) + \frac{5}{23}(.5) + \frac{103}{230}(.80)} \approx .028.$$

51. a. Let A, B, C, D, and E denote the events that a respondent's annual household income is less than 15,000, 15,000–29,999, 30,000–49,999, 50,000–74,999, and 75,000 and higher, respectively; and let R, M, and P denote the probabilities that a person considers himself rich, middle class, and poor, respectively. The probability that a respondent chosen at random calls himself or herself middle class is thus

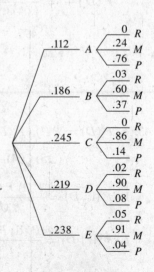

$(.112)(.24) + (.186)(.60) + (.245)(.86) + (.219)(.90) + (.238)(.91)$
$$\approx .763.$$

b. $P(C \mid M) = \dfrac{.245(.86)}{.112(.24) + .186(.60) + .245(.86) + .219(.9) + .238(.91)}$

$\approx .276.$

c. Using the results of part (b), the required probability is $1 - .276 = .724$.

53. Denote the categories professional, white collar, blue collar, unskilled, farmers, and housewives by A, B, C, D, E, and F, respectively, and let V denote the event that a person voted in the election. The required probability is

$$P(F \mid V) = \frac{.14\,(.66)}{\left[\begin{array}{l} .12\,(.84) + .24\,(.73) + .32\,(.66) \\ \qquad + .10\,(.57) + .08\,(.68) + .14\,(.66) \end{array}\right]}$$

$$\approx .1337.$$

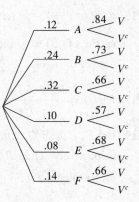

CHAPTER 7   Concept Review Questions   page 452

1. experiment, sample, space, event

3. uniform, $1/n$

5. independent

CHAPTER 7   Review Exercises   page 452

1. a. $P(E \cap F) = 0$ because E and F are mutually exclusive.

b. $P(E \cup F) = P(E) + P(F) - P(E \cap F) = .4 + .2 = .6$.

c. $P(E^c) = 1 - P(E) = 1 - .4 = .6$.

d. $P(E^c \cap F^c) = P(E \cup F)^c = 1 - P(E \cup F) = 1 - .6 = .4$.

e. $P(E^c \cup F^c) = P(E \cap F)^c = 1 - P(E \cap F) = 1 - 0 = 1$.

3. a. $P(F^c) = 1 - P(F) = 1 - .47 = .53$.

b. Because $E \cap F = \varnothing$, $P(E \cap F^c) = P(E) = .35$.

c. $P(E \cup F) = P(E) + P(F) = .35 + .47 = .82$.

d. $P(E^c \cap F^c) = P\left[(E \cup F)^c\right] = 1 - P(E \cup F) = 1 - .82 = .18$.

5. The required probability is
$$P(R \cap B) + P(B \cap R)$$
$$= \frac{6}{15} \cdot \frac{5}{14} + \frac{5}{15} \cdot \frac{6}{14} = \frac{2}{7}.$$

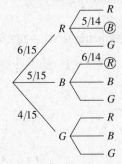

7. $P(E \mid F) = \dfrac{P(E \cap F)}{P(F)} = \dfrac{P(E) + P(F) - P(E \cup F)}{P(F)}$

$$= \frac{.35 + .55 - .70}{.55} = .364.$$

9. Because E and F are independent, $P(E \cap F) = P(E)\,P(F)$, and

so $P(F) = \dfrac{P(E \cap F)}{P(E)} = \dfrac{.16}{.32} = .5$.

11. $P(B \cap E) = (.5)(.5) = .25$.

13. $P(E) = .18 + .25 + .06 = .49$.

15. a. $P(A) = 1 - P(A^c) = 1 - \frac{1}{8} = \frac{7}{8}$. **b.** $P(B) = 1 - P(B^c) = 1 - \frac{1}{8} = \frac{7}{8}$.

   c. $P(A \cap B) = \frac{7}{8}$ and $P(A) \cdot P(B) = \frac{7}{8} \cdot \frac{7}{8} = \frac{49}{64}$. Because $P(A \cap B) \neq P(A) \cdot P(B)$, they are not independent events.

17. $P(E) = \dfrac{7 \cdot 6 \cdot 5 \cdot 4 \cdot 3}{7^5} \approx .150$.

19. Let E, F, and G denote the events that the first toss is even, the second toss is odd, and the third toss is a 1, respectively. Then $P(E) = \frac{1}{2}$, $P(F) = \frac{1}{2}$, and $P(G) = \frac{1}{6}$. Because the outcomes are independent, the required probability is $P(E \cap F \cap G) = P(E) P(F) P(G) = \frac{1}{2} \cdot \frac{1}{2} \cdot \frac{1}{6} = \frac{1}{24}$.

21. The probability that all three cards are aces is $\dfrac{C(4, 3)}{C(52, 3)} \approx .00018$.

23.

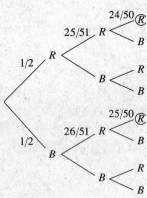

25.

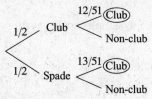

The probability that the second card is a club given that the first card was black is $\frac{1}{2} \cdot \frac{12}{51} + \frac{1}{2} \cdot \frac{13}{51} \approx .245$.

The required probability is

$\frac{1}{2} \cdot \frac{25}{51} \cdot \frac{24}{50} + \frac{1}{2} \cdot \frac{26}{51} \cdot \frac{25}{50} = \frac{1250}{5100} \approx .2451$.

27.

Answer	Falling behind	Staying even	Increasing faster	Don't know
Probability	.40	.44	.12	.04

29. $A \cap B \neq \varnothing$ since it is possible to obtain a straight and a flush—that is, a straight flush.

31. a. The probability that he or she was never late for work is $\frac{4746}{7780} \approx .61$, or approximately 61%.

   b. The probability that he or she was late once a year is $\frac{934}{7780} \approx .12$, or approximately 12%.

33. a. The probability is $\dfrac{16{,}520 + 698}{18{,}598} \approx .926$. **b.** The probability is $\dfrac{845 + 535}{18{,}598} \approx .074$.

35. a. The probability that the survey participant said that it would be the same or better is $.41 + .38 = .79$.

   b. The probability that the survey participant said that it would be the same or worse is $.41 + .18 = .59$.

37. a. The probability is $.26 + .154 + .137 + .133 + .073 = .757$.

   b. The probability is $1 - .757 = .243$.

39. a. The required probability is $.236 + .174 = .41$.

b. The probability that they were planning to use computer software to prepare their taxes or to do their taxes by hand is $.339 + .143 = .482$. The probability that they were not planning to either use computer software to prepare their taxes or do their taxes by hand is $1 - .482 = .518$.

41.

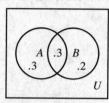

The required probability is

$$P(A \cap B^c) + P(A^c \cap B) = .3 + .2 = .5.$$

43.

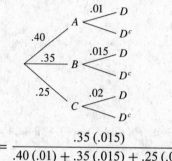

$$P(B \mid D) = \frac{.35\,(.015)}{.40\,(.01) + .35\,(.015) + .25\,(.02)}$$
$$= \frac{.35\,(.015)}{.01425} \approx .368.$$

45. Let I and II denote customers who purchased the drug in capsule and tablet form, respectively, and let E denote a customer who purchased the extra-strength dosage.

$$\text{Thus, } P(I \mid E) = \frac{.57\,(.38)}{.57\,(.38) + .43\,(.31)} = .619.$$

47. Let M and F denote the events that a respondent is male and female, respectively; and let Y, N, and X denote the events that a respondent favors the lottery, does not favor the lottery, and expresses no opinion, respectively.

a. The required probability is

$$P(F \mid Y) = \frac{.49\,(.68)}{.51\,(.62) + .49\,(.68)} \approx .513.$$

b. The required probability is

$$P(F \mid X) = \frac{.49\,(.04)}{.51\,(.06) + .49\,(.04)} \approx .390.$$

49. Let D and R denote the events that a representative is a Democrat and a Republican, respectively; and let G denote the event that the representative is a gun owner. Then the required probability is

$$P(G \mid D) = \frac{P(D)\,P(G \mid D)}{P(D)\,P(G \mid D) + P(R)\,P(G \mid R)}$$
$$= \frac{\left(\frac{200}{432}\right)\left(\frac{30}{200}\right)}{\left(\frac{200}{432}\right)\left(\frac{30}{200}\right) + \left(\frac{232}{432}\right)\left(\frac{93}{232}\right)} \approx .244.$$

CHAPTER 7   Before Moving On...   page 456

1. $P(s_1, s_3, s_6) = \frac{1}{12} + \frac{3}{12} + \frac{1}{12} = \frac{5}{12}$.

2. The number of ways of drawing a deuce or face card is 16. Therefore, the required probability is $\frac{16}{52} = \frac{4}{13}$.

3.

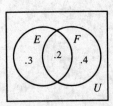

    a. $P(E \cup F) = .3 + .2 + .4 = .9$.
               b. $P(E \cap F^c) = .3$.

4. Because A and B are independent, $P(A \cap B) = P(A) \cdot P(B) = .3(.6) = .18$, so
$P(A \cup B) = P(A) + P(B) - P(A \cap B) = .3 + .6 - .18 = .72$.

5. $P(A \mid D) = \dfrac{.4(.2)}{.4(.2) + .6(.3)} \approx .308$.

PROBABILITY DISTRIBUTIONS AND STATISTICS

8.1 Distributions of Random Variables

Problem-Solving Tips

1. A **random variable** is a rule that assigns a number to each outcome of a chance experiment.

2. A **probability distribution of a random variable** gives the distinct values of the random variable X and the probabilities associated with these values.

3. A **histogram** is the graph of the probability distribution of a random variable.

Concept Questions   page 463

1. A random variable is a rule that assigns a number to each outcome of a chance experiment. Examples vary.

3. To construct a histogram for a probability distribution, follow these steps:

   a. Locate the values of the random variable on a number line.

   b. Above each such number on the number line, erect a rectangle with width 1 and height equal to the probability associated with that value of the random variable.

Exercises   page 463

1. a,b.

Outcome	GGG	GGR	GRG	RGG	GRR	RGR	RRG	RRR
Value	3	2	2	2	1	1	1	0

   c. {GGG}

3. X may assume the values in the set $S = \{1, 2, 3, ...\}$.

5. The event that the sum of the dice rolls is 7 is $E = \{(1, 6), (2, 5), (3, 4), (4, 3), (5, 2), (6, 1)\}$ and $P(E) = \frac{6}{36} = \frac{1}{6}$.

7. X may assume the value of any positive integer. The random variable is infinite discrete.

9. $\{d \mid d \geq 0\}$. The random variable is continuous.

11. X may assume the value of any positive integer. The random variable is infinite discrete.

13. No. The probability assigned to a value of the random variable X cannot be negative. Here $P(X = 0) = -.2$, which is proscribed.

15. No. The sum of the probabilities exceeds 1.

211

17. We must have $.1 + .4 + a + .1 + .2 = 1$, so $a = .2$.

19. a. $P(X = -10) = .20$

 b. $P(X \geq 5) = .1 + .25 + .1 + .15 = .60$

 c. $P(-5 \leq X \leq 5) = .15 + .05 + .1 = .30$

 d. $P(X \leq 20) = .20 + .15 + .05 + .1 + .25 + .1 + .15 = 1$

 e. $P(X < 5) = P(X = -10) + P(X = -5) + P(X = 0) = .20 + .15 + .05 = .4$

 f. $P(X = 3) = 0$

21.

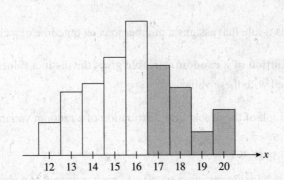

23. a.

x	1	2	3	4	5	6
$P(X = x)$	$\frac{1}{6}$	$\frac{1}{6}$	$\frac{1}{6}$	$\frac{1}{6}$	$\frac{1}{6}$	$\frac{1}{6}$

y	1	2	3	4	5	6
$P(Y = y)$	$\frac{1}{6}$	$\frac{1}{6}$	$\frac{1}{6}$	$\frac{1}{6}$	$\frac{1}{6}$	$\frac{1}{6}$

 b.

$x + y$	2	3	4	5	6	7	8	9	10	11	12
$P(X + Y = x + y)$	$\frac{1}{36}$	$\frac{2}{36}$	$\frac{3}{36}$	$\frac{4}{36}$	$\frac{5}{36}$	$\frac{6}{36}$	$\frac{5}{36}$	$\frac{4}{36}$	$\frac{3}{36}$	$\frac{2}{36}$	$\frac{1}{36}$

25. a.

x	2	2.25	2.55	2.56	2.58	2.6	2.65	2.85
$P(X = x)$	$\frac{1}{30}$	$\frac{7}{30}$	$\frac{7}{30}$	$\frac{1}{30}$	$\frac{1}{30}$	$\frac{8}{30}$	$\frac{3}{30}$	$\frac{2}{30}$

 b. $P(X = 2) + P(X = 2.25) + P(X = 2.55) = \frac{1}{30} + \frac{7}{30} + \frac{7}{30} = \frac{1}{2}$

27. a.

x	$P(X = x)$
0	.017
1	.067
2	.033
3	.117
4	.233
5	.133
6	.167
7	.1
8	.05
9	.067
10	.017

 b.

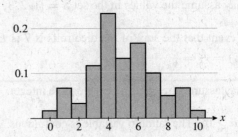

 c. $P(X = 1) + P(X = 2) + P(X = 3) \approx .067 + .033 + .117 = .217$

29. a. We divide each number in the second row by 100,091, the sum of the numbers in that row.

x	1	2	3	4
$P(X=x)$	.228	.492	.148	.132

b. $P(X=1) + P(X=2) = .228 + .492 = .72$.

31. a. We divide each number in the second row by 5765, the sum of the numbers in that row.

x	1	2	3	4	5
$P(X=x)$	.020	.110	.250	.54	.080

b. $P(X=1) + P(X=2) = .020 + .110 = .13$, or 13%.

33. a. We divide each number in the second row by 18,762, the sum of the numbers in that row.

x	1	2	3	4	5
$P(X=x)$	.131	.160	.179	.230	.300

b. $P(X=4) + P(X=5) = .230 + .300 = .530$, and $P(X=1) + P(X=2) = .131 + .160 = .291$.

35. True. This follows from the definition.

Technology Exercises   page 471

1.

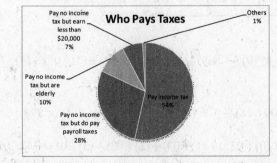

3.

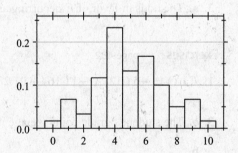

5.

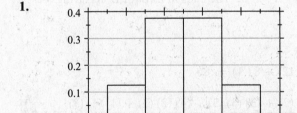

7.

9.

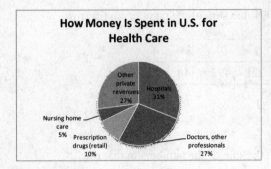

How Money Is Spent in U.S. for Health Care

Other private revenues 27%
Hospitals 31%
Nursing home care 5%
Prescription drugs (retail) 10%
Doctors, other professionals 27%

8.2   Expected Value

Problem-Solving Tips

1. The **expected value** of a random variable X is given by $E(X) = x_1 p_1 + x_2 p_2 + \cdots + x_n p_n$ where $x_1, x_2, \ldots x_n$ are the values assumed by X and $p_1, p_2, \ldots, p_n$ are the associated probabilities.

2. If $P(E)$ is the probability of an event E occurring, then the **odds in favor** of E occurring are $\dfrac{P(E)}{P(E^c)}$ and the **odds against** E occurring are $\dfrac{P(E^c)}{P(E)}$.

3. If the odds in favor of an event E occurring are a to b, then the probability of E occurring is $P(E) = \dfrac{a}{a+b}$.

Concept Questions   page 482

1. The expected value of a random variable X is given by $E(X) = x_1 p_1 + x_2 p_2 + \cdots + x_n p_n$. Examples will vary.

3. **a.** The odds in favor of E occurring are $\dfrac{P(E)}{P(E^c)}$.   **b.** The odds in favor of E occurring are $\dfrac{a}{a+b}$.

Exercises   page 482

1. $E(X) = -5(.12) + -1(.16) + 0(.28) + 1(.22) + 5(.12) + 8(.1) = .86$.

3. **a.** The student's grade point average is given by $\dfrac{(2)(4)(3) + (3)(3)(3) + (4)(2)(3) + (1)(1)(3)}{(10)(3)} = 2.6$.

b.

x	0	1	2	3	4
$P(X = x)$	0	.1	.4	.3	.2

$E(X) = 1(.1) + 2(.4) + 3(.3) + 4(.2) = 2.6$.

5. A customer entering the store is expected to buy $E(X) = (0)(.42) + (1)(.36) + (2)(.14) + (3)(.05) + (4)(.03) = .91$, or .91 DVDs.

7. $E(X) = 0(.07) + 25(.12) + 50(.17) + 75(.14) + 100(.28) + 125(.18) + 150(.04) = 78.5$, or \$78.50.

9. The expected number of accidents is given by $E(X) = (0)(.935) + (1)(.03) + (2)(.02) + (3)(.01) + (4)(.005) = .12$.

11. The expected number of machines that will break down on a given day is given by

$$E(X) = (0)(.43) + (1)(.19) + (2)(.12) + (3)(.09) + (4)(.04) + (5)(.03) + (6)(.03) + (7)(.02) + (8)(.05) = 1.73.$$

13. The associated probabilities are $\frac{3}{50}$, $\frac{8}{50}$, ..., and $\frac{5}{50}$, respectively. Therefore, the expected interest rate is

$$(2.9)\left(\frac{3}{50}\right) + 3\left(\frac{8}{50}\right) + 3.1\left(\frac{12}{50}\right) + 3.2\left(\frac{14}{50}\right) + 3.3\left(\frac{8}{50}\right) + 3.4\left(\frac{5}{50}\right) \approx 3.162, \text{ or } 3.16\%.$$

15. The expected net earnings of a person who buys one ticket are

$$-1(.997) + 24(.002) + 99(.0006) + 499(.0002) + 1999(.0002) = -.39, \text{ a loss of } \$.39 \text{ per ticket.}$$

17. The expected gain of the insurance company is given by $E(X) = .992(260) - (19,740)(.008) = 100$, or $100.

19. His expected profit is

$$E = (580,000 - 450,000)(.24) + (570,000 - 450,000)(.4) + (560,000 - 450,000)(.36) \approx 118,800, \text{ or } \$118,800.$$

21. City A: $E(X) = (10,000,000)(.2) - 250,000 = 1,750,000$, or $1.75 million. City B:
$E(X) = (7,000,000)(.3) - 200,000 = 1,900,000$, or $1.9 million. We see that the company should bid for the rights in City B.

23. The expected number of houses sold per year at Company A is given by

$$E(X) = (12)(.02) + (13)(.03) + (14)(.05) + (15)(.07) + (16)(.07) + (17)(.16) + (18)(.17)$$
$$+ (19)(.13) + (20)(.11) + (21)(.09) + (22)(.06) + (23)(.03) + (24)(.01) = 18.09.$$

The expected number of houses sold per year at Company B is given by

$$E(X) = (6)(.01) + (7)(.04) + (8)(.07) + (9)(.06) + (10)(.11) + (11)(.12) + (12)(.19)$$
$$+ (13)(.17) + (14)(.13) + (15)(.04) + (16)(.03) + (17)(.02) + (18)(.01) = 11.77.$$

Thus, Sally's expected commission at Company A is $(.03)(308,000)(18.09) = 167,151.60$, or $167,151.60. Her expected commission at Company B is $(.03)(474,000)(11.77) = 167,369.40$, or $167,369.40. Based on these expectations, she should accept the job offer from Company B.

25. Maria expects her business to grow at the rate of $(5)(.12) + (4.5)(.24) + (3)(.4) + (0)(.2) + (-.5)(.04) = 2.86$, or 2.86%/year during the upcoming year.

27. The expected value of the winnings on a $1 bet placed on a split is $E(X) = 17 \cdot \frac{2}{38} + (-1) \cdot \frac{36}{38} \approx -.0526$, a loss of 5.3 cents.

29. The expected value of a player's winnings are $(1)\left(\frac{18}{37}\right) + (-1)\left(\frac{19}{37}\right) = -\frac{1}{37} \approx -.027$, a loss of 2.7 cents per bet.

31. We take $x_1 = (0 + 10)/2 = 5$, $x_2 = (10 + 15)/2 = 12.5$, $x_3 = 17.5$, ..., $x_9 = 67.5$. Also, $p_1 = \frac{9049}{62,407}$,
$p_2 = \frac{9611}{62,407}$, $p_3 = \frac{10,172}{62,407}$, $p_4 = \frac{9423}{62,407}$, $p_5 = \frac{3807}{62,407}$, $p_6 = \frac{8175}{62,407}$, $p_7 = \frac{3807}{62,407}$, $p_8 = \frac{4244}{62,407}$, and $p_9 = \frac{4119}{62,407}$. Then

$$E(X) = x_1 p_1 + x_2 p_2 + \cdots + x_9 p_9 = (5)\left(\frac{9049}{62,407}\right) + (12.5)\left(\frac{9611}{62,407}\right) + \cdots + (67.5)\left(\frac{4119}{62,407}\right)$$
$$= \frac{1}{62,407}[5(9049) + 12.5(9611) + 17.5(10,172) + 22.5(9423) + 27.5(3807)$$
$$+ 32.5(8175) + 40(3807) + 52.5(4244) + 67.5(4119)]$$
$$\approx 25.30.$$

Thus, the average time is approximately 25.3 minutes.

33. We take $x_1 = (0+5)/2 = 2.5$, $x_2 = (5+14)/2 = 9.5$, $x_3 = 19.5$, $x_4 = 29.5, \ldots, x_{11} = 99.5$. Also, $p_1 = \frac{2531}{37,253}$, $p_2 = \frac{5097}{37,253}, \ldots, p_{11} = \frac{46}{37,253}$. Then

$$E(X) = x_1 p_1 + x_2 p_2 + \cdots + x_{11} p_{11} = (2.5)\left(\frac{2531}{37,253}\right) + (9.5)\left(\frac{5097}{37,253}\right) + \cdots + (99.5)\left(\frac{46}{37,253}\right)$$

$$= \frac{1}{37,253}[2.5(2531) + 9.5(5097) + 19.5(5590) + 29.5(5318) + 39.5(5183)$$

$$+ 49.5(5252) + 59.5(4036) + 69.5(2275) + 79.5(1370) + 89.5(555) + 99.5(46)]$$

$$\approx 36.152.$$

Thus, the average age is approximately 36.2.

35. The odds in favor of E occurring are $\dfrac{P(E)}{P(E^c)} = \dfrac{.4}{.6}$, or 2 to 3. The odds against E occurring are 3 to 2.

37. The probability of E not occurring is given by $P(E) = \frac{2}{3+2} = \frac{2}{5} = .4$.

39. The probability that she will win her match is $P(E) = \frac{7}{7+5} = \frac{7}{12} \approx .583$.

41. The probability that the business deal will not go through is $P(E) = \frac{5}{5+9} = \frac{5}{14} \approx .357$.

43. Let X and Y be random variables that assume values $x_1, x_2, \ldots, x_n$ and $y_1, y_2, \ldots, y_n$ with probabilities $p_1, p_2, \ldots, p_n$, respectively.

a. $E(c) = cp_1 + cp_2 + \cdots + cp_n = c(1) = c$

b. $E(cX) = cx_1 p_1 + cx_2 p_2 + \cdots + cx_n p_n = c(x_1 p_1 + x_2 p_2 + \cdots + x_n p_n) = cE(X)$

c. $E(X+Y) = (x_1 + y_1)p_1 + (x_2 + y_2)p_2 + \cdots + (x_n + y_n)p_n$

$$= (x_1 p_1 + x_2 p_2 + \cdots + x_n p_n) + (y_1 p_1 + y_2 p_2 + \cdots + y_n p_n) = E(X) + E(Y)$$

d. $E(X-Y) = (x_1 - y_1)p_1 + (x_2 - y_2)p_2 + \cdots + (x_n - y_n)p_n$

$$= (x_1 p_1 + x_2 p_2 + \cdots + x_n p_n) - (y_1 p_1 + y_2 p_2 + \cdots + y_n p_n) = E(X) - E(Y)$$

45. a. The mean is $\dfrac{40 + 45 + 2(50) + 55 + 2(60) + 2(75) + 2(80) + 4(85) + 2(90) + 2(95) + 100}{20} = 74$, the mode is 85 (the value that appears most frequently), and the median is 80 (the middle value).

b. The mode is the least representative of this set of test scores.

47. We first arrange the numbers in increasing order:

$$0, 0, \underbrace{1, 1, \ldots, 1}_{9 \text{ times}}, \underbrace{2, 2, \ldots, 2}_{16 \text{ times}}, \underbrace{3, 3, \ldots, 3}_{12 \text{ times}}, \underbrace{4, 4, \ldots, 4}_{8 \text{ times}}, \underbrace{5, 5, \ldots, 5}_{6 \text{ times}}, 6, 6, 6, 6, 7, 7, 8$$

There are 60 numbers, so the median is 3. This is close to the mean of 3.1 obtained in Example 1.

49. The average is $\frac{1}{10}(16.1 + 16 + 15.8 + 16 + 15.9 + 16.1 + 15.9 + 16 + 16 + 16.2) = 16$, or 16 oz. Next, we arrange the numbers in increasing order: 15.8, 15.9, 15.9, 16, 16, 16, 16, 16.1, 16.1, 16.2, so the median is $\frac{16+16}{2} = 16$, or 16 oz. The mode is also 16 oz.

51. The average concentration of type A^+ blood is
$\frac{1}{12}(8.0 + 8.0 + 7.2 + 7.0 + 7.0 + 7.0 + 7.0 + 7.0 + 7.0 + 7.0 + 7.0 + 6.6) = 7.15$. The median is $\frac{7+7}{2} = 7$, and the mode is 7, occurring eight times.

53. True. This follows from the definition.

8.3 Variance and Standard Deviation

Problem-Solving Tips

1. The **variance** of a random variable X is a measure of the spread of a probability distribution about its mean. The variance of a random variable X is given by $\text{Var}\,(X) = p_1\,(x_1 - \mu)^2 + p_2\,(x_2 - \mu)^2 + \cdots + p_n\,(x_n - \mu)^2$, where $x_1, x_2, \ldots, x_n$ are the values assumed by X and $p_1 = P\,(X = x_1)$, $p_2 = P\,(X = x_2)$, $\ldots$, $p_n = P\,(X = x_n)$. The **standard deviation** of a random variable X is $\sigma = \sqrt{\text{Var}\,(X)}$.

2. **Chebychev's Inequality** gives the proportion of values of a random variable X lying within k standard deviations of the expected value of X. The probability that a randomly chosen outcome of the experiment lies between
$$\mu - k\sigma \text{ and } \mu + k\sigma \text{ is } P\,(\mu - k\sigma \le X \le \mu + k\sigma) \ge 1 - \frac{1}{k^2}.$$

Concept Questions   page 494

1. If a random variable has the probability distribution shown at right and expected value $E\,(X) = \mu$, then the variance of the random variable X is

x	x_1	x_2	x_3	$\cdots$	x_n
$P\,(X = x)$	p_1	p_2	p_3	$\cdots$	p_n

$\text{Var}\,(X) = p_1\,(x_1 - \mu)^2 + p_2\,(x_2 - \mu)^2 + \cdots + p_n\,(x_n - \mu)^2$ and the standard variation of the random variable X is given by $\sigma = \sqrt{\text{Var}\,(X)}$.

Exercises   page 494

1. $\mu = (1)\,(.4) + (2)\,(.3) + 3\,(.2) + (4)\,(.1) = 2$,
$\text{Var}\,(X) = (.4)\,(1 - 2)^2 + (.3)\,(2 - 2)^2 + (.2)\,(3 - 2)^2 + (.1)\,(4 - 2)^2 = .4 + 0 + .2 + .4 = 1$, and $\sigma = \sqrt{1} = 1$.

3. $\mu = -2\left(\frac{1}{16}\right) - 1\left(\frac{4}{16}\right) + 0\left(\frac{6}{16}\right) + 1\left(\frac{4}{16}\right) + 2\left(\frac{1}{16}\right) = \frac{0}{16} = 0$,
$\text{Var}\,(X) = \frac{1}{16}\,(-2 - 0)^2 + \frac{4}{16}\,(-1 - 0)^2 + \frac{6}{16}\,(0 - 0)^2 + \frac{4}{16}\,(1 - 0)^2 + \frac{1}{16}\,(2 - 0)^2 = 1$, and $\sigma = \sqrt{1} = 1$.

5. $\mu = .1\,(430) + (.2)\,(480) + (.4)\,(520) + (.2)\,(565) + (.1)\,(580) = 518$,
$\text{Var}\,(X) = .1\,(430 - 518)^2 + (.2)\,(480 - 518)^2 + (.4)\,(520 - 518)^2 + (.2)\,(565 - 518)^2 + (.1)\,(580 - 518)^2 = 1891$,
and $\sigma = \sqrt{1891} \approx 43.5$.

7. The mean of the histogram in Figure (b) is more concentrated about its mean than the histogram in Figure (a). Therefore, the histogram in Figure (a) has the larger variance.

9. $E\,(X) = 1\,(.1) + 2\,(.2) + 3\,(.3) + 4\,(.2) + 5\,(.2) = 3.2$, so
$\text{Var}\,(X) = (.1)\,(1 - 3.2)^2 + (.2)\,(2 - 3.2)^2 + (.3)\,(3 - 3.2)^2 + (.2)\,(4 - 3.2)^2 + (.2)\,(5 - 3.2)^2 = 1.56$.

11. $\mu = \frac{1 + 2 + 3 + 4 + 5 + 6 + 7 + 8}{8} = 4.5$, so
$V\,(X) = \frac{1}{8}\,(1 - 4.5)^2 + \frac{1}{8}\,(2 - 4.5)^2 + \frac{1}{8}\,(3 - 4.5)^2 + \frac{1}{8}\,(4 - 4.5)^2$
$$+ \frac{1}{8}\,(5 - 4.5)^2 + \frac{1}{8}\,(6 - 4.5)^2 + \frac{1}{8}\,(7 - 4.5)^2 + \frac{1}{8}\,(8 - 4.5)^2 = 5.25.$$

13. a. Let X be the annual birth rate during the years 2003–2012.

b.

x	13.7	13.8	14.0	14.2	14.7
$P(X = x)$	.1	.3	.3	.2	.1

c. $E(X) = (.1)(13.7) + (.3)(13.8) + (.3)(14.0) + (.2)(14.2) + (.1)(14.7) = 14.02$,

$\text{Var}(X) = (.1)(13.7 - 14.02)^2 + (.3)(13.8 - 14.02)^2 + (.3)(14.0 - 14.02)^2 + (.2)(14.2 - 14.02)^2 +$
$(.1)(14.7 - 14.02)^2 = 0.0776$,

and $\sigma = \sqrt{.0776} \approx .2786$.

15. a. For Mutual Fund A, $\mu_X = (.2)(-4) + (.5)(8) + (.3)(10) = 6.2$, or \$620, and

$\text{Var}(X) = (.2)(-4 - 6.2)^2 + (.5)(8 - 6.2)^2 + (.3)(10 - 6.2)^2 = 26.76$, or \$267,600.

For Mutual Fund B, $\mu_X = (.2)(-2) + (.4)(6) + (.4)(8) = 5.2$, or \$520, and

$\text{Var}(X) = (.2)(-2 - 5.2)^2 + (.4)(6 - 5.2)^2 + (.4)(8 - 5.2)^2 = 13.76$, or \$137,600.

b. Mutual Fund A

c. Mutual Fund B

17. $\text{Var}(X) = (.4)(1)^2 + (.3)(2)^2 + (.2)(3)^2 + (.1)(4)^2 - (2)^2 = 1$.

19. $\mu_X = \frac{1}{6}(70 + 55 + 57 + 62 + 53 + 56) \approx 58.8333$,

$\text{Var}(X) \approx \frac{1}{6}\left[(70 - 58.8333)^2 + (55 - 58.8333)^2 + (57 - 58.8333)^2 \right.$
$\left. + (62 - 58.8333)^2 + (53 - 58.8333)^2 + (56 - 58.8333)^2\right] \approx 32.4722$,

and $\sigma_X \approx \sqrt{32.4722} \approx 5.70$.

21. $\mu_X = \frac{1}{6}(81 + 91 + 82 + 75 + 82 + 95) \approx 84.3333$,

$\text{Var}(X) \approx \frac{1}{6}\left[(81 - 84.3333)^2 + (91 - 84.3333)^2 + (82 - 84.3333)^2 \right.$
$\left. + (75 - 84.3333)^2 + (82 - 84.3333)^2 + (95 - 84.3333)^2\right] \approx 44.5556$,

and $\sigma_X \approx \sqrt{44.5556} \approx 6.67$.

23.

x	1342	1428	1545	1707	1807	1815
Relative Frequency	1	1	1	1	1	1
$P(X = x)$	$\frac{1}{6}$	$\frac{1}{6}$	$\frac{1}{6}$	$\frac{1}{6}$	$\frac{1}{6}$	$\frac{1}{6}$

$\mu_x = \frac{1}{6}(1342 + 1428 + 1545 + 1707 + 1807 + 1815) = 1607.33$,

$\text{Var}(X) \approx \frac{1}{6}\left[(1342 - 1607.33)^2 + (1428 - 1607.33)^2 + (1545 - 1607.33)^2 \right.$
$\left. + (1707 - 1607.33)^2 + (1807 - 1607.33)^2 + (1815 - 1607.33)^2\right] \approx 33,228.889$,

and $\sigma_X = \sqrt{32,228.889} \approx 182.2879$, so the average number of hours worked is approximately 1607 and the standard deviation is approximately 182 hours.

25. $\mu_X = \frac{1}{7}(16.3 + 15.4 + 22.2 + 17.2 + 23.2 + 30.4 + 26.4) \approx 21.59$,

$\text{Var}(X) \approx \frac{1}{7}\left[(16.3 - 21.59)^2 + (15.4 - 21.59)^2 + (22.2 - 21.59)^2 + (17.2 - 21.59)^2 \right.$
$\left. + (23.2 - 21.59)^2 + (30.4 - 21.59)^2 + (26.4 - 21.59)^2\right] \approx 27.04$,

and $\sigma_X \approx \sqrt{27.04} = 5.2$.

27. $\mu_X = \frac{1}{8}(50 + 47 + 47 + 45 + 45 + 42 + 42 + 40) = 44.75$,

$\text{Var}(X) = \frac{1}{8}\left[(50 - 44.75)^2 + 2(47 - 44.75)^2 + 2(45 - 44.75)^2 + 2(42 - 44.75)^2 + (40 - 44.75)^2\right] = 9.4375$,

and $\sigma_X \approx 3.07$.

29. $\mu_X = \frac{1}{9}(67 + 72 + 83 + 88 + 89 + 91 + 113 + 121 + 127) \approx 94.5556$,

$\text{Var}(X) \approx \frac{1}{9}\left[(67 - 94.5556)^2 + (72 - 94.5556)^2 + (83 - 94.5556)^2 + (88 - 94.5556)^2 + (89 - 94.5556)^2\right.$

$\left. + (91 - 94.5556)^2 + (113 - 94.5556)^2 + (121 - 94.5556)^2 + (127 - 94.5556)^2\right] \approx 397.8025$,

and $\sigma_X \approx \sqrt{397.8025} \approx 19.94$.

31. $\mu_X = \frac{1}{10}(2.38 + 3.03 + 1.89 + 3.47 + 3.23 + 1.81 + 3.89 + 4.35 + 1.24 + 1.93) = 2.722$,

$\text{Var}(X) = \frac{1}{10}\left[(2.38 - 2.722)^2 + (3.03 - 2.722)^2 + (1.89 - 2.722)^2 + \cdots + (1.93 - 2.722)^2\right] \approx 0.939156$, and

$\sigma_X \approx 0.969$.

33. $\mu_X = \frac{1}{10}(3.76 + 3.7 + 3.66 + 3.5 + 3.46 + 3.46 + 3.43 + 3.2 + 2.88 + 2.19) = 3.324$,

$\text{Var}(X) = \frac{1}{10}\left[(3.76 - 3.324)^2 + (3.7 - 3.324)^2 + (3.66 - 3.324)^2 + (3.5 - 3.324)^2 + 2(3.46 - 3.324)^2\right.$

$\left. + (3.43 - 3.324)^2 + (3.2 - 3.324)^2 + (2.88 - 3.324)^2 + (2.19 - 3.324)^2\right] \approx .202204$,

and $\sigma_X \approx \sqrt{.202204} \approx .4497$.

35. $\mu_X = \frac{1}{13}(2.4 + 4.8 + 4.4 + 3.8 + 7.0 + 5.7 + 4.9 + 5.2 + 4.7 + 5.7 + 5.4 + 5.3 + 5.9) \approx 5.0154$,

$\text{Var}(X) = \frac{1}{13}\left[(2.4 - 5.0154)^2 + (4.8 - 5.0154)^2 + (4.4 - 5.0154)^2 + \cdots + (5.9 - 5.0154)^2\right] \approx 1.1367$, and

$\sigma_X \approx 1.07$.

37. a. $\mu_X = \frac{1}{12}(7.7 + 8.3 + 8.3 + 8.9 + 8.9 + 9.0 + 8.9 + 8.1 + 7.9 + 7.4 + 7.1 + 6.2) \approx 8.0583$,

$\text{Var}(X) = \frac{1}{12}\left[(7.7 - 8.0583)^2 + (8.3 - 8.0583)^2 + (8.3 - 8.0583)^2 + \cdots + (6.2 - 8.0583)^2\right] \approx 0.6741$, and

$\sigma_X \approx 0.82$.

b. $\mu_X = \frac{1}{12}[3(6.1) + 2(6.4) + 6.5 + 6.3 + 6.0 + 5.5 + 5.3 + 4.8 + 4.4] = 5.825$,

$\text{Var}(X) = \frac{1}{12}\left[3(6.1 - 5.825)^2 + 2(6.4 - 5.825)^2 + (6.5 - 5.825)^2 + \cdots + (4.4 - 5.825)^2\right] \approx 0.4219$, and

$\sigma_X \approx 0.65$.

c. The average monthly supply of single-family homes for sale dropped from 2011 to 2012 as the economy recovered from the Great Recession.

39. $\mu_X = 22\left(\frac{1182}{26,415}\right) + 27\left(\frac{3810}{26,415}\right) + 32\left(\frac{6104}{26,415}\right) + 37\left(\frac{7124}{26,415}\right) + 42\left(\frac{8195}{26,415}\right) \approx 35.2822$, or 35.28,

$\text{Var}(X) = \left(\frac{1182}{26,415}\right)(22 - 35.2822)^2 + \left(\frac{3810}{26,415}\right)(27 - 35.2822)^2 + \left(\frac{6104}{26,415}\right)(32 - 35.2822)^2$

$+ \left(\frac{7124}{26,415}\right)(37 - 35.2822)^2 + \left(\frac{8195}{26,415}\right)(42 - 35.2822)^2 \approx 35.074$,

and $\sigma_X \approx \sqrt{35.074} \approx 5.922$, or approximately 5.92.

41. a. Using Chebychev's inequality $P(\mu - k\sigma \leq X \leq \mu + k\sigma) \geq 1 - 1/k^2$ with $\mu - k\sigma = 42 - k(2) = 38$ and

$k = 2$, we have $P(\mu - k\sigma \leq X \leq \mu + k\sigma) \geq 1 - \frac{1}{2^2} = 1 - \frac{1}{4} = \frac{3}{4}$, or at least .75.

b. Using Chebychev's inequality with $\mu - k\sigma = 42 - k(2) = 32$ and $k = 5$, we have

$P(\mu - k\sigma \leq X \leq \mu + k\sigma) \geq 1 - \frac{1}{5^2} = 1 - \frac{1}{25} = \frac{24}{25}$, or at least .96.

43. Here $\mu = 50$ and $\sigma = 1.4$. We require that $c = k\sigma$, so $k = \dfrac{c}{1.4}$. Next, we solve $.96 = 1 - \left(\dfrac{1.4}{c}\right)^2$, obtaining

$\dfrac{1.96}{c^2} = .04$, $c^2 = \dfrac{1.96}{.04} = 49$, and $c = 7$.

45. Using Chebychev's inequality with $\mu - k\sigma = 24 - k\,(3) = 20$ and $k = \frac{4}{3}$, we have

$P\,(\mu - k\sigma \le X \le \mu + k\sigma) \ge 1 - \dfrac{1}{(4/3)^2} = 1 - \dfrac{9}{16} = \dfrac{7}{16}$, or at least $.4375$.

47. Using Chebychev's inequality with $\mu - k\sigma = 58{,}000 - k\,(500) = 56{,}000$ and $k = 4$, we have

$P\,(\mu - k\sigma \le X \le \mu + k\sigma) \ge 1 - \dfrac{1}{4^2} = 1 - \dfrac{1}{16} = \dfrac{15}{16}$, or at least $.9375$.

49. True. This follows from the definition.

Technology Exercises   page 500

1. a.

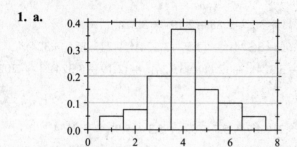

b. $\mu = 4$ and $\sigma \approx 1.40$.

3. a.

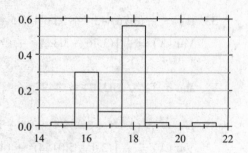

b. $\mu = 17.34$ and $\sigma \approx 1.11$.

5. a. Let X denote the random variable that gives the weight of a carton of sugar.

b.

x	4.96	4.97	4.98	4.99	5.00	5.01	5.02	5.03	5.04	5.05	5.06
$P\,(X = x)$	$\frac{3}{30}$	$\frac{4}{30}$	$\frac{4}{30}$	$\frac{1}{30}$	$\frac{1}{30}$	$\frac{5}{30}$	$\frac{3}{30}$	$\frac{3}{30}$	$\frac{4}{30}$	$\frac{1}{30}$	$\frac{1}{30}$

c. $\mu = 5.00467 \approx 5.00$, $\mathrm{Var}\,(X) = .0009$, and $\sigma = \sqrt{.0009} = .03$.

7. a.

b. $\mu = 65.875$ and $\sigma = 1.73$.

8.4   The Binomial Distribution

Problem-Solving Tips

1. A **binomial experiment** has the following properties:

   a. The number of trials is fixed.

   b. There are two possible outcomes.

   c. The probability of success in each trial is the same.

   d. The trials are independent of each other.

2. The **probability of exactly x successes in n independent trials**, in a binomial experiment in which the probability of success in any trial is p and the probability of failure in any trial is q, is $C(n, x)\, p^x q^{n-x}$.

3. The mean, variance, and standard deviation of a random variable X associated with a binomial experiment in which the probability of success is p and the probability of failure is q are $\mu = E(X) = np$, $\mathrm{Var}(X) = npq$, and $\sigma_X = \sqrt{npq}$.

Concept Questions   page 509

1. **a.** There are two outcomes in each trial.   **b.** The number is fixed.

   c. They are independent.   **d.** The probability is $C(n, x)\, p^x q^{n-x}$.

Exercises   page 509

1. Yes. The number of trials is fixed, there are two outcomes of the experiment, the probability in each trial is fixed ($p = \frac{1}{6}$), and the trials are independent of each other.

3. No. There are more than two outcomes in each trial.

5. No. There are more than two outcomes in each trial and the probability of success (an accident) in each trial is not the same.

7. $C(4, 2)\left(\frac{1}{3}\right)^2\left(\frac{2}{3}\right)^2 = \frac{4!}{2!\,2!}\left(\frac{4}{81}\right) \approx .296.$ 9. $C(5, 3)(.2)^3(.8)^2 = \left(\frac{5!}{2!\,3!}\right)(.2)^3(.8)^2 \approx .051.$

11. The required probability is $P(X = 0) = C(5, 0)\left(\frac{1}{3}\right)^0\left(\frac{2}{3}\right)^5 \approx .132.$

13. The required probability is
$$P(X \ge 3) = C(6, 3)\left(\frac{1}{2}\right)^3\left(\frac{1}{2}\right)^{6-3} + C(6, 4)\left(\frac{1}{2}\right)^4\left(\frac{1}{2}\right)^{6-4} + C(6, 5)\left(\frac{1}{2}\right)^5\left(\frac{1}{2}\right)^{6-5} + C(6, 6)\left(\frac{1}{2}\right)^6\left(\frac{1}{2}\right)^{6-6}$$
$$= \frac{6!}{3!\,3!}\left(\frac{1}{2}\right)^6 + \frac{6!}{4!\,2!}\left(\frac{1}{2}\right)^6 + \frac{6!}{5!\,1!}\left(\frac{1}{2}\right)^6 + \frac{6!}{6!\,0!}\left(\frac{1}{2}\right)^6 = \frac{1}{64}(20 + 15 + 6 + 1) = \frac{21}{32} \approx .656.$$

15. The probability of no failure or (equivalently) five successes is $P(X = 5) = C(5, 5)\left(\frac{1}{3}\right)^5\left(\frac{2}{3}\right)^{5-5} = \frac{1}{243} \approx .004.$

17. Here $n = 4$ and $p = \frac{1}{6}$, so $P(X = 2) = C(4, 2)\left(\frac{1}{6}\right)^2\left(\frac{5}{6}\right)^2 = \frac{25}{216} \approx .116.$

19. Here $n = 5$ and $p = .4$, so $q = 1 - .4 = .6$.

a. $P(X = 0) = C(5, 0)(.4)^0(.6)^5 \approx .078,$

$P(X = 1) = C(5, 1)(.4)(.6)^4 \approx .259,$

$P(X = 2) = C(5, 2)(.4)^2(.6)^3 \approx .346,$

$P(X = 3) = C(5, 3)(.4)^3(.6)^3 \approx .230,$

$P(X = 4) = C(5, 4)(.4)^4(.6)^2 \approx .077,$ and

$P(X = 5) = C(5, 5)(.4)^5(.6)^0 \approx .010.$

b.

x	0	1	2	3	4	5
$P(X = x)$	.078	.259	.346	.230	.077	.010

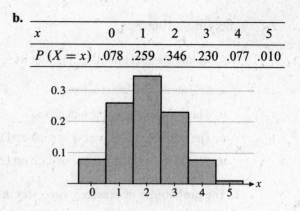

c. $\mu = np = 5(.4) = 2$ and $\sigma = \sqrt{npq} = \sqrt{5(.4)(.6)} = \sqrt{1.2} \approx 1.095.$

21. Here $1 - p = \frac{1}{50}$, so $p = \frac{49}{50}$. Thus, the probability of obtaining 49 or 50 nondefective fuses is

$P(X = 49) + P(X = 50) = C(50, 49)\left(\frac{49}{50}\right)^{49}\left(\frac{1}{50}\right) + C(50, 50)\left(\frac{49}{50}\right)^{50}\left(\frac{1}{50}\right)^0 \approx .74.$ This is also the probability of at most one defective fuse, so the inference is incorrect.

23. The probability that she will serve no ace or one ace is

$P(X = 0) + P(X = 1) = C(5, 0)(.15)^0(.85)^5 + C(5, 1)(.15)^1(.85)^4 = \frac{5!}{0!\,5!}(.85)^5 + \frac{5!}{1!\,4!}(.15)(.85)^4 \approx .83521.$ Therefore, the probability that she will serve at least two aces is $1 - P(X = 0) - P(X = 1) = 1 - .83521 = .16479,$ or approximately .165.

25. The required probability is $P(X = 6) = C(6, 6)\left(\frac{1}{4}\right)^6\left(\frac{3}{4}\right)^0 \approx .0002.$

27. a. The probability that a restaurant is in violation of the health code is $p = \frac{3}{10}$ and so the probability that a restaurant is not in violation of the health code is $q = \frac{7}{10}$. Let X denote the number of restaurants in violation of the health code. Then the probability that none of the restaurants is in violation of the health code is

$P(X = 0) = C(7, 0)\left(\frac{3}{10}\right)^0\left(\frac{7}{10}\right)^5 \approx .168.$

b. The probability that just one of the restaurants is in violation of the health code is

$P(X = 1) = C(5, 1)\left(\frac{3}{10}\right)^1\left(\frac{7}{10}\right)^4 \approx .360.$

c. The probability that at least two of the restaurants are in violation of the health code is

$1 - P(X \leq 1) = 1 - (.168 + .360) = .472.$

29. In order to obtain a score of at least 90%, the student needs to answer three or four of the remaining questions correctly. The probability of doing this is $P(X \geq 3) = C(4, 3)(.5)^3(.5) + C(4, 4)(.5)^4(.5)^0 \approx .313.$

31. The required probability is

$P(X \geq 3) = C(6, 3)(.78)^3(.22)^3 + C(6, 4)(.78)^4(.22)^2 + C(6, 5)(.78)^5(.22)^1 + C(6, 6)(.78)^6(.22)^0$

$= \frac{6!}{3!\,3!}(.78)^3(.22)^3 + \frac{6!}{4!\,2!}(.78)^4(.22)^2 + \frac{6!}{5!\,1!}(.78)^5(.22)^1 + \frac{6!}{6!\,0!}(.78)^6(1) \approx .976.$

33. The probability of success is $\frac{1}{5}$. The probability that he will pass the examination through random guesses is

$$P\left(X \geq 5\right) = C\left(8, 5\right)\left(.2\right)^5\left(.8\right)^3 + C\left(8, 6\right)\left(.2\right)^6\left(.8\right)^2 + C\left(8, 7\right)\left(.2\right)^7\left(.8\right)^1 + C\left(8, 8\right)\left(.2\right)^8\left(.8\right)^0$$

$$\approx .009 + .001 + .0000 + .0000 \approx .01.$$

35. This is a binomial experiment with $n = 9$, $p = \frac{1}{3}$, and $q = \frac{2}{3}$.

a. The probability is $P\left(X = 3\right) = C\left(9, 3\right)\left(\frac{1}{3}\right)^3\left(\frac{2}{3}\right)^6 = \frac{9!}{6!\,3!}\left(\frac{1}{3}\right)^3\left(\frac{2}{3}\right)^6 \approx .273.$

b. The probability is

$$P\left(X = 0\right) + P\left(X = 1\right) + P\left(X = 2\right) + P\left(X = 3\right)$$

$$= C\left(9, 0\right)\left(\frac{1}{3}\right)^0\left(\frac{2}{3}\right)^9 + C\left(9, 1\right)\left(\frac{1}{3}\right)\left(\frac{2}{3}\right)^8 + C\left(9, 2\right)\left(\frac{1}{3}\right)^2\left(\frac{2}{3}\right)^7 + C\left(9, 3\right)\left(\frac{1}{3}\right)^3\left(\frac{2}{3}\right)^6$$

$$\approx .026 + .117 + .234 + .273 \approx .650.$$

37. This is a binomial experiment with $n = 10$, $p = .02$, and $q = .98$.

a. The probability that the sample contains no defectives is given by $P\left(X = 0\right) = C\left(10, 0\right)\left(.02\right)^0\left(.98\right)^{10} \approx .817.$

b. The probability that the sample contains at most two defectives is given by

$$P\left(X \leq 2\right) = C\left(10, 0\right)\left(.02\right)^0\left(.98\right)^{10} + C\left(10, 1\right)\left(.02\right)\left(.98\right)^9 + C\left(10, 2\right)\left(.02\right)^2\left(.98\right)^8 \approx .817 + .167 + .015$$

$$= .999.$$

39. The required probability is

$$P\left(X = 9\right) + P\left(X = 10\right) = C\left(10, 9\right)\left(.58\right)^9\left(.42\right) + C\left(10, 10\right)\left(.58\right)^{10}\left(.42\right)^0$$

$$= \frac{10!}{9!\,1!}\left(.58\right)^9\left(.42\right) + \frac{10!}{10!\,0!}\left(.58\right)^{10} \approx .0355.$$

41. The required probability is the complement of the probability that there are at most 4 females. The latter probability is

$$P\left(X \leq 4\right) = P\left(X = 0\right) + P\left(X = 1\right) + P\left(X = 2\right) + P\left(X = 3\right) + P\left(X = 4\right)$$

$$= C\left(10, 0\right)\left(.23\right)^0\left(.77\right)^{10} + C\left(10, 1\right)\left(.23\right)^1\left(.77\right)^9 + C\left(10, 2\right)\left(.23\right)^2\left(.77\right)^8$$

$$+ C\left(10, 3\right)\left(.23\right)^3\left(.77\right)^7 + C\left(10, 4\right)\left(.23\right)^4\left(.77\right)^6$$

$$= \frac{10!}{0!\,10!}\left(.77\right)^{10} + \frac{10!}{1!\,9!}\left(.23\right)^1\left(.77\right)^9 + \frac{10!}{2!\,8!}\left(.23\right)^2\left(.77\right)^8 + \frac{10!}{3!\,7!}\left(.23\right)^3\left(.77\right)^7 + \frac{10!}{4!\,6!}\left(.23\right)^4\left(.77\right)^6 \approx .9431.$$

Thus, the required probability is $1 - 0.9431 \approx 0.0569.$

43. The probability that all 10 of them believed that engaging in internal politics is the way to get ahead at work is
$$P\left(X = 10\right) = C\left(10, 10\right)\left(0.51\right)^{10}\left(0.49\right)^0 = \frac{10!}{0!\,10!}\left(0.51\right)^{10} \approx .0012.$$

45. a. The required probability is $P\left(X = 2\right) = C\left(10, 2\right)\left(.05\right)^2\left(.95\right)^8 \approx .075.$

b. The required probability is

$$P\left(X \geq 2\right) = 1 - P\left(X \leq 2\right) = 1 - \left[C\left(10, 2\right)\left(.05\right)^2\left(.95\right)^8 + C\left(10, 1\right)\left(.05\right)\left(.95\right)^9 + C\left(10, 0\right)\left(.05\right)^0\left(.95\right)^{10}\right]$$

$$\approx 1 - \left(.0746 + .3151 + .5987\right) \approx 0.012.$$

47. $E\left(X\right) = 45$, implying that $np = 45$, and $\sigma_X = 6$, implying that $\sqrt{npq} = 6$. The second equation gives $npq = 36$, and since $np = 45$, we have $45q = 36$, $45\left(1 - p\right) = 36$, $45 - 45p = 36$, $45p = 9$, and $p = \frac{1}{5}$. This implies that $n\left(\frac{1}{5}\right) = 45$, so $n = 225$.

49. The mean number of defective panels per sample is $\mu = \frac{1}{40}[33 \cdot 0 + 3 \cdot 1 + 2 \cdot 2 + 1 \cdot 3 + 1 \cdot 4] = .35$. Using Equation (15b) with $n = 10$, we find $0.35 = 10p$, so $p = .035$. Thus, approximately 3.5% of the day's production of solar panels is defective.

51. a. The required probability is $P(X = 0) = C(20, 0)(.1)^0 (.9)^{20} \approx .122$.

   b. The required probability is $P(X = 0) = C(20, 0)(.05)^0 (.95)^{20} \approx .358$.

53. The required probability is $P(X = 0) = C(10, 0)(.1)^0 (.9)^{10} \approx .349$.

55. Take $p = \frac{1}{2}$. The probability of obtaining no heads in n tosses is $P(X = n) = C(n, n)\left(\frac{1}{2}\right)^n \left(\frac{1}{2}\right)^0 = \left(\frac{1}{2}\right)^n$, and the probability of obtaining at least one head is $1 - \left(\frac{1}{2}\right)^n$. We want this to exceed .99. Thus, $1 - \left(\frac{1}{2}\right)^n \geq .99$, giving $\frac{1}{2^n} \geq .01$, $2^n \geq 100$, and $n \geq \frac{\ln 100}{\ln 2} \approx 6.64$. Therefore, one must toss the coin at least 7 times.

57. The mean number of people for whom the drug is effective is $\mu = np = (500)(.75) = 375$. The standard deviation of the number of people for whom the drug can be expected to be effective is $\sigma = \sqrt{npq} = \sqrt{(500)(.75)(.25)} \approx 9.68$.

59. False. There are exactly two outcomes.

61. False. Here $p = \frac{1}{4}$ and $q = 1 - \frac{1}{4} = \frac{3}{4}$, so the probability that the batter will get a hit if he bats four times is
$1 - P(X = 0) = 1 - C(4, 0)\left(\frac{1}{4}\right)^0 \left(\frac{3}{4}\right)^4 = 1 - \frac{4!}{4!\,0!} \cdot 1 \cdot \frac{81}{256} = 1 - \frac{81}{256} = \frac{175}{256} \approx .68$, which is far from guaranteeing that he gets a hit.

8.5   The Normal Distribution

Problem-Solving Tips

1. **Normal distributions** are a special class of continuous probability distributions. A normal curve is the bell-shaped graph of a normal distribution. The standard normal curve has mean $\mu = 0$ and standard deviation $\sigma = 1$. The random variable associated with a standard normal distribution is called a **standard normal variable** and is denoted by Z. The areas under the standard normal curve to the left of the number z, corresponding to the probabilities $P(Z < z)$ or $P(Z \leq z)$, are given in Table 2, Appendix D.

2. The area of the region under the normal curve between $x = a$ and $x = b$ is equal to the area of the region under the standard normal curve between $z = \dfrac{a - \mu}{\sigma}$ and $z = \dfrac{b - \mu}{\sigma}$. The probability of the random variable X associated with this area is $P(a < X < b) = \left(\dfrac{a - \mu}{\sigma} < Z < \dfrac{b - \mu}{\sigma}\right)$.

Concept Questions   page 520

1. a. The curve peaks at $x = \mu$.

   b. The curve is symmetrical with respect to the vertical line $x = \mu$.

   c. The curve lies above the x-axis and approaches 0 as x extends indefinitely in either direction.

d. The value of the area under the normal curve is 1.

e. For any normal curve, 68.27% of the area under the curve lies within one standard deviation of the mean.

 Exercises   page 520

1. $P(Z < 1.45) = .9265$.

3. $P(Z < -1.75) = .0401$.

5. $P(-1.32 < Z < 1.74) = P(Z < 1.74) - P(Z < -1.32) = .9591 - .0934 = .8657$.

7. a.

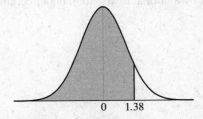

b. $P(Z < 1.38) = .9162$.

9. a.

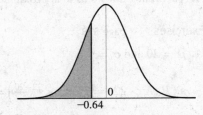

b. $P(Z < -.64) = .2611$.

11. a.

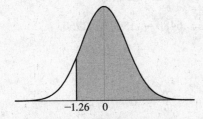

b. $P(Z > -1.26) = 1 - P(Z < -1.26)$

$= 1 - .1038 = .8962$.

13. a.

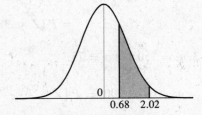

b. $P(.68 < Z < 2.02)$

$= P(Z < 2.02) - P(Z < .68)$

$= .9783 - .7517 = .2266$.

15. a. Referring to Table 2 in Appendix D, we see that $P(Z < z) = .8907$ implies that $z = 1.23$.

b. Referring to Table 2 in Appendix D, we see that $P(Z < z) = .2090$ implies that $z = -.81$.

17. a. $P(Z > -z) = 1 - P(Z < -z) = 1 - .9713 = .0287$ implies $z = 1.9$.

b. $P(Z < -z) = .9713$ implies that $z = -1.9$.

19. a. $P(X < 60) = P\left(Z < \frac{60-50}{5}\right) = P(Z < 2) = .9772$.

b. $P(X > 43) = P\left(Z > \frac{43-50}{5}\right) = P(Z > -1.4) = P(Z < 1.4) = .9192$.

c. $P(46 < X < 58) = P\left(\frac{46-50}{5} < Z < \frac{58-50}{5}\right) = P(-.8 < Z < 1.6) = P(Z < 1.6) - P(Z < -.8)$

$= .9452 - .2119 = .7333$.

8.6   Applications of the Normal Distribution

Problem-Solving Tips

A binomial distribution associated with a binomial experiment involving n trials with probability of success p and probability of failure q may be approximated by a normal distribution (if n is large and p is not close to 0 or 1) with $\mu = np$ and $\sigma = \sqrt{npq}$.

Concept Questions   page 527

1. Theorem 1 allows us to approximate a binomial distribution by a normal distribution under certain conditions.

Exercises   page 528

1. $\mu = 20$ and $\sigma = 2.6$.

   a. $P(X > 22) = P\left(Z > \frac{22-20}{2.6}\right) = P(Z > .7692) = P(Z < -.7692) \approx .2206.$

   b. $P(X < 18) = P\left(Z < \frac{18-20}{2.6}\right) = P(Z < -.7692) \approx .2206.$

   c. $P(19 < X < 21) = P\left(\frac{19-20}{2.6} < Z < \frac{21-20}{2.6}\right) = P(-.3846 < Z < .3846)$

$$= P(Z < .3846) - P(Z < -.3846) = .6480 - .3520 = .2960.$$

3. $\mu = 750$ and $\sigma = 75$.

   a. $P(X > 900) = P\left(Z > \frac{900-750}{75}\right) = P(Z > 2) = P(Z < -2) = .0228.$

   b. $P(X < 600) = P\left(Z < \frac{600-750}{75}\right) = P(Z < -2) = .0228.$

   c. $P(750 < X < 900) = P\left(Z < \frac{750-750}{75} < Z < \frac{900-750}{75}\right) = P(0 < Z < 2) = P(Z < 2) - P(Z < 0)$

$$= .9772 - .5000 = .4772.$$

   d. $P(600 < X < 800) = P\left(\frac{600-750}{75} < Z < \frac{800-750}{75}\right) = P(-2 < Z < .667) = P(Z < .667) - P(Z < -2)$

$$= .7486 - .0228 = .7258.$$

5. $\mu = 100$ and $\sigma = 15$.

   a. $P(X > 140) = P\left(Z > \frac{140-100}{15}\right) = P(Z > 2.667) = P(Z < -2.667) = .0038.$

   b. $P(X > 120) = P\left(Z > \frac{120-100}{15}\right) = P(Z > 1.33) = P(Z < -1.33) = .0918.$

   c. $P(100 < X < 120) = P\left(\frac{100-100}{15} < Z < \frac{120-100}{15}\right) = P(0 < Z < 1.333) = P(Z < 1.333) - P(Z < 0)$

$$= .9082 - .5000 = .4082.$$

   d. $P(X < 90) = P\left(Z < \frac{90-100}{15}\right) = P(Z < -.667) = .2514.$

7. Here $\mu = 675$ and $\sigma = 50$.

$$P(650 < X < 750) = P\left(\frac{650-675}{50} < Z < \frac{750-675}{50}\right) = P(-.5 < Z < 1.5) = P(Z < 1.5) - P(Z < -.5)$$

$$= .9332 - .3085 = .6247.$$

9. Here $\mu = 22$ and $\sigma = 4$. $P(X < 12) = P\left(Z < \frac{12-22}{4}\right) = P(Z < -2.5) = .0062$, or 0.62%.

11. $\mu = 70$ and $\sigma = 10$. To find the cutoff point for an A, we solve $P(Y < y) = .85$ for y, obtaining

$$P(Y < y) = P\left(Z < \frac{y-70}{10}\right) = .85 \text{ and so } \frac{y-70}{10} = 1.04 \text{ and } y = 80.4 \approx 80.$$

For a B: $P(Y < y) = P\left(Z < \frac{y-70}{10}\right) \approx .60$, so $\frac{y-70}{10} = .25$ and $y \approx 73$.

For a C: $P(Y < y) = P\left(Z < \frac{y-70}{10}\right) \approx .2$, so $\frac{y-70}{10} = -.84$ and $y \approx 62$.

For a D: $P(Y < y) = P\left(Z < \frac{y-70}{10}\right) \approx .05$, so $\frac{y-70}{10} = -1.64$ and $y \approx 54$.

13. Let X denote the number of heads in 25 tosses of the coin. Then X is a binomial random variable with $n = 25$, $p = .4$, and $q = .6$, so $\mu = (25)(.4) = 10$ and $\sigma = \sqrt{(25)(.4)(.6)} \approx 2.45$. We approximate the binomial distribution by a normal distribution with a mean of 10 and a standard deviation of 2.45, with Y as the associated normal random variable.

a. $P(X < 10) \approx P(Y < 9.5) = P\left(Z < \frac{9.5-10}{2.45}\right) = P(Z < -.20) \approx .4207$.

b. $P(10 \le X \le 12) \approx P(9.5 < Y < 12.5) = P\left(\frac{9.5-10}{2.45} < Z < \frac{12.5-10}{2.45}\right) = P(Z < 1.02) - P(Z < -.20)$

$$= P(Z < 1.02) - P(Z < -.20) \approx .8461 - .4207 = .4254.$$

c. $P(X > 15) \approx P(Y \ge 15) = P\left(Z > \frac{15.5-10}{2.45}\right) = P(Z > 2.24) = P(Z < -2.24) \approx .0125$.

15. Let X denote the number of times the marksman hits his target. Then X has a binomial distribution with $n = 30$, $p = .6$, and $q = .4$. Therefore, $\mu = (30)(.6) = 18$ and $\sigma = \sqrt{(30)(.6)(.4)} = 2.68$.

a. $P(X \ge 20) \approx P(Y \ge 19.5) = P\left(Z > \frac{19.5-18}{2.68}\right) = P(Z > .56) = P(Z < -.56) \approx .2877$.

b. $P(X < 10) \approx P(Y < 9.5) = P\left(Z < \frac{9.5-18}{2.68}\right) = P(Z < -3.17) \approx .0008$.

c. $P(15 \le X \le 20) \approx P(14.5 < Y < 20.5) = P\left(\frac{14.5-18}{2.68} < Z < \frac{20.5-18}{2.68}\right) = P(Z < .93) - P(Z < -1.31)$

$$= .8238 - .0951 = .7287.$$

17. Let X denote the number of seconds. Then X has a binomial distribution with $n = 200$, $p = .03$, and $q = .97$. Thus, $\mu = (200)(.03) = 6$, $\sigma = \sqrt{(200)(.03)(.97)} \approx 2.41$, and

$$P(X < 10) \approx P(Y < 9.5) = P\left(Z < \frac{9.5-6}{2.41}\right) = P(Z < 1.45) \approx .9265.$$

19. Let X denote the number of workers who meet with an accident during a one-year period. Then $\mu = (800)(.1) = 80$, $\sigma = \sqrt{(800)(.1)(.9)} \approx 8.49$, and

$$P(X > 70) \approx P(Y > 70.5) = P\left(Z > \frac{70.5-80}{8.49}\right) = P(Z > -1.12) = P(Z < 1.12) \approx .8686.$$

21. a. Let X denote the number of mice that recovered from the disease. Then X has a binomial distribution with $n = 50$, $p = .5$, and $q = .5$, so $\mu = (50)(.5) = 25$ and $\sigma = \sqrt{(50)(.5)(.5)} \approx 3.54$. Approximating the binomial distribution by a normal distribution with a mean of 25 and a standard deviation of 3.54, we find that the probability that 35 or more of the mice would recover from the disease without benefit of the drug is

$$P(X \ge 35) \approx P(Y > 34.5) = P\left(Z > \frac{34.5-25}{3.54}\right) = P(Z > 2.68) = P(Z < -2.68) \approx .0037.$$

b. The drug is effective.

23. Let n denote the number of reservations the company should accept. Then we need to find

$$P(X \geq 2000) \approx P(Y > 1999.5) = .01 \text{ or, equivalently, } P\left(Z \geq \frac{1999.5 - np}{\sqrt{npq}}\right) = P\left(Z \leq \frac{np - 1999.5}{\sqrt{npq}}\right) = .01$$

with $p = .92$ and $q = .08$; that is, $P\left(Z \leq \frac{.92n - 1999.5}{\sqrt{.0736n}}\right) = .01$. We simplify $\frac{.92n - 1999.5}{\sqrt{.0736n}} = -2.33$,

obtaining $(.92n - 1999.5)^2 = (-2.33)^2(.0736n)$, $.8464n^2 - 3679.08n + 3,998,000.25 = .39956704n$,

and $.8464n^2 - 3679.479567n + 3,998,000.25 = 0$. Using the quadratic formula, we obtain

$$n = \frac{3679.479567 \pm \sqrt{2940.2376}}{1.6928} \approx 2142 \text{ or } 2206. \text{ We discard } n \approx 2206 \text{ because it does not satisfy the original}$$

equation. Therefore, the company should accept no more than 2142 reservations.

CHAPTER 8   Concept Review Questions   page 531

1. random

3. sum, $\left(\frac{1}{2}\right)(-2) + \left(\frac{1}{4}\right)(3) + \left(\frac{1}{4}\right)(4) = \frac{3}{4}$

5. $p_1(x_1 - \mu)^2 + p_2(x_2 - \mu)^2 + \cdots + p_n(x_n - \mu)^2$, $\sqrt{\text{Var}(X)}$

7. continuous, probability density function, set

CHAPTER 8   Review Exercises   page 531

1. a. $S = \{WWW, BWW, WBW, WWB, BBW, BWB, WBB, BBB\}$

d.

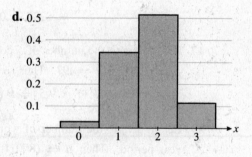

b.

Outcome	WWW	BWW	WBW	WWB	BBW	BWB	WBB	BBB
Value	0	1	1	1	2	2	2	3

c.

x	0	1	2	3
$P(X = x)$	$\frac{1}{35}$	$\frac{12}{35}$	$\frac{18}{35}$	$\frac{4}{35}$

3. a. $P(1 \leq X \leq 4) = .1 + .2 + .3 + .2 = .8$.

b. $\mu = 0(.1) + 1(.1) + 2(.2) + 3(.3) + 4(.2) + 5(.1) = 2.7$,

$\text{Var}(X) = .1(0 - 2.7)^2 + .1(1 - 2.7)^2 + .2(2 - 2.7)^2 + .3(3 - 2.7)^2 + .2(4 - 2.7)^2 + .1(5 - 2.7)^2 \approx 2.01$,

and $\sigma \approx \sqrt{2.01} \approx 1.42$.

5. $P(Z < .5) \approx .6915$

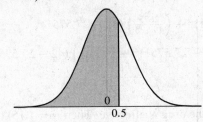

7. $P(-.75 < Z < .5)$

$= P(Z < .5) - P(Z < -.75)$

$\approx .6915 - .2266 = .4649$

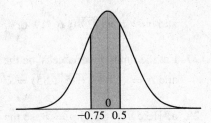

9. If $P(Z < z) = .9922$, then $z \approx 2.42$.

11. If $P(Z > z) = .9788$, then $P(Z < -z) = .9788$, $-z \approx 2.03$, and $z \approx -2.03$.

13. $P(X < 11) = P\left(Z < \frac{11-10}{2}\right) = P(Z < .5) \approx .6915$.

15. $P(7 < X < 9) = P\left(\frac{7-10}{2} < Z < \frac{9-10}{2}\right) = P(-1.5 < Z < -.5) = P(Z < -.5) - P(Z < -1.5)$

$\approx .3085 - .0668 = .2417$.

17. a. X gives the minimum age requirement for a regular driver's license.

b.

x	15	16	17	18	19	21
$P(X=x)$	.02	.30	.08	.56	.02	.02

c. $\mu = E(X) = (.02)(15) + (.3)(16) + (.08)(17) + (.56)(18) + (.02)(19) + (.02)(21) = 17.34$,

$\text{Var}(X) = (.02)(15 - 17.34)^2 + (.3)(16 - 17.34)^2 + (.08)(17 - 17.34)^2$
$+ (.56)(18 - 17.34)^2 + (.02)(19 - 17.34)^2 + (.02)(21 - 17.34)^2 = 1.2244$,

and $\sigma = \sqrt{1.2244} \approx 1.11$.

19. Let X denote the speed of a vehicle. Then the average vehicle speed is

$E(X) = (32)(.07) + (37)(.28) + (42)(.42) + (47)(.18) + (52)(.05) \approx 41.3$, or 41.3 mi/h.

21. This is a binomial experiment with $p = .7$ and $q = .3$. The probability that he will get exactly two strikes in four attempts is given by $P(X = 2) = C(4, 2)(.7)^2(.3)^2 \approx .2646$. The probability that he will get at least two strikes in four attempts is given by

$P(X = 2) + P(X = 3) + P(X = 4) = C(4, 2)(.7)^2(.3)^2 + C(4, 3)(.7)^3(.3) + C(4, 4)(.7)^4(.3)^0$

$\approx .2646 + .4116 + .2401 = .9163$.

23. Here $\mu = 64.5$ and $\sigma = 2.5$. Thus, $64.5 - 2.5k = 59.5$, $64.5 + 2.5k = 69.5$, and $k = 2$. Therefore, the required probability is $P(59.5 \le X \le 69.5) \ge 1 - \frac{1}{2^2} = .75$.

25. $\mu_X = 22 \left(\frac{8296}{19,198}\right) + 27 \left(\frac{5026}{19,198}\right) + 32 \left(\frac{2678}{19,198}\right) + 37 \left(\frac{1768}{19,198}\right) + 42 \left(\frac{1430}{19,198}\right) \approx 27.5751$, or approximately 27.58;

$\text{Var}(X) = \left(\frac{8296}{19,198}\right)(22 - 27.5751)^2 + \left(\frac{5026}{19,198}\right)(27 - 27.5751)^2 + \left(\frac{2678}{19,198}\right)(32 - 27.5751)^2$

$$+ \left(\frac{1768}{19,198}\right)(37 - 27.5751)^2 + \left(\frac{1430}{19,198}\right)(42 - 27.5751)^2 \approx 39.929;$$

and $\sigma \approx \sqrt{39.929} \approx 6.319$, or approximately 6.32.

27. Let the random variable X be the number of people for whom the drug is effective. Then $\mu = (.15)(800) = 120$ and $\sigma = \sqrt{(800)(.15)(.85)} = \sqrt{102} \approx 10.1$.

29. a. Here $n = 6$ and $p = .8$, so the probability is $P(X = 4) = C(6, 4)(.8)^4(.2)^2 \approx .246$.

b. The probability is

$$P(X = 4) + P(X = 5) + P(X = 6) = C(6, 4)(.8)^4(.2)^2 + C(6, 5)(.8)^5(.2) + C(6, 6)(.8)^6(.2)^0$$

$$\approx .901.$$

31. Here $\mu = (.6)(100) = 60$ and $\sigma = \sqrt{100(.6)(.4)} = 4.899$. Thus,

$$P(X > 50) \approx P(Y > 50.5) = P\left(Z > \frac{50.5 - 60}{4.899}\right) = P(Z > -1.94) = P(Z < 1.94) \approx .9738.$$

CHAPTER 8   **Before Moving On...**   page 533

1.

x	-3	-2	0	1	2	3
$P(X = x)$	.05	.1	.25	.3	.2	.1

2. a. $P(X \le 0) = P(X = 0) + P(X = -1) + P(X = -3) + P(X = -4) = .28 + .32 + .14 + .06 = .8$.

b. $P(-4 \le X \le 1) = 1 - P(X = 3) = 1 - .08 = .92$.

3. $\mu_X = (-3)(.08) + (-1)(.24) + 0(.32) + 1(.16) + 3(.12) + 5(.08) = .44$, $\text{Var}(X) = .08(-3 - .44)^2 +$ $.24(-1 - .44)^2 + .32(0 - .44)^2 + .16(1 - .44)^2 + .12(3 - .44)^2 + .08(5 - .44)^2 \approx 4.0064$, and $\sigma_X \approx \sqrt{4.0064} \approx 2.0016$.

4. a. $P(X = 0) = C(4, 0)(.3)^0(.7)^4 = \frac{4!}{0!\,4!} \cdot 1 \cdot (.7)^4 \approx .2401$,

$P(X = 1) = C(4, 1)(.3)^1(.7)^3 = \frac{4!}{1!\,3!}(.3)(.7)^3 \approx .4116$,

$P(X = 2) = C(4, 2)(.3)^2(.7)^2 = \frac{4!}{2!\,2!}(.3)^2(.7)^2 \approx .2646$,

$P(X = 3) = C(4, 3)(.3)^3(.7)^1 = \frac{4!}{3!\,1!}(.3)^3(.7) \approx .0756$, and

$P(X = 4) = C(4, 4)(.3)^4(.7)^0 = \frac{4!}{4!\,0!}(.3)^4 \cdot 1 \approx .0081$.

b. $\mu_X = E(X) = np = 4(.3) = 1.2$ and $\sigma_X = \sqrt{npq} = \sqrt{4(.3)(.7)} \approx .917$.

5. Here $\mu = 60$ and $\sigma = 5$.

a. $P(X < 70) = P\left(Z < \frac{70 - 60}{5}\right) = P(Z < 2) \approx .9772$.

b. $P(X > 50) = P\left(Z > \frac{50 - 60}{5}\right) = P(Z > -2) = P(Z < 2) \approx .9772$.

c. $P(50 < X < 70) = P\left(\frac{50-60}{5} < Z < \frac{70-60}{5}\right) = P(-2 < Z < 2) = P(Z < 2) - P(Z < -2)$

$\approx .9772 - .0228 = .9544.$

6. Here $n = 30$, $p = .5$, and $q = .5$. Thus, $\mu = np = 30(.5) = 15$ and $\sigma = \sqrt{npq} = \sqrt{(30)(.5)(.5)} \approx 2.7386$, or approximately 2.74.

a. $P(X < 10) \approx P(Y < 9.5) = P\left(Z < \frac{9.5-15}{2.74}\right) = P(Z < -2.01) \approx .0222.$

b. $P(12 \le X \le 16) \approx P(12 < Y < 16) = P\left(\frac{11.5-15}{2.7386} < Z < \frac{16.5-15}{2.7386}\right) = P(-1.28 < Z < .55)$

$= P(Z < .55) - P(Z < -1.28) \approx .7088 - .1003 = .6085.$

c. $P(X > 20) \approx P(Y > 20) = P\left(Z > \frac{20.5-15}{2.74}\right) = P(Z > 2.01) = P(Z < -2.01) \approx .0222.$

9 MARKOV CHAINS AND THE THEORY OF GAMES

9.1 Markov Chains

Problem-Solving Tips

1. A **transition matrix** associated with a Markov chain with n states is an $n \times n$ matrix T with entries a_{ij}, $1 \le i \le n$, $1 \le j \le n$. The **transition probability** associated with the transition from state i to state j is represented by the entry a_{ij}.

2. Each entry in the transition matrix is nonnegative. The sum of the entries in each column of T is 1.

3. If T is the $n \times n$ transition matrix associated with a Markov process, then the probability distribution of the system after m observations is $X_m = T^m X_0$.

Concept Questions   page 542

1. A *finite stochastic process* is an experiment consisting of a finite number of stages in which the outcomes and associated probabilities at each stage depend on the outcomes and associated probabilities of the *preceding stages*. In a *Markov chain*, the probabilities associated with the outcomes at any stage of the experiment depend only on the outcomes of the *preceding stage*.

3. **a.** $n \times n$ **b.** $a_{ij} = P$ (state i |state j), no   **c.** 1

Exercises   page 542

1. Yes. All entries are nonnegative and the sum of the entries in each column is equal to 1.

3. Yes.

5. No. The sum of the entries in the third column is not 1.

7. Yes.

9. No. It is not a square $n \times n$ matrix.

11. **a.** The conditional probability that the outcome state 1 will occur given that the outcome state 1 has occurred is .3.

     b. .7

     c. We compute $X_1 = TX_0 = \begin{bmatrix} .3 & .6 \\ .7 & .4 \end{bmatrix} \begin{bmatrix} .4 \\ .6 \end{bmatrix} = \begin{bmatrix} .48 \\ .52 \end{bmatrix}$.

233

13. We compute $TX_0 = \begin{bmatrix} .6 & .2 \\ .4 & .8 \end{bmatrix}\begin{bmatrix} .5 \\ .5 \end{bmatrix} = \begin{bmatrix} .4 \\ .6 \end{bmatrix}$ Thus, after 1 stage of the

experiment, the probability of state 1 occurring is .4 and the probability of

state 2 occurring is .6. Using the tree diagram, we see that the probabilities

of state 1 and state 2 occurring in the next stage of the experiment are

given by $P(S_1) = (.5)(.6) + (.5)(.2) = .4$ and

$P(S_2) = (.5)(.4) + (.5)(.8) = .6$, respectively. Observe that these

probabilities are precisely those represented in the probability distribution

vector TX_0.

Current state   Next state

15. $X_1 = TX_0 = \begin{bmatrix} .4 & .8 \\ .6 & .2 \end{bmatrix}\begin{bmatrix} .6 \\ .4 \end{bmatrix} = \begin{bmatrix} .56 \\ .44 \end{bmatrix}$ and $X_2 = TX_1 = \begin{bmatrix} .4 & .8 \\ .6 & .2 \end{bmatrix}\begin{bmatrix} .56 \\ .44 \end{bmatrix} = \begin{bmatrix} .576 \\ .424 \end{bmatrix}$.

17. $X_1 = TX_0 = \begin{bmatrix} \frac{1}{4} & \frac{1}{4} & \frac{1}{2} \\ \frac{1}{4} & \frac{1}{2} & \frac{1}{2} \\ \frac{1}{2} & \frac{1}{4} & 0 \end{bmatrix}\begin{bmatrix} \frac{1}{4} \\ \frac{1}{2} \\ \frac{1}{4} \end{bmatrix} = \begin{bmatrix} \frac{5}{16} \\ \frac{7}{16} \\ \frac{1}{4} \end{bmatrix}$ and $X_2 = TX_0 = \begin{bmatrix} \frac{1}{4} & \frac{1}{4} & \frac{1}{2} \\ \frac{1}{4} & \frac{1}{2} & \frac{1}{2} \\ \frac{1}{2} & \frac{1}{4} & 0 \end{bmatrix}\begin{bmatrix} \frac{5}{16} \\ \frac{7}{16} \\ \frac{1}{4} \end{bmatrix} = \begin{bmatrix} \frac{5}{16} \\ \frac{27}{64} \\ \frac{17}{64} \end{bmatrix}$.

19. a.   Current state   Next state

b. $T = \begin{matrix} L \\ R \end{matrix}\begin{matrix} L \quad R \\ \begin{bmatrix} .8 & .9 \\ .2 & .1 \end{bmatrix} \end{matrix}$

c. $X_0 = \begin{matrix} L \\ R \end{matrix}\begin{bmatrix} .5 \\ .5 \end{bmatrix}$

d. $X_1 = \begin{matrix} L \\ R \end{matrix}\begin{matrix} L \quad R \\ \begin{bmatrix} .8 & .9 \\ .2 & .1 \end{bmatrix} \end{matrix}\begin{bmatrix} .5 \\ .5 \end{bmatrix} = \begin{matrix} L \\ R \end{matrix}\begin{bmatrix} .85 \\ .15 \end{bmatrix}$, so the probability that the

mouse will turn left on the second trial is .85.

21. a. $X_1 = TX_0 = \begin{matrix} R \\ D \end{matrix}\begin{matrix} R \quad D \\ \begin{bmatrix} .7 & .2 \\ .3 & .8 \end{bmatrix} \end{matrix}\begin{bmatrix} .6 \\ .4 \end{bmatrix} = \begin{matrix} R \\ D \end{matrix}\begin{bmatrix} .5 \\ .5 \end{bmatrix}$ so if the election were held now, it would be a tie.

b. $X_1 = TX_0 = \begin{matrix} R \\ D \end{matrix}\begin{matrix} R \quad D \\ \begin{bmatrix} .7 & .2 \\ .3 & .8 \end{bmatrix} \end{matrix}\begin{bmatrix} .5 \\ .5 \end{bmatrix} = \begin{matrix} R \\ D \end{matrix}\begin{bmatrix} .45 \\ .55 \end{bmatrix}$ so the Democratic candidate would win.

23. $X_1 = TX_0 = \begin{bmatrix} .97 & .06 \\ .03 & .94 \end{bmatrix}\begin{bmatrix} .80 \\ .20 \end{bmatrix} = \begin{bmatrix} .788 \\ .212 \end{bmatrix}$ and $X_2 = TX_1 = \begin{bmatrix} .97 & .06 \\ .03 & .94 \end{bmatrix}\begin{bmatrix} .788 \\ .212 \end{bmatrix} = \begin{bmatrix} .777 \\ .223 \end{bmatrix}$, so after one

year, 78.8% of the population will be in the city and 21.2% in the suburbs. After two years, 77.7% of the population

will be in the city and 22.3% in the suburbs.

25. The expected distribution is given by $X_1 = TX_0 = \begin{bmatrix} .80 & .10 & .05 \\ .10 & .75 & .05 \\ .10 & .15 & .90 \end{bmatrix} \begin{bmatrix} .4 \\ .4 \\ .2 \end{bmatrix} = \begin{bmatrix} .37 \\ .35 \\ .28 \end{bmatrix}$, and we

conclude that at the beginning of the second quarter, the University Bookstore will have 37% of the market, the Campus Bookstore will have 35%, and the Book Mart will have 28%. Similarly,

$X_2 = TX_2 = \begin{bmatrix} .80 & .10 & .05 \\ .10 & .75 & .05 \\ .10 & .15 & .90 \end{bmatrix} \begin{bmatrix} .37 \\ .35 \\ .28 \end{bmatrix} = \begin{bmatrix} .3450 \\ .3135 \\ .3415 \end{bmatrix}$, so at the beginning of the third quarter, the University

Bookstore will have 34.5% of the market, the Campus Bookstore will have 31.35%, and the Book Mart will have 34.15%.

27. $X_1 = TX_0 = \begin{bmatrix} .80 & .10 & .20 & .10 \\ .10 & .70 & .10 & .05 \\ .05 & .10 & .60 & .05 \\ .05 & .10 & .10 & .80 \end{bmatrix} \begin{bmatrix} .3 \\ .3 \\ .2 \\ .2 \end{bmatrix} = \begin{bmatrix} .33 \\ .27 \\ .175 \\ .225 \end{bmatrix}$,

$X_2 = TX_1 = \begin{bmatrix} .80 & .10 & .20 & .10 \\ .10 & .70 & .10 & .05 \\ .05 & .10 & .60 & .05 \\ .05 & .10 & .10 & .80 \end{bmatrix} \begin{bmatrix} .33 \\ .27 \\ .175 \\ .225 \end{bmatrix} = \begin{bmatrix} .3485 \\ .25075 \\ .15975 \\ .241 \end{bmatrix}$, and

$X_3 = TX_2 = \begin{bmatrix} .80 & .10 & .20 & .10 \\ .10 & .70 & .10 & .05 \\ .05 & .10 & .60 & .05 \\ .05 & .10 & .10 & .80 \end{bmatrix} \begin{bmatrix} .3485 \\ .25075 \\ .15975 \\ .241 \end{bmatrix} \approx \begin{bmatrix} .3599 \\ .2384 \\ .1504 \\ .2513 \end{bmatrix}$. Assuming that the present trend continues, 36.0%

of students in their senior year will major in business, 23.8% will major in the humanities, 15.0% will major in education, and 25.1% will major in the natural sciences.

29. False. In a Markov chain, an outcome depends only on the preceding stage.

Technology Exercises   page 546

1. $X_5 = \begin{bmatrix} .204489 \\ .131869 \\ .261028 \\ .186814 \\ .215800 \end{bmatrix}$

3. Manufacturer A will have 23.95% of the market, manufacturer B will have 49.71%, and manufacturer C will have 26.34%.

9.2   Regular Markov Chains

Problem-Solving Tips

1. A stochastic matrix T is **regular** if and only if some power of T has entries that are all positive.

2. To find the **steady-state distribution vector** X for a transition matrix T, solve the vector equation $TX = X$ together with the condition that the sum of the elements of the vector X is 1.

Concept Questions   page 552

1. a. Let T be an $n \times n$ transition matrix and let X_0 be an $n \times 1$ initial distribution vector. If the sequence of vectors $X_1, X_2, \ldots, X_n, \ldots$, defined for $i = 1, 2, 3, \ldots$ by $X_i = TX_{i-1}$ approaches a vector X as n gets larger and larger, then T is called the *steady-state distribution vector.*

   b. If the sequence $T, T^2, T^3, \ldots, T^m$ approaches a matrix L as m increases, then L is called the *steady-state matrix.*

   c. A stochastic matrix T is a regular Markov chain if the sequence $T, T^2, T^3, \ldots$ approaches a steady-state matrix in which the rows of the limiting matrix are all equal and all the entries are all positive.

Exercises   page 552

1. Because all entries in the matrix are positive, it is regular.

3. $T^2 = \begin{bmatrix} 1 & .8 \\ 0 & .2 \end{bmatrix} \begin{bmatrix} 1 & .8 \\ 0 & .2 \end{bmatrix} = \begin{bmatrix} 1 & .96 \\ 0 & .04 \end{bmatrix}$ and $T^3 = \begin{bmatrix} 1 & .96 \\ 0 & .04 \end{bmatrix} \begin{bmatrix} 1 & .8 \\ 0 & .2 \end{bmatrix} = \begin{bmatrix} 1 & .992 \\ 0 & .008 \end{bmatrix}$, so we see that the a_{21} entry will always be 0. Thus, T is not regular.

5. $T^2 = \begin{bmatrix} \frac{1}{2} & \frac{3}{4} & 0 \\ \frac{1}{2} & 0 & \frac{1}{2} \\ 0 & \frac{1}{4} & \frac{1}{2} \end{bmatrix} \begin{bmatrix} \frac{1}{2} & \frac{3}{4} & 0 \\ \frac{1}{2} & 0 & \frac{1}{2} \\ 0 & \frac{1}{4} & \frac{1}{2} \end{bmatrix} = \begin{bmatrix} \frac{5}{8} & \frac{3}{8} & \frac{3}{8} \\ \frac{1}{4} & \frac{1}{2} & \frac{1}{4} \\ \frac{1}{8} & \frac{1}{8} & \frac{3}{8} \end{bmatrix}$, and so T is regular.

7. $T^2 = \begin{bmatrix} .7 & .2 & .3 \\ .3 & .8 & .3 \\ 0 & 0 & .4 \end{bmatrix} \begin{bmatrix} .7 & .2 & .3 \\ .3 & .8 & .3 \\ 0 & 0 & .4 \end{bmatrix} = \begin{bmatrix} .55 & .3 & .39 \\ .45 & .7 & .45 \\ 0 & 0 & .16 \end{bmatrix}$, and so forth. Continuing, we see that the a_{31} and a_{32} entries of $T^3, T^4, \ldots$ will be 0, and so T is not regular.

9. We have $x + y = 1$ along with the matrix equation $\begin{bmatrix} \frac{1}{3} & \frac{1}{4} \\ \frac{2}{3} & \frac{3}{4} \end{bmatrix} \begin{bmatrix} x \\ y \end{bmatrix} = \begin{bmatrix} x \\ y \end{bmatrix}$. Solving the equivalent system of

equations $\begin{cases} \frac{1}{3}x + \frac{1}{4}y = x \\ \frac{2}{3}x + \frac{3}{4}y = y \\ x + y = 1 \end{cases}$ we find the required vector to be $\begin{bmatrix} \frac{3}{11} \\ \frac{8}{11} \end{bmatrix}$.

11. We have $TX = X$, that is, $\begin{bmatrix} .5 & .2 \\ .5 & .8 \end{bmatrix} \begin{bmatrix} x \\ y \end{bmatrix} = \begin{bmatrix} x \\ y \end{bmatrix}$, or equivalently the system of equations

$\begin{cases} .5x + .2y = x \\ .5x + .8y = y \end{cases}$ This system is equivalent to the single equation $.5x - .2y = 0$. We must also have $x + y = 1$,

so we have the system $\begin{cases} .5x - .2y = x \\ x + y = 1 \end{cases}$ The second equation gives $y = 1 - x$, and substituting this value into

the first equation yields $.5x - .2(1 - x) = 0$, $0.7x - .2 = 0$, and $x = \frac{2}{7}$. Therefore, $y = \frac{5}{7}$ and the steady-state

distribution vector is $\begin{bmatrix} \frac{2}{7} \\ \frac{5}{7} \end{bmatrix}$.

13. We solve the system $\begin{bmatrix} 0 & \frac{1}{8} & 1 \\ 1 & \frac{5}{8} & 0 \\ 0 & \frac{1}{4} & 0 \end{bmatrix} \begin{bmatrix} x \\ y \\ z \end{bmatrix} = \begin{bmatrix} x \\ y \\ z \end{bmatrix}$ together with the equation $x + y + z = 1$; that is, the system

$\begin{cases} -x + \frac{1}{8}y + z = 0 \\ x - \frac{3}{8}y = 0 \\ \frac{1}{4}y - z = 0 \\ x + y + z = 1 \end{cases}$ Using the Gauss-Jordan method, we find that the required steady-state vector is $\begin{bmatrix} \frac{3}{13} \\ \frac{8}{13} \\ \frac{2}{13} \end{bmatrix}$.

15. We solve the system $\begin{bmatrix} .2 & 0 & .3 \\ 0 & .6 & .4 \\ .8 & .4 & .3 \end{bmatrix} \begin{bmatrix} x \\ y \\ z \end{bmatrix} = \begin{bmatrix} x \\ y \\ z \end{bmatrix}$ together with the equation $x + y + z = 1$; that is, the system

$\begin{cases} -.8x + .3z = 0 \\ -.4y + .4z = 0 \\ .8x + .4y - .7z = 0 \\ x + y + z = 1 \end{cases}$ Using the Gauss-Jordan method, we find that the required steady-state vector is $\begin{bmatrix} \frac{3}{19} \\ \frac{8}{19} \\ \frac{8}{19} \end{bmatrix}$.

17. We want to solve $\begin{bmatrix} .8 & .9 \\ .2 & .1 \end{bmatrix} \begin{bmatrix} x \\ y \end{bmatrix} = \begin{bmatrix} x \\ y \end{bmatrix}$, or equivalently $\begin{cases} .2x + .9y = 0 \\ .2x - .9y = 0 \\ x + y = 1 \end{cases}$ We find that the required

steady-state vector is $\begin{bmatrix} \frac{2}{11} \\ \frac{9}{11} \end{bmatrix}$ and conclude that in the long run the mouse will turn left 81.8% of the time.

19. We compute $X_1 = \begin{bmatrix} .72 & .12 \\ .28 & .88 \end{bmatrix} \begin{bmatrix} .48 \\ .52 \end{bmatrix} = \begin{bmatrix} .408 \\ .592 \end{bmatrix}$ and conclude that ten years from now, there will be 40.8% of

families will have one wage earner and 59.2% will have two wage earners.

To find out what happens in the long run, we solve the system $\begin{bmatrix} .72 & .12 \\ .28 & .88 \end{bmatrix} \begin{bmatrix} x \\ y \end{bmatrix} = \begin{bmatrix} x \\ y \end{bmatrix}$ together with the

equation $x + y = 1$; that is, the system $\begin{cases} -.28x + .12y = 0 \\ .28x - .12y = 0 \\ x + y = 1 \end{cases}$ We find that $x = .3$ and $y = .7$, and conclude that in

the long run, 30% of families will have one wage earner and 70% will have two wage earners.

21. If the trend continues, the percentage of homeowners in the city who will own single-family homes or condominiums

two decades from now will be $X_2 = TX_1$. We calculate $X_1 = TX_0 = \begin{bmatrix} .85 & .35 \\ .15 & .65 \end{bmatrix} \begin{bmatrix} .8 \\ .2 \end{bmatrix} = \begin{bmatrix} .75 \\ .25 \end{bmatrix}$ and

$X_2 = TX_1 = \begin{bmatrix} .85 & .35 \\ .15 & .65 \end{bmatrix} \begin{bmatrix} .75 \\ .25 \end{bmatrix} = \begin{bmatrix} .725 \\ .275 \end{bmatrix}$, and conclude that 72.5% of homeowners will own single-family

homes and 27.5% will own condominiums two decades from now.

To find out what happens in the long run, we solve the system $\begin{bmatrix} .85 & .35 \\ .15 & .65 \end{bmatrix} \begin{bmatrix} x \\ y \end{bmatrix} = \begin{bmatrix} x \\ y \end{bmatrix}$ together with the

equation $x + y = 1$; that is, $\begin{cases} -.15x + .35y = 0 \\ .15x - .35y = 0 \\ x + y = 1 \end{cases}$ We find that $x = .7$ and $y = .3$, and conclude that in the long

run 70% of homeowners will own single family homes and 30% will own condominiums.

23. a. $X_1 = TX_0 = \begin{bmatrix} .8 & .1 & .1 \\ .1 & .85 & .05 \\ .1 & .05 & .85 \end{bmatrix} \begin{bmatrix} .3 \\ .4 \\ .3 \end{bmatrix} = \begin{bmatrix} .31 \\ .385 \\ .305 \end{bmatrix}$ and $X_2 = TX_1 = \begin{bmatrix} .8 & .1 & .1 \\ .1 & .85 & .05 \\ .1 & .05 & .85 \end{bmatrix} \begin{bmatrix} .31 \\ .385 \\ .305 \end{bmatrix} = \begin{bmatrix} .317 \\ .3735 \\ .3095 \end{bmatrix}$.

From our computations, we conclude that after two weeks, 31.7% of the viewers will watch the ABC news, 37.35% will watch the CBS news, and 30.95% will watch the NBC news.

b. We solve the system $\begin{bmatrix} .8 & .1 & .1 \\ .1 & .85 & .05 \\ .1 & .05 & .85 \end{bmatrix} \begin{bmatrix} x \\ y \\ z \end{bmatrix} = \begin{bmatrix} x \\ y \\ z \end{bmatrix}$ together with the equation $x + y + z = 1$; that is, the

system $\begin{cases} -.2x + .1y + .1z = 0 \\ .1x - .15y + .05z = 0 \\ .1x + .05y - .15z = 0 \\ x + y + z = 1 \end{cases}$ Using the Gauss-Jordan elimination method, we find that the required

steady-state vector is $\begin{bmatrix} \frac{1}{3} \\ \frac{1}{3} \\ \frac{1}{3} \end{bmatrix}$, and conclude that each network will command one-third of the audience in the long

run.

25. We wish to solve $\begin{bmatrix} \frac{1}{2} & \frac{1}{4} & 0 \\ \frac{1}{2} & \frac{1}{2} & \frac{1}{2} \\ 0 & \frac{1}{4} & \frac{1}{2} \end{bmatrix} \begin{bmatrix} x \\ y \\ z \end{bmatrix} = \begin{bmatrix} x \\ y \\ z \end{bmatrix}$ together with the equation $x + y + z = 1$; that is, the system

$$\begin{cases} -\frac{1}{2}x + \frac{1}{4}y & = 0 \\ \frac{1}{2}x - \frac{1}{2}y + \frac{1}{2}z = 0 \\ \frac{1}{4}y - \frac{1}{2}z = 0 \\ x + y + z = 1 \end{cases}$$

We find that $x = \frac{1}{4}$, $y = \frac{1}{2}$, and $z = \frac{1}{4}$, so in the long run 25% of the plants will have red flowers, 50% will have pink flowers, and 25% will have white flowers.

27. False. All entries of the limiting matrix must be positive as well.

29. Let T be a regular stochastic matrix and X the steady-state distribution vector that satisfies the equation $TX = T$, and assume that the sum of the elements of X are equal to 1. Then $TX = X$ implies that $TX = T^2X$, so $X = T^2X$. Thus, we have $X = T^nX$. Next, let L be the steady-state distribution vector. When m is large, $L \approx X_m \approx T^mX_0$, so $L \approx T^mX_0$. Multiplying both sides of this last equation by T, we obtain $TL = T^{m+1}X_0$. But if m is large, $T^{m+1}X_0 \approx L$. Thus, L also satisfies $TL = L$, together with the condition that the sum of the elements in L is equal to 1. Because the matrix equation $TX = X$ has a unique solution, we conclude that $X = L$.

Technology Exercises   page 556

1. $\begin{bmatrix} .2045 \\ .1319 \\ .2610 \\ .1868 \\ .2158 \end{bmatrix}$

9.3   Absorbing Markov Chains

Problem-Solving Tips

1. An **absorbing stochastic matrix** has at least one absorbing state and it is possible to go from any nonabsorbing state to an absorbing state in one or more stages.

2. To find the steady-state matrix of an absorbing stochastic matrix A, partition the matrix A into submatrices $\begin{bmatrix} I & S \\ \hline O & R \end{bmatrix}$. Then the steady-state matrix of A is given by $\begin{bmatrix} I & S(I-R)^{-1} \\ \hline O & O \end{bmatrix}$, where the order of I in the expression $(I-R)^{-1}$ is the same as the order of R.

Concept Questions   page 562

1. An absorbing stochastic matrix has the following properties: (1) There is at least one absorbing state. (2) It is possible to go from any nonabsorbing state to an absorbing state in one or more stages.

Exercises   page 562

1. For the matrix $\begin{array}{c} \\ 1 \\ 2 \end{array} \begin{array}{cc} 1 & 2 \\ \left[\begin{array}{cc} \frac{2}{5} & 0 \\ \frac{3}{5} & 1 \end{array} \right] \end{array}$, state 2 is an absorbing state. State 1 is nonabsorbing, but an object in this state has a

probability of $\frac{3}{5}$ of going to the absorbing state 2. Thus, the matrix is an absorbing matrix.

3. For the matrix $\begin{array}{c} \\ 1 \\ 2 \\ 3 \end{array} \begin{array}{ccc} 1 & 2 & 3 \\ \left[\begin{array}{ccc} 1 & .5 & 0 \\ 0 & 0 & 1 \\ 0 & .5 & 0 \end{array} \right] \end{array}$, states 1 and 3 are absorbing states. State 2 is not absorbing, but an object in this state

has probability .5 of going to the absorbing state 1 and probability .5 of going to the absorbing state 3. Thus, the matrix is an absorbing matrix.

5. Yes. This is an absorbing stochastic matrix because it is possible to go from state 1 to the absorbing states 2 and 3.

7. For the matrix $\begin{array}{c} \\ 1 \\ 2 \\ 3 \\ 4 \end{array} \begin{array}{cccc} 1 & 2 & 3 & 4 \\ \left[\begin{array}{cccc} 1 & 0 & .3 & 0 \\ 0 & 1 & .2 & 0 \\ 0 & 0 & .1 & .5 \\ 0 & 0 & .4 & .5 \end{array} \right] \end{array}$, states 1 and 2 are absorbing states and states 3 and 4 are not. However, it is

possible for an object to go from state 3 to state 1 with probability .3, and it is also possible for an object to go from the non-absorbing state 4 to an absorbing state (via state 3, for example). Therefore, the given matrix is an absorbing matrix.

9. The required matrix is $\begin{array}{c} \\ 2 \\ 1 \end{array} \begin{array}{cc} 2 & 1 \\ \left[\begin{array}{c|c} 1 & .4 \\ \hline 0 & .6 \end{array} \right] \end{array}$, where $S = \left[\begin{array}{c} .4 \end{array} \right]$ and $R = \left[\begin{array}{c} .6 \end{array} \right]$.

11. $\begin{array}{c} \\ 3 \\ 2 \\ 1 \end{array} \begin{array}{ccc} 3 & 2 & 1 \\ \left[\begin{array}{c|cc} 1 & .4 & .5 \\ \hline 0 & .4 & .5 \\ 0 & .2 & 0 \end{array} \right] \end{array}$, where $S = \left[\begin{array}{cc} .4 & .5 \end{array} \right]$ and $R = \left[\begin{array}{cc} .4 & .5 \\ .2 & 0 \end{array} \right]$, or $\begin{array}{c} \\ 3 \\ 1 \\ 2 \end{array} \begin{array}{ccc} 3 & 2 & 1 \\ \left[\begin{array}{c|cc} 1 & .5 & .4 \\ \hline 0 & 0 & .2 \\ 0 & .5 & .4 \end{array} \right] \end{array}$, where $S = \left[\begin{array}{cc} .5 & .4 \end{array} \right]$ and

$R = \left[\begin{array}{cc} 0 & .2 \\ .5 & .4 \end{array} \right]$.

13. Possible matrices are $\begin{bmatrix} 1 & 0 & .2 & .4 \\ 0 & 1 & .3 & 0 \\ \hline 0 & 0 & .3 & .2 \\ 0 & 0 & .2 & .4 \end{bmatrix}$, where $S = \begin{bmatrix} .2 & .4 \\ .3 & 0 \end{bmatrix}$ and $R = \begin{bmatrix} .3 & .2 \\ .2 & .4 \end{bmatrix}$; $\begin{bmatrix} 1 & 0 & .4 & .2 \\ 0 & 1 & 0 & .3 \\ \hline 0 & 0 & .4 & .2 \\ 0 & 0 & .2 & .3 \end{bmatrix}$, where

$S = \begin{bmatrix} .4 & .2 \\ 0 & .3 \end{bmatrix}$ and $R = \begin{bmatrix} .4 & .2 \\ .2 & .3 \end{bmatrix}$; and so forth.

15. Rewriting the matrix so that the absorbing states appear first, we have $\begin{array}{cc} & \begin{array}{cc} 2 & 1 \end{array} \\ \begin{array}{c} 2 \\ 1 \end{array} & \begin{bmatrix} 1 & .45 \\ \hline 0 & .55 \end{bmatrix} \end{array}$, where $S = \begin{bmatrix} .45 \end{bmatrix}$. and

$R = \begin{bmatrix} .55 \end{bmatrix}$. Then $(I - R) = \begin{bmatrix} .45 \end{bmatrix}$ and $(I - R)^{-1} = \begin{bmatrix} \frac{1}{.45} \end{bmatrix}$, so $S(I - R)^{-1} = \begin{bmatrix} .45 \end{bmatrix}\begin{bmatrix} \frac{1}{.45} \end{bmatrix} = \begin{bmatrix} 1 \end{bmatrix}$.

Therefore, the steady-state matrix is $\begin{array}{cc} & \begin{array}{cc} 2 & 1 \end{array} \\ \begin{array}{c} 2 \\ 1 \end{array} & \begin{bmatrix} 1 & 1 \\ \hline 0 & 0 \end{bmatrix} \end{array}$.

17. Here we have $\begin{bmatrix} 1 & .2 & .3 \\ \hline 0 & .4 & .2 \\ 0 & .4 & .5 \end{bmatrix}$, where $S = \begin{bmatrix} .2 & .3 \end{bmatrix}$ and $R = \begin{bmatrix} .4 & .2 \\ .4 & .5 \end{bmatrix}$. Next,

$I - R = \begin{bmatrix} 1 & 0 \\ 0 & 1 \end{bmatrix} - \begin{bmatrix} .4 & .2 \\ .4 & .5 \end{bmatrix} = \begin{bmatrix} .6 & -.2 \\ -.4 & .5 \end{bmatrix}$. Using the formula for the inverse of a 2 × 2 matrix, we have

$(I - R)^{-1} = \begin{bmatrix} 2.27 & .91 \\ 1.82 & 2.73 \end{bmatrix}$, so $S(I - R)^{-1} = \begin{bmatrix} .2 & .3 \end{bmatrix}\begin{bmatrix} 2.27 & .91 \\ 1.82 & 2.73 \end{bmatrix} = \begin{bmatrix} .994 & 1 \end{bmatrix} \approx \begin{bmatrix} 1 & 1 \end{bmatrix}$ We conclude

that the steady-state matrix is $\begin{bmatrix} 1 & 1 & 1 \\ \hline 0 & 0 & 0 \\ 0 & 0 & 0 \end{bmatrix}$.

19. Upon rewriting the given matrix so that the absorbing states appear first, we have
$$\begin{array}{cc} & \begin{matrix} 2 & 4 & 1 & 3 \end{matrix} \\ \begin{matrix} 2 \\ 4 \\ 1 \\ 3 \end{matrix} & \left[\begin{array}{cc|cc} 1 & 0 & \frac{1}{2} & 0 \\ 0 & 1 & 0 & 0 \\ \hline 0 & 0 & \frac{1}{2} & \frac{1}{3} \\ 0 & 0 & 0 & \frac{2}{3} \end{array}\right] \end{array}$$, where

$S = \begin{bmatrix} \frac{1}{2} & 0 \\ 0 & 0 \end{bmatrix}$ and $R = \begin{bmatrix} \frac{1}{2} & \frac{1}{3} \\ 0 & \frac{2}{3} \end{bmatrix}$. Next, we compute $I - R = \begin{bmatrix} 1 & 0 \\ 0 & 1 \end{bmatrix} - \begin{bmatrix} \frac{1}{2} & \frac{1}{3} \\ 0 & \frac{2}{3} \end{bmatrix} = \begin{bmatrix} \frac{1}{2} & -\frac{1}{3} \\ 0 & \frac{1}{3} \end{bmatrix}$. Using the

formula for the inverse of a 2×2 matrix, we have $(I - R)^{-1} = \begin{bmatrix} 2 & 2 \\ 0 & 3 \end{bmatrix}$, and so

$S(I-R)^{-1} = \begin{bmatrix} \frac{1}{2} & 0 \\ 0 & 0 \end{bmatrix}\begin{bmatrix} 2 & 2 \\ 0 & 3 \end{bmatrix} = \begin{bmatrix} 1 & 1 \\ 0 & 0 \end{bmatrix}$. Therefore, the steady-state matrix is $\left[\begin{array}{cc|cc} 1 & 0 & 1 & 1 \\ 0 & 1 & 0 & 0 \\ \hline 0 & 0 & 0 & 0 \\ 0 & 0 & 0 & 0 \end{array}\right]$.

21. Here $\left[\begin{array}{cc|cc} 1 & 0 & \frac{1}{4} & \frac{1}{3} \\ 0 & 1 & \frac{1}{4} & \frac{1}{3} \\ \hline 0 & 0 & \frac{1}{2} & 0 \\ 0 & 0 & 0 & \frac{1}{3} \end{array}\right]$, $S = \begin{bmatrix} \frac{1}{4} & \frac{1}{3} \\ \frac{1}{4} & \frac{1}{3} \end{bmatrix}$, $R = \begin{bmatrix} \frac{1}{2} & 0 \\ 0 & \frac{1}{3} \end{bmatrix}$ and $I - R = \begin{bmatrix} 1 & 0 \\ 0 & 1 \end{bmatrix} - \begin{bmatrix} \frac{1}{2} & 0 \\ 0 & \frac{1}{3} \end{bmatrix} = \begin{bmatrix} \frac{1}{2} & 0 \\ 0 & \frac{2}{3} \end{bmatrix}$.

Using the formula for the inverse of a 2×2 matrix, we find $(I - R)^{-1} = \begin{bmatrix} 2 & 0 \\ 0 & \frac{3}{2} \end{bmatrix}$ and so

$S(I-R)^{-1} = \begin{bmatrix} \frac{1}{4} & \frac{1}{3} \\ \frac{1}{4} & \frac{1}{3} \end{bmatrix}\begin{bmatrix} 2 & 0 \\ 0 & \frac{3}{2} \end{bmatrix} = \begin{bmatrix} \frac{1}{2} & \frac{1}{2} \\ \frac{1}{2} & \frac{1}{2} \end{bmatrix}$. The steady-state matrix is given by $\left[\begin{array}{cc|cc} 1 & 0 & \frac{1}{2} & \frac{1}{2} \\ 0 & 1 & \frac{1}{2} & \frac{1}{2} \\ \hline 0 & 0 & 0 & 0 \\ 0 & 0 & 0 & 0 \end{array}\right]$.

23. The absorbing states already appear first in the matrix, so the matrix need not be rewritten. Next,

$R = \begin{bmatrix} .2 & .2 \\ .2 & .4 \end{bmatrix}$, $S = \begin{bmatrix} .2 & .1 \\ .1 & .2 \\ .3 & .1 \end{bmatrix}$, and $I - R = \begin{bmatrix} .8 & -.2 \\ -.2 & .6 \end{bmatrix}$ so $(I - R)^{-1} = \begin{bmatrix} \frac{15}{11} & \frac{5}{11} \\ \frac{5}{11} & \frac{20}{11} \end{bmatrix}$ and

$S(I-R)^{-1} = \begin{bmatrix} \frac{2}{10} & \frac{1}{10} \\ \frac{1}{10} & \frac{2}{10} \\ \frac{3}{10} & \frac{1}{10} \end{bmatrix}\begin{bmatrix} \frac{15}{11} & \frac{5}{11} \\ \frac{5}{11} & \frac{20}{11} \end{bmatrix} = \begin{bmatrix} \frac{7}{22} & \frac{3}{11} \\ \frac{5}{22} & \frac{9}{22} \\ \frac{5}{11} & \frac{7}{22} \end{bmatrix}$. Therefore, the steady-state matrix is $\left[\begin{array}{ccc|cc} 1 & 0 & 0 & \frac{7}{22} & \frac{3}{11} \\ 0 & 1 & 0 & \frac{5}{22} & \frac{9}{22} \\ 0 & 0 & 1 & \frac{5}{11} & \frac{7}{22} \\ \hline 0 & 0 & 0 & 0 & 0 \\ 0 & 0 & 0 & 0 & 0 \end{array}\right]$.

25. a. State 2 is absorbing. State 1 is not absorbing, but it is possible for an object to go from state 1 to state 2 with

probability .8. Therefore, the matrix is absorbing. Rewriting, we obtain $\begin{array}{c} \\ 2 \\ 1 \end{array}\begin{array}{c} 2\quad 1 \\ \left[\begin{array}{c|c} 1 & .2 \\ \hline 0 & .8 \end{array}\right] \end{array}$, where $S = \left[\begin{array}{c} .2 \end{array}\right]$ and

$R = \left[\begin{array}{c} .8 \end{array}\right]$.

b. We compute $I - R = \left[\begin{array}{c} 1 \end{array}\right] - \left[\begin{array}{c} .8 \end{array}\right] = \left[\begin{array}{c} .2 \end{array}\right]$, so $(I - R)^{-1} = \left[\begin{array}{c} 5 \end{array}\right]$. Therefore,

$S(I - R)^{-1} = \left[\begin{array}{c} .2 \end{array}\right]\left[\begin{array}{c} 5 \end{array}\right] = \left[\begin{array}{c} 1 \end{array}\right]$ and the steady state matrix is $\left[\begin{array}{c|c} 1 & 1 \\ \hline 0 & 0 \end{array}\right]$. This result tells us that in the long

run only broadband Internet service will be used.

27. Here
$$
\begin{array}{c}
 \\
\$0 \\
\$4 \\
\$1 \\
\$2 \\
\$3
\end{array}
\begin{array}{c}
\$0\ \ \$4\ \ \$1\ \ \$2\ \ \$3 \\
\left[\begin{array}{cc|ccc}
1 & 0 & \frac{1}{2} & 0 & 0 \\
0 & 1 & 0 & 0 & \frac{1}{2} \\
\hline
0 & 0 & 0 & \frac{1}{2} & 0 \\
0 & 0 & \frac{1}{2} & 0 & \frac{1}{2} \\
0 & 0 & 0 & \frac{1}{2} & 0
\end{array}\right]
\end{array}
$$
, where $S = \left[\begin{array}{ccc} \frac{1}{2} & 0 & 0 \\ 0 & 0 & \frac{1}{2} \end{array}\right]$ and $R = \left[\begin{array}{ccc} 0 & \frac{1}{2} & 0 \\ \frac{1}{2} & 0 & \frac{1}{2} \\ 0 & \frac{1}{2} & 0 \end{array}\right]$. Next,

$I - R = \left[\begin{array}{ccc} 1 & -\frac{1}{2} & 0 \\ -\frac{1}{2} & 1 & -\frac{1}{2} \\ 0 & -\frac{1}{2} & 1 \end{array}\right]$, $(I - R)^{-1} = \left[\begin{array}{ccc} \frac{3}{2} & 1 & \frac{1}{2} \\ 1 & 2 & 1 \\ \frac{1}{2} & 1 & \frac{3}{2} \end{array}\right]$, and $S(I - R)^{-1} = \left[\begin{array}{ccc} \frac{3}{4} & \frac{1}{2} & \frac{1}{4} \\ \frac{1}{4} & \frac{1}{2} & \frac{3}{4} \end{array}\right]$. Therefore, the

steady-state matrix is given by
$$
\begin{array}{c}
\$0 \\
\$4 \\
\$1 \\
\$2 \\
\$3
\end{array}
\begin{array}{c}
\$0\ \ \$4\ \ \$1\ \ \$2\ \ \$3 \\
\left[\begin{array}{cc|ccc}
1 & 0 & \frac{3}{4} & \frac{1}{2} & \frac{1}{4} \\
0 & 1 & \frac{1}{4} & \frac{1}{2} & \frac{3}{4} \\
\hline
0 & 0 & 0 & 0 & 0 \\
0 & 0 & 0 & 0 & 0 \\
0 & 0 & 0 & 0 & 0
\end{array}\right]
\end{array}
$$
We conclude that if Diane started out with $1, the

probability that she would leave the game a winner is $\frac{1}{4}$. Similarly, if she started out with $2, the probability that she would leave the game a winner is $\frac{1}{2}$, and if she started out with $3, the probability that she would leave as a winner is $\frac{3}{4}$.

29. a. $\left[\begin{array}{cc|cc} 1 & 0 & .25 & .1 \\ 0 & 1 & 0 & .9 \\ \hline 0 & 0 & 0 & 0 \\ 0 & 0 & .75 & 0 \end{array}\right]$

b. $I - R = \begin{bmatrix} 1 & 0 \\ -.75 & 1 \end{bmatrix}$, $(I - R)^{-1} = \begin{bmatrix} 1 & 0 \\ .75 & 1 \end{bmatrix}$, and $S(I - R)^{-1} = \begin{bmatrix} .25 & .1 \\ 0 & .9 \end{bmatrix} \begin{bmatrix} 1 & 0 \\ .75 & 1 \end{bmatrix} = \begin{bmatrix} .325 & .1 \\ .675 & .9 \end{bmatrix}$.

Therefore, the steady-state matrix is $\begin{bmatrix} 1 & 0 & .325 & .1 \\ 0 & 1 & .675 & .9 \\ 0 & 0 & 0 & 0 \\ 0 & 0 & 0 & 0 \end{bmatrix}$.

c. From the steady-state matrix, we see that the probability that a beginning student enrolled in the program will compete the course successfully is .675.

31. The transition matrix is $T = \begin{array}{c} \\ aa \\ Aa \\ AA \end{array} \begin{array}{c} \begin{array}{ccc} aa & Aa & AA \end{array} \\ \begin{bmatrix} 1 & \frac{1}{2} & 0 \\ 0 & \frac{1}{2} & 1 \\ 0 & 0 & 0 \end{bmatrix} \end{array}$. Because the entries in T are exactly the same as those in

Example 4, the steady-state matrix is $\begin{array}{c} \\ aa \\ Aa \\ AA \end{array} \begin{array}{c} \begin{array}{ccc} aa & Aa & AA \end{array} \\ \begin{bmatrix} 1 & 1 & 1 \\ 0 & 0 & 0 \\ 0 & 0 & 0 \end{bmatrix} \end{array}$. Interpreting the steady-state matrix, we see that in the

long run all the flowers produced by the plants will be white.

33. True. See Section 9.3.

9.4 Game Theory and Strictly Determined Games

Problem-Solving Tips

1. A **zero-sum game** is a game in which the payoff to one party results in an equal loss to the other.

2. The **maximin strategy** for the row player is to find the *smallest* entry in each row and then to choose the row that has the *largest* entry among these entries—thus "maximizing the minima".

3. The **minimax strategy** for the column player is to find the *largest* entry in each column and then to choose the column that has the *smallest* entry among these entries—thus "minimizing the maxima".

4. A **strictly determined game** has a **saddle point** that is simultaneously the smallest entry in its row and the largest entry in its column. The optimal strategy for the row player is to play the row containing the saddle point and that for the column player is to play the column containing the saddle point.

Concept Questions   page 572

1. **a.** To follow the maximin strategy, (1) find the smallest entry in each row of the payoff matrix and (2) choose the row for which the entry found in step 1 is as large as possible. This row represents R's best move.

  b. To follow the minimax strategy, (1) find the largest entry in each column of the payoff matrix and (2) choose the column for which the entry found in step 1 is as small as possible. This column represents C's best move.

Exercises   page 572

1. We first determine the minimum of each row and the maxima of each column of the payoff matrix. Next, we find the larger of the row minima and the smaller of the column maxima, as shown below.

C's move

$$
\begin{array}{c}
 & \begin{array}{ccc} C_1 & C_2 & \text{Row minima} \end{array} \\
R\text{'s move} \quad \begin{array}{c} R_1 \\ R_2 \end{array} & \left[\begin{array}{cc} 2 & 3 \\ 4 & 1 \end{array} \right] \quad \begin{array}{c} ② \leftarrow \text{Larger} \\ 1 \end{array}
\end{array}
$$

Column maxima   4   ③
                        ↑
                     Smaller

From these results, we conclude that the row player's maximum strategy is to play row 1, whereas the column player's minimax strategy is to play column 2.

3.

C's move

$$
\begin{array}{c}
 & \begin{array}{ccc} C_1 & C_2 & C_3 \end{array} \quad \begin{array}{c} \text{Row minima} \end{array} \\
R\text{'s move} \quad \begin{array}{c} R_1 \\ R_2 \end{array} & \left[\begin{array}{ccc} 1 & 3 & 2 \\ 0 & -1 & 4 \end{array} \right] \quad \begin{array}{c} ① \leftarrow \text{Larger} \\ -1 \end{array}
\end{array}
$$

Column maxima   ①   3   4
                  ↑
               Smallest

From the payoff matrix, we conclude that the row player's maximum strategy is to play row 1, whereas the column player's minimax strategy is to play column 1.

5.

C's move

$$
\begin{array}{c}
 & \begin{array}{ccc} C_1 & C_2 & C_3 \end{array} \quad \begin{array}{c} \text{Row minima} \end{array} \\
R\text{'s move} \quad \begin{array}{c} R_1 \\ R_2 \\ R_2 \end{array} & \left[\begin{array}{ccc} 3 & 2 & 1 \\ 1 & -2 & 3 \\ 6 & 4 & 1 \end{array} \right] \quad \begin{array}{c} ① \\ -2 \\ ① \end{array} \quad \text{Largest}
\end{array}
$$

Column maxima   6   4   ③
                          ↑
                       Smallest

From the payoff matrix, we conclude that the row player's maximin strategy is to play either row 1 or row 3, whereas the column player's minimax strategy is to play column 3.

7.

C's move

$$
\begin{array}{c}
 & \begin{array}{ccc} C_1 & C_2 & C_3 \end{array} \quad \begin{array}{c} \text{Row minima} \end{array} \\
R\text{'s move} \quad \begin{array}{c} R_1 \\ R_2 \\ R_3 \end{array} & \left[\begin{array}{ccc} 4 & 2 & 1 \\ 0 & 0 & -1 \\ 2 & 1 & 3 \end{array} \right] \quad \begin{array}{c} ① \\ -1 \\ ① \end{array} \quad \text{Largest}
\end{array}
$$

Column maxima   4   ②   3
                       ↑
                    Smallest

From the payoff matrix, we conclude that the row player's maximin strategy is to play either row 1 or row 3, whereas the column player's minimax strategy is to play column 2.

a. The saddle point is 2.

b. The optimum strategy for the row player is to play row 1 and the optimum strategy for the column player is to play column 1.

c. The value of the game is 2.

d. The game favors the row player.

9.

C's move

$$
\begin{array}{c}
 & \begin{array}{cc} C_1 & C_2 \end{array} \quad \begin{array}{c} \text{Row minima} \end{array} \\
R\text{'s move} \quad \begin{array}{c} R_1 \\ R_2 \end{array} & \left[\begin{array}{cc} ② & 3 \\ 1 & -4 \end{array} \right] \quad \begin{array}{c} 2 \leftarrow \text{Larger} \\ -4 \end{array}
\end{array}
$$

Column maxima   2   3
                 ↑
              Smaller

From the payoff matrix, we see that the game is strictly determined.

11.

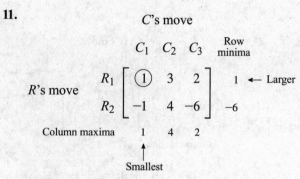

C's move

$$\begin{array}{c} & \begin{array}{ccc} C_1 & C_2 & C_3 \end{array} & \begin{array}{c} \text{Row} \\ \text{minima} \end{array} \\ R\text{'s move} \quad \begin{array}{c} R_1 \\ R_2 \end{array} & \left[\begin{array}{ccc} ① & 3 & 2 \\ -1 & 4 & -6 \end{array}\right] & \begin{array}{c} 1 \ \leftarrow \text{Larger} \\ -6 \end{array} \end{array}$$

Column maxima   1   4   2

↑
Smallest

From the payoff matrix, we see that the game is strictly determined.

a. The saddle point is 1.

b. The optimum strategy for the row player is to play row 1, whereas, the optimum strategy for the column player is to play column 1.

c. The value of the game is 1.

d. The game favors the row player.

13.

C's move

$$\begin{array}{c} & \begin{array}{cccc} C_1 & C_2 & C_3 & C_4 \end{array} & \begin{array}{c} \text{Row} \\ \text{minima} \end{array} \\ \begin{array}{c} R_1 \\ R\text{'s move} \ R_2 \\ R_3 \end{array} & \left[\begin{array}{cccc} ① & 3 & 4 & 2 \\ 0 & 2 & 6 & -4 \\ -1 & -3 & -2 & 1 \end{array}\right] & \begin{array}{c} 1 \ \leftarrow \text{Largest} \\ -4 \\ -3 \end{array} \end{array}$$

Column maxima   1   3   6   2

↑
Smallest

From the payoff matrix, we see that the game is strictly determined.

a. The saddle point is 1.

b. The optimum strategy for the row player is to play row 1 and the optimum strategy for the column player is to play column 1.

c. The value of the game is 1.

d. The game favors the row player.

15.

C's move

$$\begin{array}{c} & \begin{array}{cc} C_1 & C_2 \end{array} & \begin{array}{c} \text{Row} \\ \text{minima} \end{array} \\ \begin{array}{c} R_1 \\ R_2 \\ R\text{'s move} \quad R_3 \\ R_4 \end{array} & \left[\begin{array}{cc} 1 & 2 \\ 0 & 3 \\ -1 & 2 \\ 2 & -2 \end{array}\right] & \begin{array}{c} ① \ \leftarrow \text{Largest} \\ 0 \\ -1 \\ -2 \end{array} \end{array}$$

Column maxima   ②   3

↑
Smaller

From the payoff matrix, we see that the game is not strictly determined and consequently there is no saddle point.

17.

C's move

	C_1	C_2	C_3	C_4	Row minima
R_1	1	−1	3	2	−1
R_2	1	0	2	2	⓪ ← Largest
R_3	−2	2	3	−1	−2

R's move

Column maxima ① 2 3 2

↑ Smallest

From the payoff matrix, we see that the game is not strictly determined and consequently there is no saddle point.

19. a.

	1	2	3
1	2	−3	4
2	−3	4	−5
3	4	−5	6

c. The game is not strictly determined.

d. The game is not strictly determined.

b.

C's move

	C_1	C_2	C_3	Row minima
R_1	2	−3	4	−3 ← Largest
R_2	−3	4	−5	−5
R_3	4	−5	6	−5

R's move

Column maxima 4 4 6

↑ ↑ Smallest

From the payoff matrix, we see that the maximin strategy for Robin is to play row 1 (extend one finger), whereas the minimax strategy for Cathy is to play column 1 or column 2 (extend one or two fingers).

21. a.

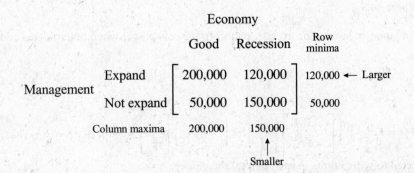

Economy

	Good	Recession	Row minima
Expand	200,000	120,000	120,000 ← Larger
Not expand	50,000	150,000	50,000

Management

Column maxima 200,000 150,000

↑ Smaller

b. The row player's (management's) minimax strategy is to play row 1, that is, to expand its line of conventional speakers.

23. a.

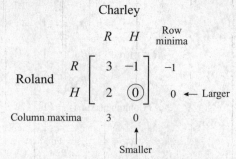

Charley

$$
\begin{array}{c}
 & R \quad H \quad L \\
\begin{array}{c} R \\ \text{Roland} \quad H \\ L \end{array}
\left[
\begin{array}{ccc}
3 & -1 & -3 \\
2 & 0 & -2 \\
5 & 2 & \boxed{1}
\end{array}
\right]
\begin{array}{c} \text{Row} \\ \text{minima} \\ -3 \\ -2 \\ 1 \leftarrow \text{Largest} \end{array}
\end{array}
$$

Column maxima   5   2   1

↑

Smallest

b. From the payoff matrix, we see that the game is strictly determined.

c. If neither party were willing to lower their price, the payoff matrix would be as follows.

Charley

$$
\begin{array}{c}
 & R \quad H \\
\text{Roland} \quad
\begin{array}{c} R \\ H \end{array}
\left[
\begin{array}{cc}
3 & -1 \\
2 & \boxed{0}
\end{array}
\right]
\begin{array}{c} \text{Row} \\ \text{minima} \\ -1 \\ 0 \leftarrow \text{Larger} \end{array}
\end{array}
$$

Column maxima   3   0

↑

Smaller

In this case, we see that the game is strictly determined, so that the optimal strategy for each barber is to charge his current price for a haircut.

25. True. This follows from the definition.

9.5   Games with Mixed Strategies

Problem-Solving Tips

1. The **expected value** of a game gives the average payoff to the row player when both players adopt a particular set of mixed strategies. If P and Q are the vectors representing the mixed strategies for the row player R and the column player C in a game with an $m \times n$ payoff matrix, then the expected value of the game is given by $E = PAQ$.

2. The **optimal strategy** for the row player in a nonstrictly determined game with payoff matrix $A = \begin{bmatrix} a & b \\ c & d \end{bmatrix}$ is

$P = \begin{bmatrix} p_1 & p_2 \end{bmatrix}$, where $p_1 = \dfrac{d - c}{a + d - b - c}$ and $p_2 = 1 - p_1$. The optimal strategy for the column player

is $Q = \begin{bmatrix} q_1 \\ q_2 \end{bmatrix}$, where $q_1 = \dfrac{d - b}{a + d - b - c}$ and $q_2 = 1 - q_1$. The expected value of the game is given by

$$E = PAQ = \dfrac{ad - bc}{a + d - b - c}.$$

Concept Questions   page 582

1. The *expected value of a game* measures the average payoff to the row player when both players adopt a particular set of mixed strategies.

Exercises   page 583

1. We compute $E = PAQ = \begin{bmatrix} \frac{1}{2} & \frac{1}{2} \end{bmatrix} \begin{bmatrix} 3 & 1 \\ -4 & 2 \end{bmatrix} \begin{bmatrix} \frac{3}{5} \\ \frac{2}{5} \end{bmatrix} = \begin{bmatrix} -\frac{1}{2} & \frac{3}{2} \end{bmatrix} \begin{bmatrix} \frac{3}{5} \\ \frac{2}{5} \end{bmatrix} = -\frac{3}{10} + \frac{6}{10} = \frac{3}{10}$. Thus, in the long run,

the row player is expected to win .3 units in each play of the game.

3. We compute $E = PAQ = \begin{bmatrix} \frac{1}{3} & \frac{2}{3} \end{bmatrix} \begin{bmatrix} -4 & 3 \\ 2 & 1 \end{bmatrix} \begin{bmatrix} \frac{3}{4} \\ \frac{1}{4} \end{bmatrix} = \begin{bmatrix} 0 & \frac{5}{3} \end{bmatrix} \begin{bmatrix} \frac{3}{4} \\ \frac{1}{4} \end{bmatrix} = \frac{5}{12}$. Thus, in the long run, the row player

may be expected to win .4167 units in each play of the game.

5. We compute $E = PAQ = \begin{bmatrix} .2 & .6 & .2 \end{bmatrix} \begin{bmatrix} 2 & 0 & -2 \\ 1 & -1 & 3 \\ 2 & 1 & -4 \end{bmatrix} \begin{bmatrix} .2 \\ .6 \\ .2 \end{bmatrix} = \begin{bmatrix} 1.4 & -.4 & .6 \end{bmatrix} \begin{bmatrix} .2 \\ .6 \\ .2 \end{bmatrix} = .16.$

7. a. $E = PAQ = \begin{bmatrix} 1 & 0 \end{bmatrix} \begin{bmatrix} 1 & -2 \\ -2 & 3 \end{bmatrix} \begin{bmatrix} 1 \\ 0 \end{bmatrix} = \begin{bmatrix} 1 & -2 \end{bmatrix} \begin{bmatrix} 1 \\ 0 \end{bmatrix} = 1.$

b. $E = PAQ = \begin{bmatrix} 0 & 1 \end{bmatrix} \begin{bmatrix} 1 & -2 \\ -2 & 3 \end{bmatrix} \begin{bmatrix} 1 \\ 0 \end{bmatrix} = \begin{bmatrix} -2 & 3 \end{bmatrix} \begin{bmatrix} 1 \\ 0 \end{bmatrix} = -2.$

c. $E = PAQ = \begin{bmatrix} \frac{1}{2} & \frac{1}{2} \end{bmatrix} \begin{bmatrix} 1 & -2 \\ -2 & 3 \end{bmatrix} \begin{bmatrix} \frac{1}{2} \\ \frac{1}{2} \end{bmatrix} = \begin{bmatrix} -\frac{1}{2} & \frac{1}{2} \end{bmatrix} \begin{bmatrix} \frac{1}{2} \\ \frac{1}{2} \end{bmatrix} = 0.$

d. $E = PAQ = \begin{bmatrix} .5 & .5 \end{bmatrix} \begin{bmatrix} 1 & -2 \\ -2 & 3 \end{bmatrix} \begin{bmatrix} .8 \\ .2 \end{bmatrix} = \begin{bmatrix} -.5 & .5 \end{bmatrix} \begin{bmatrix} .8 \\ .2 \end{bmatrix} = -.3.$

Thus, the strategy in part (a) is the most advantageous to R.

9. a.

$$\begin{array}{c} \\ \\ \\ \text{Column} \\ \text{maxima} \end{array} \begin{bmatrix} -3 & 3 & 2 \\ -3 & 1 & 1 \\ 1 & -2 & 1 \end{bmatrix} \begin{array}{c} \text{Row} \\ \text{minima} \\ -3 \\ -3 \\ \boxed{-2} \leftarrow \text{Largest} \end{array}$$

Column maxima: $\boxed{1} \quad 3 \quad 2$

↑ Smallest

From the payoff matrix, we see that the expected payoff to a row player using the minimax strategy is 1.

b. The expected payoff is given by

$$E = PAQ = \begin{bmatrix} .25 & .25 & .5 \end{bmatrix} \begin{bmatrix} -3 & 3 & 2 \\ -3 & 1 & 1 \\ 1 & -2 & 1 \end{bmatrix} \begin{bmatrix} .6 \\ .2 \\ .2 \end{bmatrix}$$

$$= -.35.$$

c. The minimax strategy [part (a)] is the better strategy for the row player.

11. The game under consideration has no saddle point and is accordingly nonstrictly determined. Using the formulas for determining the optimal mixed strategies for a 2×2 game with $a = 4$, $b = 1$, $c = 2$, and $d = 3$, we find that

$$p_1 = \frac{d - c}{a + d - b - c} = \frac{3 - 2}{4 + 3 - 1 - 2} = \frac{1}{4} \text{ and } p_2 = 1 - p_1 = 1 - \frac{1}{4} = \frac{3}{4}, \text{ so that the row player's optimal}$$

mixed strategy is given by $P = \begin{bmatrix} \frac{1}{4} & \frac{3}{4} \end{bmatrix}$. Next, we compute $q_1 = \frac{d - b}{a + d - b - c} = \frac{3 - 1}{4 + 3 - 1 - 2} = \frac{2}{4} = \frac{1}{2}$

and $q_2 = 1 - q_1 = 1 - \frac{1}{2} = \frac{1}{2}$. Thus, the optimal strategy for the column player is $Q = \begin{bmatrix} \frac{1}{2} \\ \frac{1}{2} \end{bmatrix}$. To

determine whether the game favors one player over the other, we compute the expected value of the game:

$$E = \frac{ad - bc}{a + d - b - c} = \frac{(4)(3) - (1)(2)}{4 + 3 - 1 - 2} = \frac{10}{4} = \frac{5}{2}, \text{ or } 2.5 \text{ units for each play of the game. These results imply}$$
that the game favors the row player.

13. Because the game is not strictly determined, we use the formulas for determining the optimal mixed strategies for a 2×2 game with $a = -1$, $b = 2$, $c = 1$, and $d = -3$. We find that $p_1 = \frac{d - c}{a + d - b - c} = \frac{-3 - 1}{-1 - 3 - 2 - 1} = \frac{4}{7}$

and $p_2 = 1 - p_1 = 1 - \frac{4}{7} = \frac{3}{7}$. Thus, the optimal mixed strategy for the row player is given by $P = \begin{bmatrix} \frac{4}{7} & \frac{3}{7} \end{bmatrix}$. To

find the optimal mixed strategy for the column player, we compute $q_1 = \frac{d - b}{a + d - b - c} = \frac{-3 - 2}{-1 - 3 - 2 - 1} = \frac{5}{7}$

and $q_2 = 1 - q_1 = 1 - \frac{5}{7} = \frac{2}{7}$, so $Q = \begin{bmatrix} \frac{5}{7} \\ \frac{2}{7} \end{bmatrix}$. The expected value of the game is given by

$$E = \frac{ad - bc}{a + d - b - c} = \frac{(-1)(-3) - (2)(1)}{-1 - 3 - 2 - 1} = -\frac{1}{7}. \text{ Because the value of the game is negative, we conclude that}$$
the game favors the column player.

15. Because the game is not strictly determined, we use the formulas for determining the optimal mixed strategies for a 2×2 game with $a = -2$, $b = -6$, $c = -8$, and $d = -4$. Then $p_1 = \frac{d - c}{a + d - b - c} = \frac{-4 + 8}{-2 - 4 + 6 + 8} = \frac{1}{2}$,

$p_2 = 1 - p_1 = \frac{1}{2}$, $q_1 = \frac{d - b}{a + d - b - c} = \frac{-4 + 6}{-2 - 4 + 6 + 8} = \frac{1}{4}$, and $q_2 = 1 - q_1 = \frac{3}{4}$, so $P = \begin{bmatrix} \frac{1}{2} & \frac{1}{2} \end{bmatrix}$,

$Q = \begin{bmatrix} \frac{1}{4} \\ \frac{3}{4} \end{bmatrix}$, and $E = \frac{ad - bc}{a + d - b - c} = \frac{(-2)(-4) - (-6)(-8)}{-2 - 4 + 6 + 8} = -5$. We conclude that the game favors the

column player.

17. a. Because the game is not strictly determined, we employ the formulas for determining the optimal mixed strategies for a 2×2 game with $a = 4$, $b = -2$, $c = -2$, and $d = 1$. We find that

$$p_1 = \frac{d - c}{a + d - b - c} = \frac{1 - (-2)}{4 + 1 - (-2) - (-2)} = \frac{3}{9} = \frac{1}{3} \text{ and } p_2 = 1 - p_1 = 1 - \frac{1}{3} = \frac{2}{3}, \text{ so Richie's optimal}$$

mixed strategy is $P = \begin{bmatrix} \frac{1}{3} & \frac{2}{3} \end{bmatrix}$. Next, we compute $q_1 = \frac{d - b}{a + d - b - c} = \frac{1 - (-2)}{4 + 1 - (-2) - (-2)} = \frac{3}{9} = \frac{1}{3}$ and

$q_2 = 1 - q_1 = 1 - \frac{1}{3} = \frac{2}{3}$. Thus, Chuck's optimal strategy is $Q = \begin{bmatrix} \frac{1}{3} \\ \frac{2}{3} \end{bmatrix}$.

b. The expected value of the game is given by $E = \frac{ad - bc}{a + d - b - c} = \frac{(4)(1) - (-2)(-2)}{4 + 1 - (-2) - (-2)} = 0$ and conclude that

in the long run the game will end in a draw.

19. a.

	Expanding economy	Economic recession
Hotel stocks	25	−5
Brewery stocks	10	15

From the payoff matrix, we see that the game is not strictly determined, so we use the formulas for finding the optimal mixed strategies for a 2×2 nonstrictly determined game with $a = 25$, $b = -5$, $c = 10$, and $d = 15$:

$$p_1 = \frac{d - c}{a + d - b - c} = \frac{15 - 10}{25 + 15 + 5 - 10} = \frac{5}{35} = \frac{1}{7} \text{ and}$$

$p_2 = 1 - p_2 = 1 - \frac{1}{7} = \frac{6}{7}$, so the Maxwells optimal mixed strategy is $P = \begin{bmatrix} \frac{1}{7} & \frac{6}{7} \end{bmatrix}$ Thus, the Maxwells should invest $\frac{1}{7}$ (40,000) ≈ \$5714 in hotel stocks and $\frac{6}{7}$ (40,000) ≈ \$34,286 in brewery stocks.

b. The profit that the Maxwells expect to make is given by

$$E = \frac{ad - bc}{a + d - b - c} = \frac{(25)(15) - (-5)(10)}{35} = \frac{425}{35} = 12.1429.$$ We conclude that the Maxwells will realize a profit of (.12143) (\$40,000) ≈ \$4857 by employing their optimal mixed strategy.

21. a.

	C		
	N	F	Row minima
N	.48	.65	(.48) ← Larger
F	.50	.45	.45
Column maxima	(.50)	.65	

Smaller ↑ (.50)

From the payoff matrix, we see that there is no saddle point, and thus the game is not strictly determined.

b. Employing the formulas for finding the optimal mixed strategies for a 2×2 nonstrictly determined game, we find that $p_1 = \dfrac{d - c}{a + d - b - c} = \dfrac{.45 - .50}{.48 + .45 - .65 - .50} \approx .227$,

$p_2 = 1 - p_1 = 1 - .227 = .773$,

$q_1 = \dfrac{d - b}{a + d - b - c} = \dfrac{.45 - .65}{.48 + .45 - .65 - .50} \approx .909$, and

$q_2 = 1 - q_1 \approx 1 - .909 = .091$.

We conclude that Dr. Russell's strategy is $P = \begin{bmatrix} .227 & .773 \end{bmatrix}$ and Dr. Carlton's strategy is $Q = \begin{bmatrix} .91 \\ .09 \end{bmatrix}$. Thus, Dr. Russell should place 22.7% of his advertisements in the local newspaper and 77.3% in fliers, whereas Dr. Carlton should place 91% of his advertisements in the local newspaper and 9% in fliers.

23. The optimal strategies for the row and column players are $P = \begin{bmatrix} p_1 & p_2 \end{bmatrix}$, where $p_1 = \dfrac{d - c}{a + d - b - c}$ and

$p_2 = 1 - p_1 = 1 - \dfrac{d - c}{a + d - b - c} = \dfrac{a - b}{a + d - b - c}$, and $Q = \begin{bmatrix} q_1 \\ q_2 \end{bmatrix}$, where $q_1 = \dfrac{d - b}{a + d - b - c}$ and

$q_2 = 1 - q_1 = 1 - \dfrac{d - b}{a + d - b - c} = \dfrac{a - c}{a + d - b - c}$. Therefore, the expected value of the game is

$$E = PAQ = \begin{bmatrix} p_1 & p_2 \end{bmatrix} \begin{bmatrix} a & b \\ c & d \end{bmatrix} \begin{bmatrix} q_1 \\ q_2 \end{bmatrix} = \begin{bmatrix} ap_1 + cp_2 & bp_1 + dp_2 \end{bmatrix} \begin{bmatrix} q_1 \\ q_2 \end{bmatrix} = (ap_1 + cp_2)q_1 + (bp_1 + dp_2)q_2$$

$$= \left[\frac{a(d - c)}{a + d - b - c} + \frac{c(a - b)}{a + d - b - c} \right] \frac{d - b}{a + d - b - c} + \left[\frac{b(d - c)}{a + d - b - c} + \frac{d(a - b)}{a + d - b - c} \right] \frac{a - c}{a + d - b - c}$$

$$= \frac{(ad - bc)(d - b)}{(a + d - b - c)^2} + \frac{(ad - bc)(a - c)}{(a + d - b - c)^2} = \frac{(ad - bc)(a + d - b - c)}{(a + d - b - c)^2} = \frac{ad - bc}{a + d - b - c},$$

as was to be shown.

1. probabilities, preceding   **3.** transition   **5.** distribution, steady-state

7. absorbing, leave, steps

9. optimal

CHAPTER 9   Review Exercises   page 587

1. Because the entries $a_{12} = -2$ and $a_{22} = -8$ are negative, the given matrix is not stochastic and is hence not a regular stochastic matrix.

3. $T^2 = \begin{bmatrix} \frac{1}{2} & 0 & \frac{1}{3} \\ 0 & 0 & \frac{1}{3} \\ \frac{1}{2} & 1 & \frac{1}{3} \end{bmatrix} \begin{bmatrix} \frac{1}{2} & 0 & \frac{1}{3} \\ 0 & 0 & \frac{1}{3} \\ \frac{1}{2} & 1 & \frac{1}{3} \end{bmatrix} = \begin{bmatrix} \frac{5}{12} & \frac{1}{3} & \frac{5}{18} \\ \frac{1}{6} & \frac{1}{3} & \frac{1}{9} \\ \frac{5}{12} & \frac{1}{3} & \frac{11}{18} \end{bmatrix}$ and so the matrix is regular.

5. $X_1 = \begin{bmatrix} 0 & \frac{1}{4} & \frac{3}{5} \\ \frac{2}{5} & \frac{1}{2} & \frac{1}{5} \\ \frac{3}{5} & \frac{1}{4} & \frac{1}{5} \end{bmatrix} \begin{bmatrix} \frac{1}{2} \\ \frac{1}{2} \\ 0 \end{bmatrix} = \begin{bmatrix} \frac{1}{8} \\ \frac{9}{20} \\ \frac{17}{40} \end{bmatrix}$ and $X_2 = \begin{bmatrix} 0 & \frac{1}{4} & \frac{3}{5} \\ \frac{2}{5} & \frac{1}{2} & \frac{1}{5} \\ \frac{3}{5} & \frac{1}{4} & \frac{1}{5} \end{bmatrix} \begin{bmatrix} \frac{1}{8} \\ \frac{9}{20} \\ \frac{17}{40} \end{bmatrix} = \begin{bmatrix} \frac{147}{400} \\ \frac{9}{25} \\ \frac{109}{400} \end{bmatrix} = \begin{bmatrix} .3675 \\ .36 \\ .2725 \end{bmatrix}$.

7. This is an absorbing matrix because state 1 is an absorbing state and it is possible to go from any nonabsorbing state to state 1.

9. This is not an absorbing stochastic matrix because there is no absorbing state.

11. We solve the matrix equation $\begin{bmatrix} .6 & .3 \\ .4 & .7 \end{bmatrix} \begin{bmatrix} x \\ y \end{bmatrix} = \begin{bmatrix} x \\ y \end{bmatrix}$ together with $x + y = 1$; that is, the system

$\begin{cases} -.4x + .3y = 0 \\ .4x - .3y = 0 \\ x + y = 1 \end{cases}$ We find the steady-state distribution vector to be $\begin{bmatrix} \frac{3}{7} \\ \frac{4}{7} \end{bmatrix}$ and the steady-state matrix to be

$\begin{bmatrix} \frac{3}{7} & \frac{3}{7} \\ \frac{4}{7} & \frac{4}{7} \end{bmatrix}$.

13. We solve the matrix equation $\begin{bmatrix} .6 & .4 & .3 \\ .2 & .2 & .2 \\ .2 & .4 & .5 \end{bmatrix} \begin{bmatrix} x \\ y \\ z \end{bmatrix} = \begin{bmatrix} x \\ y \\ z \end{bmatrix}$ together with $x + y + z = 1$; that is, the system

$$\begin{cases} -.4x + .4y + .3z = 0 \\ .2x - .8y + .2z = 0 \\ .2x + .4y - .5z = 0 \\ x + y + z = 1 \end{cases}$$ We find the solution $x = .457$, $y = .20$, $z = .343$, so the steady-state distribution vector

is $\begin{bmatrix} .457 \\ .200 \\ .343 \end{bmatrix}$ and the steady-state matrix is $\begin{bmatrix} .457 & .457 & .457 \\ .200 & .200 & .200 \\ .343 & .343 & .343 \end{bmatrix}$.

15. a. The transition matrix for the Markov chain is $T = \begin{array}{c} \\ A \\ U \\ N \end{array} \begin{array}{c} A \quad U \quad N \\ \begin{bmatrix} .85 & 0 & .10 \\ .10 & .95 & .05 \\ .05 & .05 & .85 \end{bmatrix} \end{array}$.

b. The probability vector describing the distribution of land 10 years ago is given by $\begin{array}{c} A \\ U \\ N \end{array} \begin{bmatrix} .50 \\ .15 \\ .35 \end{bmatrix}$.

To find the required probability vector, we compute

$$TX_0 = \begin{bmatrix} .85 & 0 & .10 \\ .10 & .95 & .05 \\ .05 & .05 & .85 \end{bmatrix} \begin{bmatrix} .50 \\ .15 \\ .35 \end{bmatrix} = \begin{bmatrix} .46 \\ .21 \\ .33 \end{bmatrix} \text{ and } TX_1 = \begin{bmatrix} .85 & 0 & .10 \\ .10 & .95 & .05 \\ .05 & .05 & .85 \end{bmatrix} \begin{bmatrix} .46 \\ .21 \\ .33 \end{bmatrix} = \begin{bmatrix} .424 \\ .262 \\ .314 \end{bmatrix}.$$ Thus, the

probability vector describing the distribution of land 10 years from now is $\begin{bmatrix} .424 \\ .262 \\ .314 \end{bmatrix}$.

17.

C's move

		C_1	C_2	Row minima
	R_1	1	2	1
R's move	R_2	3	5	3
	R_3	④	6	4 ← Largest
Column maxima		4	6	

↑
Smaller

From the payoff matrix, we see that the game is strictly determined. The saddle point is 4, the optimum strategy for the row player is to play row 3 and the optimum strategy for the column player is to play column 1, the value of the game is 4, and the game favors the row player.

19.

C's move

	C_1	C_2	C_3	Row minima
R_1	①	3	6	1 ← Largest
R_2	−2	4	3	−2
R_3	−5	−4	−2	−5

R's move

Column maxima   1   4   6
↑
Smallest

From the payoff matrix, we see that $a_{11} = 1$ is the saddle point of the game, and we conclude that the game is strictly determined. The row player's optimal strategy is to play row 1 and the column player's optimal strategy is to play column 1. The value of the game is 1, and the game favors the row player.

21. We compute $E = PAQ = \begin{bmatrix} \frac{1}{2} & \frac{1}{2} \end{bmatrix} \begin{bmatrix} 4 & 8 \\ 6 & -12 \end{bmatrix} \begin{bmatrix} \frac{1}{4} \\ \frac{3}{4} \end{bmatrix} = -\frac{1}{4}$.

23. We compute $E = PAQ = \begin{bmatrix} .2 & .4 & .4 \end{bmatrix} \begin{bmatrix} 3 & -1 & 2 \\ 1 & 2 & 4 \\ -2 & 3 & 6 \end{bmatrix} \begin{bmatrix} .2 \\ .6 \\ .2 \end{bmatrix} = \begin{bmatrix} .2 & 1.8 & 4. 4 \end{bmatrix} \begin{bmatrix} .2 \\ .6 \\ .2 \end{bmatrix} = 2$. The expected

payoff for the game is 2.

25. The game under consideration has no saddle point and is accordingly nonstrictly determined. Using the formulas for determining the optimal mixed strategies for a 2 × 2 game with $a = 1$, $b = -2$, $c = 0$, and $d = 3$, we find

that $p_1 = \dfrac{d-c}{a+d-b-c} = \dfrac{3-0}{1+3+2-0} = \dfrac{3}{6} = \dfrac{1}{2}$ and $p_2 = 1 - p_1 = 1 - \frac{1}{2} = \frac{1}{2}$, so the row player's

optimal mixed strategy is $P = \begin{bmatrix} \frac{1}{2} & \frac{1}{2} \end{bmatrix}$. Next, we compute $q_1 = \dfrac{d-b}{a+d-b-c} = \dfrac{3+2}{1+3+2-0} = \dfrac{5}{6}$

and $q_2 = 1 - q_1 = 1 - \frac{5}{6} = \frac{1}{6}$, so the optimal strategy for the column player is $Q = \begin{bmatrix} \frac{5}{6} \\ \frac{1}{6} \end{bmatrix}$. To

determine whether the game favors one player over the other, we compute the expected value of the game:

$E = \dfrac{ad-bc}{a+d-b-c} = \dfrac{(1)(3) - (-2)(0)}{1+3+2-0} = \dfrac{3}{6} = \dfrac{1}{2}$. Because the expected value is positive, the game favors the

row player.

27. Using the formulas for the optimal strategies in a 2 × 2 nonstrictly determined game with $a = 3$, $b = -6$,

$c = 1$, and $d = 2$, we find that $p_1 = \dfrac{d-c}{a+d-b-c} = \dfrac{2-1}{3+2+6-1} = \dfrac{1}{10}$, $p_2 = 1 - p_1 = \frac{9}{10}$,

$q_1 = \dfrac{d-b}{a+d-b-c} = \dfrac{2+6}{3+2+6-1} = \dfrac{4}{5}$, and $q_2 = 1 - q_1 = \frac{1}{5}$. Thus, the optimal mixed strategy for the row

player is $P = \begin{bmatrix} \frac{1}{10} & \frac{9}{10} \end{bmatrix}$ and the optimal mixed strategy for the column player is $Q = \begin{bmatrix} \frac{4}{5} \\ \frac{1}{5} \end{bmatrix}$. The value of the

game is $E = \dfrac{ad-bc}{a+d-b-c} = \dfrac{3(2) - (-6)(1)}{3+2-(-6)-1} = \dfrac{6}{5}$ and the game favors the row player.

29. a.

Record World

$7 $8 Row minima

Disco-Mart $7 $\begin{bmatrix} \textcircled{.5} & .7 \\ .4 & .5 \end{bmatrix}$ .5 ← Larger

$8 .4

Column maxima .5 .7

↑ Smaller

b. Upon finding the larger of the row minima and the smaller of the column maxima, we see that the entry $a_{11} = .5$ is a saddle point, and , consequently, that the game is strictly determined. Thus, the optimal price at which both companies should sell the compact disc label is $7.

CHAPTER 9   Before Moving On...   page 589

1. $X_1 = TX_0 = \begin{bmatrix} .3 & .4 \\ .7 & .6 \end{bmatrix} \begin{bmatrix} .6 \\ .4 \end{bmatrix} = \begin{bmatrix} .34 \\ .66 \end{bmatrix}$ and $X_2 = TX_1 = \begin{bmatrix} .3 & .4 \\ .7 & .6 \end{bmatrix} \begin{bmatrix} .34 \\ .66 \end{bmatrix} = \begin{bmatrix} .366 \\ .634 \end{bmatrix}$.

2. We solve the equation $TX = X$, or $\begin{bmatrix} \frac{1}{3} & \frac{1}{4} \\ \frac{2}{3} & \frac{3}{4} \end{bmatrix} \begin{bmatrix} x \\ y \end{bmatrix} = \begin{bmatrix} x \\ y \end{bmatrix}$, along with the system $\begin{cases} -\frac{2}{3}x + \frac{1}{4}y = 0 \\ \frac{2}{3}x - \frac{1}{4}y = 0 \\ x + y = 1 \end{cases}$ We find

that $x = \frac{3}{11}$ and $y = \frac{8}{11}$. Therefore, the steady-state distribution vector is $\begin{bmatrix} \frac{3}{11} \\ \frac{8}{11} \end{bmatrix}$.

3. Rewriting the matrix so that the absorbing state appears first, we have $\begin{matrix} & \begin{matrix} 2 & 3 & 1 \end{matrix} \\ \begin{matrix} 2 \\ 3 \\ 1 \end{matrix} & \begin{bmatrix} 1 & \frac{1}{4} & 0 \\ 0 & \frac{3}{4} & \frac{2}{3} \\ 0 & 0 & \frac{1}{3} \end{bmatrix} \end{matrix}$. We see that $S = \begin{bmatrix} \frac{1}{4} & 0 \end{bmatrix}$

and $R = \begin{bmatrix} \frac{3}{4} & \frac{2}{3} \\ 0 & \frac{1}{3} \end{bmatrix}$, so $I - R = \begin{bmatrix} 1 & 0 \\ 0 & 1 \end{bmatrix} - \begin{bmatrix} \frac{3}{4} & \frac{2}{3} \\ 0 & \frac{1}{3} \end{bmatrix} = \begin{bmatrix} \frac{1}{4} & -\frac{2}{3} \\ 0 & \frac{2}{3} \end{bmatrix}$, $(I - R)^{-1} = \begin{bmatrix} 4 & 4 \\ 0 & \frac{3}{2} \end{bmatrix}$, and

$S(I - R)^{-1} = \begin{bmatrix} \frac{1}{4} & 0 \end{bmatrix} \begin{bmatrix} 4 & 4 \\ 0 & \frac{3}{2} \end{bmatrix} = \begin{bmatrix} 1 & 1 \end{bmatrix}$. Therefore, the steady-state matrix of T is $\begin{bmatrix} 1 & 1 & 1 \\ 0 & 0 & 0 \\ 0 & 0 & 0 \end{bmatrix}$.

4. a.

C's move

C_1 C_2 C_3 Row minima

R's move $\begin{matrix} R_1 \\ R_2 \\ R_3 \end{matrix}$ $\begin{bmatrix} 2 & 3 & \textcircled{-1} \\ -1 & 2 & -3 \\ 3 & 4 & -2 \end{bmatrix}$ $\begin{matrix} -1 & \text{← Largest} \\ -3 \\ -2 \end{matrix}$

Column maxima 3 4 −1

↑ Smallest

From the payoff matrix, we see that −1 is a saddle point.

b. The optimal strategy for the row player is to make the move represented by the first row. The optimal strategy for the column player is to make the move represented by the third column.

c. The value of the game is −1. The game favors the column player.

5. a.

C's move

$$
\begin{array}{c}
 & \begin{array}{cc} C_1 & C_2 \end{array} & \begin{array}{c}\text{Row}\\ \text{minima}\end{array} \\
\begin{array}{c} R_1 \\ R\text{'s move } R_2 \\ R_2 \end{array}
\begin{bmatrix} 2 & -1 \\ 3 & 2 \\ -3 & 4 \end{bmatrix}
\begin{array}{c} -1 \\ ② \leftarrow \text{Largest} \\ -3 \end{array}
\end{array}
$$

Column maxima   ③   4
                  ↑
              Smallest

From the payoff matrix, we see that R's optimal pure strategy is to choose row 2 and C's optimal pure strategy is to chose column 1. If both players use their optimal strategy, then the expected payoff to the row player is 3 units.

b. $P = \begin{bmatrix} .3 & .4 & .3 \end{bmatrix}$ and $Q = \begin{bmatrix} .6 \\ .4 \end{bmatrix}$, so

$$
E = PAQ = \begin{bmatrix} .3 & .4 & .3 \end{bmatrix} \begin{bmatrix} 2 & -1 \\ 3 & 2 \\ -3 & 4 \end{bmatrix} \begin{bmatrix} .6 \\ .4 \end{bmatrix}
$$

$$
= \begin{bmatrix} .3 & .4 & .3 \end{bmatrix} \begin{bmatrix} .8 \\ 2.6 \\ -.2 \end{bmatrix} = 1.22.
$$

6. a. Using the formulas for the optimal mixed strategies for a 2×2 game with $a = 3$, $b = 1$, $c = -2$, and $d = 2$, we find that $p_1 = \dfrac{d-c}{a+d-b-c} = \dfrac{2-(-2)}{3+2-1-(-2)} = \dfrac{4}{6} = \dfrac{2}{3}$ and $p_2 = 1 - p_1 = \frac{1}{3}$, so $P = \begin{bmatrix} \frac{2}{3} & \frac{1}{3} \end{bmatrix}$; and

$q_1 = \dfrac{d-b}{a+d-b-c} = \dfrac{2-1}{6} = \dfrac{1}{6}$ and $q_2 = 1 - q_1 = \frac{5}{6}$, so $Q = \begin{bmatrix} \frac{1}{6} \\ \frac{5}{6} \end{bmatrix}$.

b. $E = \dfrac{ad-bc}{a+d-b-c} = \dfrac{(3)(2)-(1)(-2)}{6} = \dfrac{8}{6} = \dfrac{4}{3}$. Because the value of the game is positive, it favors the row player.

A INTRODUCTION TO LOGIC

A.1 Propositions and Connectives

Exercises   page 595

1. Yes   **3.** Yes   **5.** No   **7.** Yes

9. Yes   **11.** No   **13.** No   **15.** Negation

17. Conjunction   **19.** Conjunction

21. New orders for manufactured goods did not fall last month.

23. Drinking during pregnancy does not affect both the size and weight of babies.

25. The commuter airline industry is not now undergoing a shakeup.

27. a. Domestic car sales increased over the past year, or foreign car sales decreased over the past year, or both.

    b. Domestic car sales increased over the past year, and foreign car sales decreased over the past year.

    c. Either domestic car sales increased over the past year or foreign car sales decreased over the past year.

    d. Domestic car sales did not increase over the past year.

    e. Domestic car sales did not increase over the past year, or foreign car sales decreased over the past year, or both.

    f. Domestic car sales did not increase over the past year, or foreign car sales did not decrease over the past year, or both.

29. a. Either the doctor recommended surgery to treat Sam's hyperthyroidism or the doctor recommended radioactive iodine to treat Sam's hyperthyroidism.

    b. The doctor recommended surgery to treat Sam's hyperthyroidism, or the doctor recommended radioactive iodine to treat Sam's hyperthyroidism, or both.

31. a. $p \wedge q$ **b.** $p \veebar q$ **c.** $\sim p \wedge \sim q$ **d.** $\sim(\sim q)$

33. a. The popularity of neither prime-time soaps nor prime-time situation comedies increased this year.

    b. The popularity of prime-time soaps did not increase this year, or the popularity of prime-time detective shows decreased this year, or both.

    c. The popularity of prime-time detective shows decreased this year, or the popularity of prime-time situation comedies did not increase this year, or both.

    d. Either the popularity of prime-time soaps did not increase this year or the popularity of prime-time situation comedies did not increase this year.

A.2   Truth Tables

Exercises   page 598

1.

p	q	$\sim q$	$p \vee \sim q$
T	T	F	T
T	F	T	T
F	T	F	F
F	F	T	T

3.

p	$\sim p$	$\sim(\sim p)$
T	F	T
F	T	F

5.

p	$\sim p$	$p \vee \sim p$
T	F	T
F	T	T

7.

p	q	$\sim p$	$p \vee q$	$\sim p \wedge (p \vee q)$
T	T	F	T	F
T	F	F	T	F
F	T	T	T	T
F	F	T	F	F

9.

p	q	$\sim q$	$p \vee q$	$p \wedge \sim q$	$(p \vee q) \wedge (p \wedge \sim q)$
T	T	F	T	F	F
T	F	T	T	T	T
F	T	F	T	F	F
F	F	T	F	F	F

11.

p	q	$p \vee q$	$\sim(p \vee q)$	$(p \vee q) \wedge \sim(p \vee q)$
T	T	T	F	F
T	F	T	F	F
F	T	T	F	F
F	F	F	T	F

13.

p	q	r	$p \vee q$	$p \vee r$	$(p \vee q) \wedge (p \vee r)$
T	T	T	T	T	T
T	T	F	T	T	T
T	F	T	T	T	T
T	F	F	T	T	T
F	T	T	T	T	T
F	T	F	T	F	F
F	F	T	F	T	F
F	F	F	F	F	F

15.

p	q	r	$p \wedge q$	$\sim r$	$(p \wedge q) \vee \sim r$
T	T	T	T	F	T
T	T	F	T	T	T
T	F	T	F	F	F
T	F	F	F	T	T
F	T	T	F	F	F
F	T	F	F	T	T
F	F	T	F	F	F
F	F	F	F	T	T

17.

p	q	r	$\sim q$	$p \wedge \sim q$	$p \wedge r$	$(p \wedge \sim q) \vee (p \wedge r)$
T	T	T	F	F	T	T
T	T	F	F	F	F	F
T	F	T	T	T	T	T
T	F	F	T	T	F	T
F	T	T	F	F	F	F
F	T	F	F	F	F	F
F	F	T	T	F	F	F
F	F	F	T	F	F	F

19. 16 rows

A.3   The Conditional and the Biconditional Connectives

Exercises   page 602

1. $\sim q \to p, q \to \sim p, \sim p \to q$

3. $p \to q, \sim p \to \sim q, \sim q \to \sim p$

5. *Conditional:* If it is snowing, then the temperature is below freezing.
Biconditional: It is snowing if and only if the temperature is below freezing.

7. *Conditional:* If the company's union and management reach a settlement, then the workers will not strike.
Biconditional: The company's union and management will reach a settlement if and only if the workers do not strike.

9. False

11. False

13. It is false when I do not buy the house after the owner lowers the selling price.

15.

p	q	$p \to q$	$\sim(p \to q)$
T	T	T	F
T	F	F	T
F	T	T	F
F	F	T	F

17.

p	q	$p \to q$	$\sim(p \to q)$	$\sim(p \to q) \wedge p$
T	T	T	F	F
T	F	F	T	T
F	T	T	F	F
F	F	T	F	F

19.

p	q	$\sim p$	$\sim q$	$p \to \sim q$	$(p \to \sim q) \veebar \sim p$
T	T	F	F	F	F
T	F	F	T	T	T
F	T	T	F	T	F
F	F	T	T	T	F

21.

p	q	$\sim p$	$\sim q$	$p \to q$	$\sim q \to \sim p$	$(p \to q) \leftrightarrow (\sim q \to \sim p)$
T	T	F	F	T	T	T
T	F	F	T	F	F	T
F	T	T	F	T	T	T
F	F	T	T	T	T	T

23.

p	q	$p \wedge q$	$p \vee q$	$(p \wedge q) \to (p \vee q)$
T	T	T	T	T
T	F	F	T	T
F	T	F	T	T
F	F	F	F	T

25.

p	q	r	$p \vee q$	$\sim r$	$(p \vee q) \to \sim r$
T	T	T	T	F	F
T	T	F	T	T	T
T	F	T	T	F	F
T	F	F	T	T	T
F	T	T	T	F	F
F	T	F	T	T	T
F	F	T	F	F	T
F	F	F	F	T	T

27.

p	q	r	$q \vee r$	$p \to (q \vee r)$
T	T	T	T	T
T	T	F	T	T
T	F	T	T	T
T	F	F	F	F
F	T	T	T	T
F	T	F	T	T
F	F	T	T	T
F	F	F	F	T

29.

p	q	$\sim q$	$p \to q$	$\sim p \vee q$
T	T	F	T	T
T	F	F	F	F
F	T	T	T	T
F	F	T	T	T

The propositions are logically equivalent.

31.

p	q	$\sim p$	$\sim q$	$q \to p$	$\sim p \to \sim q$
T	T	F	F	T	T
T	F	F	T	T	T
F	T	T	F	F	F
F	F	T	T	T	T

The propositions are logically equivalent.

33.

p	q	$\sim q$	$p \wedge q$	$p \to \sim q$
T	T	F	T	F
T	F	T	F	T
F	T	F	F	T
F	F	T	F	T

The propositions are not logically equivalent.

35.

p	q	r	$p \to q$	$p \vee q$	$(p \to q) \to r$	$(p \vee q) \vee r$
T	T	T	T	T	T	T
T	T	F	T	T	F	T
T	F	T	F	T	T	T
T	F	F	F	T	T	T
F	T	T	T	T	T	T
F	T	F	T	T	F	T
F	F	T	T	F	T	T
F	F	F	T	F	F	F

The propositions are not logically equivalent.

37. a. $p \to \sim q$ **b.** $\sim p \to q$ **c.** $\sim q \leftrightarrow p$ **d.** $p \to \sim q$ **e.** $p \leftrightarrow \sim q$

A.4   Laws of Logic

Exercises   page 606

1.

p	p	$p \wedge p$
T	T	T
F	F	F

3.

p	q	r	$p \wedge q$	$(p \wedge q) \wedge r$	$q \wedge r$	$p \wedge (q \wedge r)$
T	T	T	T	T	T	T
T	T	F	T	F	F	F
T	F	T	F	F	F	F
T	F	F	F	F	F	F
F	T	T	F	F	T	F
F	T	F	F	F	F	F
F	F	T	F	F	F	F
F	F	F	F	F	F	F

5.

p	q	$p \wedge q$	$q \wedge p$
T	T	T	T
T	F	F	F
F	T	F	F
F	F	F	F

7.

p	q	r	$q \wedge r$	$p \vee (q \wedge r)$	$p \vee q$	$p \vee r$	$(p \vee q) \wedge (p \vee r)$
T	T	T	T	T	T	T	T
T	T	F	F	T	T	T	T
T	F	T	F	T	T	T	T
T	F	F	F	T	T	T	T
F	T	T	T	T	T	T	T
F	T	F	F	F	T	F	F
F	F	T	F	F	F	T	F
F	F	F	F	F	F	F	F

9.

p	q	$\sim p$	$p \rightarrow q$	$\sim p \vee q$	$(p \rightarrow q) \leftrightarrow (\sim p \vee q)$
T	T	F	T	T	T
T	F	F	F	F	T
F	T	T	T	T	T
F	F	T	T	T	T

Because the entries in the last column are all Ts, the proposition is a tautology.

11.

p	q	$p \vee q$	$p \rightarrow (p \vee q)$
T	T	T	T
T	F	T	T
F	T	T	T
F	F	F	T

Because the entries in the last column are all Ts, the proposition is a tautology.

13.

p	q	$\sim p$	$\sim q$	$p \rightarrow q$	$\sim q \rightarrow \sim p$	$(p \rightarrow q) \leftrightarrow (\sim q \rightarrow \sim p)$
T	T	F	F	T	T	T
T	F	F	T	F	F	T
F	T	T	F	T	T	T
F	F	T	T	T	T	T

Because the entries in the last column are all Ts, the proposition is a tautology.

15.

p	q	$\sim p$	$\sim q$	$p \rightarrow q$	$(p \rightarrow q) \wedge \sim q$	$(p \rightarrow q) \wedge (\sim q) \rightarrow (\sim p)$
T	T	F	F	T	F	T
T	F	F	T	F	F	T
F	T	T	F	T	F	T
F	F	T	T	T	T	T

Because all the entries in the last column are Ts, the proposition is a tautology.

17.

p	q	r	$p \to q$	$q \to r$	$p \to r$	$(p \to q) \vee (q \to r)$	$[(p \to q) \vee (q \to r)] \to (p \to r)$
T	T	T	T	T	T	T	T
T	T	F	T	F	F	T	F
T	F	T	F	T	T	T	T
T	F	F	F	T	F	T	F
F	T	T	T	T	T	T	T
F	T	F	T	F	T	T	T
F	F	T	T	T	T	T	T
F	F	F	T	T	T	T	T

Because the entries in the last column are not all Ts or all Fs, the proposition is neither a tautology nor a contradiction.

19. $\sim(p \wedge q)$: The candidate does not oppose changes in the Social Security system, or the candidate does not support immigration reform.

$\sim(p \vee q)$: The candidate does not oppose changes in the Social Security system, and the candidate does not support immigration reform.

21. $p \wedge (q \vee \sim q) \vee (p \wedge q)$

$\Leftrightarrow [p \wedge t \vee (p \wedge q)$ by Law 11

$\Leftrightarrow p \vee (p \wedge q)$ by Law 14

23. $(p \wedge \sim q) \vee (p \wedge \sim r)$

$\Leftrightarrow p \wedge (\sim q \vee \sim r)]$ by Law 7

25. $p \wedge \sim(q \wedge r) \Leftrightarrow p \wedge (\sim q \vee \sim r)$ by Law 10

$\Leftrightarrow (p \wedge \sim q) \vee (p \wedge \sim r)$ by Law 7

A.5   Arguments

Exercises   page 607

1. Valid by the law of syllogisms.

3. From the associated truth table, we see that the argument is valid. (There is no row in which the conclusion is false and the premises are all true.)

p	q	$p \wedge q$	$\sim p$	q
T	T	T	F	T
T	F	F	F	F
F	T	F	T	T
F	F	F	T	F

5. From the associated truth table, we see that the argument is invalid.

p	q	$p \rightarrow q$	$\sim p$	$\sim q$
T	T	T	F	F
T	F	F	F	T
F	T	T	T	F
F	F	T	T	T

7. From the associated truth table, we see that the argument is valid.

p	q	$p \leftrightarrow q$	q	p
T	T	T	T	T
T	F	F	F	T
F	T	F	T	F
F	F	T	F	F

9. From the associated truth table, we see that the argument is valid.

p	q	$p \rightarrow q$	$q \rightarrow p$	$p \leftrightarrow q$
T	T	T	T	T
T	F	F	T	F
F	T	T	F	F
F	F	T	T	T

11. From the associated truth table, we see that the argument is valid.

p	q	r	$p \leftrightarrow q$	$q \leftrightarrow r$	$p \leftrightarrow r$
T	T	T	T	T	T
T	T	F	T	F	F
T	F	T	F	F	T
T	F	F	F	T	F
F	T	T	F	T	F
F	T	F	F	F	T
F	F	T	T	F	F
F	F	F	T	T	T

13. From the associated truth table, we see that the argument is valid.

p	q	r	$p \veebar r$	$q \wedge r$	$p \rightarrow r$
T	T	T	F	T	T
T	T	F	T	F	F
T	F	T	F	F	T
T	F	F	T	F	F
F	T	T	T	T	T
F	T	F	F	F	T
F	F	T	T	F	T
F	F	F	F	F	T

15. From the associated truth table, we see that the argument is invalid.

p	q	r	$\sim p$	$\sim r$	$p \leftrightarrow q$	$q \vee r$	$\sim p$	$\sim p \rightarrow \sim r$
T	T	T	F	F	T	T	F	T
T	T	F	F	T	T	T	F	T
T	F	T	F	F	F	T	F	T
T	F	F	F	T	F	F	F	T
F	T	T	T	F	F	T	T	F
F	T	F	T	T	F	T	T	T
F	F	T	T	F	T	T	T	F
F	F	F	T	T	T	F	T	T

17. The symbolic form of the argument and its truth table are as follows.

$$p \to q$$
$$\sim p$$
$$\overline{\therefore \sim q}$$

p	q	$p \to q$	$\sim p$	$\sim q$
T	T	T	F	F
T	F	F	F	T
F	T	T	T	F
F	F	T	T	T

Because there is a row in the truth table in which the premises are all true but the conclusion is not, the argument is invalid.

19. The symbolic form of the argument and its truth table are as follows.

$$p \vee q$$
$$\sim p \to \sim q$$
$$\overline{\therefore p}$$

p	q	$\sim p$	$\sim q$	$p \vee q$	$\sim p \to \sim q$	p
T	T	F	F	T	T	T
T	F	F	T	T	T	T
F	T	T	F	T	F	F
F	F	T	T	F	T	F

Because the conclusion is true whenever the premises are all true, the argument is valid.

21. The symbolic form of the argument and its truth table are as follows.

$$p \to q$$
$$q \to r$$
$$r$$
$$\overline{\therefore p}$$

p	q	r	$p \to q$	$q \to r$	r	p
T	T	T	T	T	T	T
T	T	F	T	F	F	T
T	F	T	F	T	T	T
T	F	F	F	T	F	T
F	T	T	T	T	T	F
F	T	F	T	F	F	F
F	F	T	T	T	T	F
F	F	F	T	T	F	F

Because there are rows in the truth table in which the premises are all true but the conclusion is not, the argument is invalid.

23. The symbolic form of the argument and its truth table are as follows.

$$p \to q$$
$$r \to s$$
$$\sim s$$
$$\overline{\therefore \sim p \land r}$$

p	q	r	s	$\sim p$	$p \to q$	$r \to s$	$\sim s$	$\sim p \land \sim r$
T	T	T	T	F	T	T	F	F
T	T	T	F	F	T	F	T	F
T	T	F	T	F	T	T	F	F
T	T	F	F	F	T	T	T	F
T	F	T	T	F	F	T	F	F
T	F	T	F	F	F	F	T	F
T	F	F	T	F	F	T	F	F
T	F	F	F	F	F	T	T	F
F	T	T	T	T	T	T	F	T
F	T	T	F	T	T	F	T	T
F	T	F	T	T	T	T	F	F
F	T	F	F	T	T	T	T	F
F	F	T	T	T	T	T	F	T
F	F	T	F	T	T	F	T	T
F	F	F	T	T	T	T	F	F
F	F	F	F	T	T	T	T	F

Because there is a row in the truth table in which the premises are all true but the conclusion is not, the argument is invalid.

25. From the associated truth table, we see that the argument is valid.

p	q	$p \to q$	$\sim q$	$\sim p$
T	T	T	F	F
T	F	F	T	F
F	T	T	F	T
F	F	T	T	T

A.6   Applications of Logic to Switching Networks

Exercises   page 614

1. $p \land q \land (r \lor s)$

3. $\big[(p \land q) \lor r\big] \land (\sim r \lor p)$

5. $\big[(p \lor q) \land r\big] \lor (\sim p) \lor \big[\sim q \land (p \lor r \lor \sim r)\big]$

7.

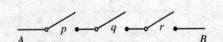

9.

11.

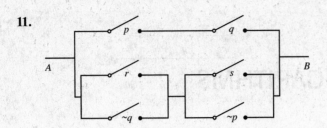

13. $p \wedge [[\sim q \vee (\sim p \wedge q)], p \wedge \sim q$

15. $p \wedge [\sim p \vee q \vee (q \wedge r)], p \wedge q$

C REVIEW OF LOGARITHMS

Exercises   page 625

1. $\log_2 64 = 6$

3. $\log_3 \frac{1}{9} = -2$

5. $\log_{32} 8 = \frac{3}{5}$

7. $\log 12 = \log(4 \cdot 3) = \log 4 + \log 3 \approx 0.6021 + 0.4771 = 1.0792$

9. $\log 16 = \log 4^2 = 2 \log 4 \approx 2(0.6021) = 1.2042$

11. $2 \ln a + 3 \ln b = \ln a^2 + \ln b^3 = \ln \left(a^2 b^3 \right)$

13. $\ln 3 + \frac{1}{2} \ln x + \ln y - \frac{1}{3} \ln z = \ln 3 + \ln x^{1/2} + \ln y - \ln z^{1/3} = \ln \dfrac{3\sqrt{x}\, y}{\sqrt[3]{z}}$

15. $\log_2 x = 3 \Leftrightarrow 2^3 = x \Leftrightarrow x = 8$

17. $\log_x 10^3 = 3 \Leftrightarrow x^3 = 10^3 \Leftrightarrow x = 10$

19. $\log_2 (2x + 5) = 3 \Leftrightarrow 2^3 = 2x + 5 \Leftrightarrow 2x + 5 = 8 \Leftrightarrow 2x = 3 \Leftrightarrow x = \frac{3}{2}$

21. $\log_5 (2x + 1) - \log_5 (x - 2) = 1 \Leftrightarrow \log_5 \dfrac{2x + 1}{x - 2} = 1 \Leftrightarrow \dfrac{2x + 1}{x - 2} = 5 \Leftrightarrow 2x + 1 = 5x - 10 \Leftrightarrow 3x = 11 \Leftrightarrow x = \frac{11}{3}$

23. $e^{0.4t} = 8 \Leftrightarrow \ln \left(e^{0.4t} \right) = \ln 8 \Leftrightarrow 0.4t = \ln 8 \Leftrightarrow t = \frac{5}{2} \ln 8 \approx 5.1986$

25. $5e^{-2t} = 6 \Leftrightarrow e^{-2t} = \frac{6}{5} \Leftrightarrow -2t = \ln \frac{6}{5} \Leftrightarrow t = -\frac{1}{2} \ln \frac{6}{5} \approx -0.0912$

27. $2e^{-0.2t} - 4 = 6 \Leftrightarrow 2e^{-0.2t} = 10 \Leftrightarrow e^{-0.2t} = 5 \Leftrightarrow -0.2t = \ln 5 \Leftrightarrow t = -5 \ln 5 \approx -8.0472$

29. $\dfrac{50}{1 + 4e^{0.2t}} = 20 \Leftrightarrow 50 = 20 + 80e^{0.2t} \Leftrightarrow e^{0.2t} = \frac{3}{8} \Leftrightarrow 0.2t = \ln \frac{3}{8} \Leftrightarrow t = 5 \ln \frac{3}{8} \approx -4.9041$